AF800124

Compendium of Plant Genomes

Series Editor

Chittaranjan Kole, Raja Ramanna Fellow, Government of India, ICAR-National Research Center on Plant Biotechnology, Pusa, New Delhi, India

Whole-genome sequencing is at the cutting edge of life sciences in the new millennium. Since the first genome sequencing of the model plant *Arabidopsis thaliana* in 2000, whole genomes of about 100 plant species have been sequenced and genome sequences of several other plants are in the pipeline. Research publications on these genome initiatives are scattered on dedicated web sites and in journals with all too brief descriptions. The individual volumes elucidate the background history of the national and international genome initiatives; public and private partners involved; strategies and genomic resources and tools utilized; enumeration on the sequences and their assembly; repetitive sequences; gene annotation and genome duplication. In addition, synteny with other sequences, comparison of gene families and most importantly potential of the genome sequence information for gene pool characterization and genetic improvement of crop plants are described.

Interested in editing a volume on a crop or model plant?

Please contact Prof. C. Kole, Series Editor, at ckoleorg@gmail.com

More information about this series at http://www.springer.com/series/11805

M. Timothy Rabanus-Wallace •
Nils Stein
Editors

The Rye Genome

Springer

Editors
M. Timothy Rabanus-Wallace
Leibniz Institute of Plant Genetics
and Crop Plant Research (IPK)
Seeland, Sachsen-Anhalt, Germany

Nils Stein
Leibniz Institute of Plant Genetics
and Crop Plant Research (IPK)
Seeland, Sachsen-Anhalt, Germany

Department of Crop Sciences
Center of integrated Breeding
Research (CiBreed)
Georg-August-University Göttingen
Göttingen, Lower Saxony, Germany

ISSN 2199-4781 ISSN 2199-479X (electronic)
Compendium of Plant Genomes
ISBN 978-3-030-83382-4 ISBN 978-3-030-83383-1 (eBook)
https://doi.org/10.1007/978-3-030-83383-1

© The Editor(s) (if applicable) and The Author(s), under exclusive license to Springer Nature Switzerland AG 2021
This work is subject to copyright. All rights are solely and exclusively licensed by the Publisher, whether the whole or part of the material is concerned, specifically the rights of translation, reprinting, reuse of illustrations, recitation, broadcasting, reproduction on microfilms or in any other physical way, and transmission or information storage and retrieval, electronic adaptation, computer software, or by similar or dissimilar methodology now known or hereafter developed.
The use of general descriptive names, registered names, trademarks, service marks, etc. in this publication does not imply, even in the absence of a specific statement, that such names are exempt from the relevant protective laws and regulations and therefore free for general use.
The publisher, the authors and the editors are safe to assume that the advice and information in this book are believed to be true and accurate at the date of publication. Neither the publisher nor the authors or the editors give a warranty, expressed or implied, with respect to the material contained herein or for any errors or omissions that may have been made. The publisher remains neutral with regard to jurisdictional claims in published maps and institutional affiliations.

This Springer imprint is published by the registered company Springer Nature Switzerland AG
The registered company address is: Gewerbestrasse 11, 6330 Cham, Switzerland

Rye (*Secale cereale* L.)—growing on a field at IPK Gatersleben in 2018 (© Nils Stein)

This book series is dedicated to my wife Phullara and our children Sourav and Devleena

Chittaranjan Kole

Preface to the Series

Genome sequencing has emerged as the leading discipline in the plant sciences coinciding with the start of the new century. For much of the twentieth century, plant geneticists were only successful in delineating putative chromosomal location, function, and changes in genes indirectly through the use of a number of "markers" physically linked to them. These included visible or morphological, cytological, protein, and molecular or DNA markers. Among them, the first DNA marker, the RFLPs, introduced a revolutionary change in plant genetics and breeding in the mid-1980s, mainly because of their infinite number and thus potential to cover maximum chromosomal regions, phenotypic neutrality, absence of epistasis, and codominant nature. An array of other hybridization-based markers, PCR-based markers, and markers based on both facilitated construction of genetic linkage maps, mapping of genes controlling simply inherited traits, and even gene clusters (QTLs) controlling polygenic traits in a large number of model and crop plants. During this period, a number of new mapping populations beyond F2 were utilized and a number of computer programs were developed for map construction, mapping of genes, and for mapping of polygenic clusters or QTLs. Molecular markers were also used in the studies of evolution and phylogenetic relationship, genetic diversity, DNA fingerprinting, and map-based cloning. Markers tightly linked to the genes were used in crop improvement employing the so-called marker-assisted selection. These strategies of molecular genetic mapping and molecular breeding made a spectacular impact during the last one and a half decades of the twentieth century. But still they remained "indirect" approaches for elucidation and utilization of plant genomes since much of the chromosomes remained unknown and the complete chemical depiction of them was yet to be unraveled.

Physical mapping of genomes was the obvious consequence that facilitated the development of the "genomic resources" including BAC and YAC libraries to develop physical maps in some plant genomes. Subsequently, integrated genetic–physical maps were also developed in many plants. This led to the concept of structural genomics. Later on, emphasis was laid on EST and transcriptome analysis to decipher the function of the active gene sequences leading to another concept defined as functional genomics. The advent of techniques of bacteriophage gene and DNA sequencing in the 1970s was extended to facilitate sequencing of these genomic resources in the last decade of the twentieth century.

As expected, sequencing of chromosomal regions would have led to too much data to store, characterize, and utilize with the then available computer software could handle. But the development of information technology made the life of biologists easier by leading to a swift and sweet marriage of biology and informatics, and a new subject was born—bioinformatics.

Thus, the evolution of the concepts, strategies, and tools of sequencing and bioinformatics reinforced the subject of genomics—structural and functional. Today, genome sequencing has traveled much beyond biology and involves biophysics, biochemistry, and bioinformatics!

Thanks to the efforts of both public and private agencies, genome sequencing strategies are evolving very fast, leading to cheaper, quicker, and automated techniques right from clone-by-clone and whole-genome shotgun approaches to a succession of second-generation sequencing methods. The development of software of different generations facilitated this genome sequencing. At the same time, newer concepts and strategies were emerging to handle sequencing of the complex genomes, particularly the polyploids.

It became a reality to chemically—and so directly—define plant genomes, popularly called whole-genome sequencing or simply genome sequencing.

The history of plant genome sequencing will always cite the sequencing of the genome of the model plant *Arabidopsis thaliana* in 2000 that was followed by sequencing the genome of the crop and model plant rice in 2002. Since then, the number of sequenced genomes of higher plants has been increasing exponentially, mainly due to the development of cheaper and quicker genomic techniques and, most importantly, the development of collaborative platforms such as national and international consortia involving partners from public and/or private agencies.

As I write this preface for the first volume of the new series "Compendium of Plant Genomes," a net search tells me that complete or nearly complete whole-genome sequencing of 45 crop plants, eight crops and model plants, eight model plants, 15 crop progenitors and relatives, and three basal plants is accomplished, the majority of which are in the public domain. This means that we nowadays know many of our model and crop plants chemically, i.e., directly, and we may depict them and utilize them precisely better than ever. Genome sequencing has covered all groups of crop plants. Hence, information on the precise depiction of plant genomes and the scope of their utilization are growing rapidly every day. However, the information is scattered in research articles and review papers in journals and dedicated Web pages of the consortia and databases. There is no compilation of plant genomes and the opportunity of using the information in sequence-assisted breeding or further genomic studies. This is the underlying rationale for starting this book series, with each volume dedicated to a particular plant.

Plant genome science has emerged as an important subject in academia, and the present compendium of plant genomes will be highly useful to both students and teaching faculties. Most importantly, research scientists involved in genomics research will have access to systematic deliberations on the plant genomes of their interest. Elucidation of plant genomes is of interest not only for the geneticists and breeders, but also for practitioners of an array of plant science disciplines, such as taxonomy, evolution, cytology, physiology, pathology, entomology, nematology, crop production, biochemistry, and obviously bioinformatics. It must be mentioned that information regarding each plant genome is ever-growing. The contents of the volumes of this compendium are, therefore, focusing on the basic aspects of the genomes and their utility. They include information on the academic and/or economic importance of the plants, description of their genomes from a molecular genetic and cytogenetic point of view, and the genomic resources developed. Detailed deliberations focus on the background history of the national and international genome initiatives, public and private partners involved, strategies and genomic resources and tools utilized, enumeration on the sequences and their assembly, repetitive sequences, gene annotation, and genome duplication. In addition, synteny with other sequences, comparison of gene families, and, most importantly, the potential of the genome sequence information for gene pool characterization through genotyping by sequencing (GBS) and genetic improvement of crop plants have been described. As expected, there is a lot of variation of these topics in the volumes based on the information available on the crop, model, or reference plants.

I must confess that as the series editor, it has been a daunting task for me to work on such a huge and broad knowledge base that spans so many diverse plant species. However, pioneering scientists with lifetime experience and expertise on the particular crops did excellent jobs editing the respective volumes. I myself have been a small science worker on plant genomes since the mid-1980s and that provided me the opportunity to personally know several stalwarts of plant genomics from all over the globe. Most, if not all, of the volume editors are my longtime friends and colleagues. It has been highly comfortable and enriching for me to work with them on this book series. To be honest, while working on this series, I have been and will remain a student first, a science worker second, and a series editor last. And I must express my gratitude to the volume editors and the chapter authors for providing me the opportunity to work with them on this compendium.

I also wish to mention here my thanks and gratitude to the Springer staff, particularly Dr. Christina Eckey and Dr. Jutta Lindenborn, for the earlier set of volumes and presently Ing. Zuzana Bernhart for all their timely help and support.

I always had to set aside additional hours to edit books beside my professional and personal commitments—hours I could and should have given to my wife, Phullara, and our kids, Sourav and Devleena. I must mention that they not only allowed me the freedom to take away those hours from them but also offered their support in the editing job itself. I am really not sure whether my dedication of this compendium to them will suffice to do justice to their sacrifices for the interest of science and the science community.

New Delhi, India Chittaranjan Kole

Preface

Cultivated rye (*Secale cereale* L.) had undeniably humble origins. Rye's wild ancestors, probably weedy highland grasses from Anatolia, were domesticated partly by mistake—a consequence of unintentional selection exerted by wheat and barley farmers, whose crops rye's ancestors invaded. Rye's domestication therefore proceeded parallel to—if somewhat later than—the North-Western/North-Eastern cultivation range expansion of wheat and barley during the origins of agriculture. Rye's ability to flourish in unforgiving environments and poor-quality soils contributed to its uptake by farmers, and its unique taste and baking qualities have insured that rye breads and fermented drinks are popular cultural staples in a broad range of cuisines across the globe. Alongside human culinary uses, rye is used as a cover crop and a source of animal feed. As of 2019, 10 million tons of rye was annually cultivated, alongside a further 10 million tons of the high-yielding and hardy wheat-rye hybrid *Triticale*. Northern Europe dominates production of both.

Rye has a large and highly repetitive diploid genome ($\sim$8 Gbp in length), and most varieties are obligate outcrossers. These factors represented significant hurdles to the development of a high-quality genome sequence, and early efforts to develop sequence-based genetic resources for cereals focused on the most widespread commercial crops, wheat and barley. While several important resources including a draft genome sequence were developed during the 2000s and 2010s, it was only in 2021 that a duo of high-quality full genome sequences for rye was tandemly published.

These achievements represent the perfect prompt to produce an up-to-date summary of the state of rye genome research. This book aims to fill that role, alongside providing much relevant background during the pre-genome-sequencing era. Those interested in learning about the fascinating progression of rye genetic research toward the genomics era would be well advised to consult Rolf Schlegel's *Rye: Genetics, Breeding, and Cultivation* (CRC Press, 2013). Topics covered here include sequencing and assembly approaches, gene prediction, chromosomal genomics, sequence diversity and structural variation, taxonomy, domestication, the repetitive genome, cytogenetics, biotic and abiotic stress responses and their genetic underpinnings, rye's self-incompatibility systems that enable efficient hybrid breeding, and the enigmatic supernumerary "B" chromosomes—researchers of which have used rye as an important model organism since 1924.

We are greatly indebted to the many authors and collaborators who contributed their expertise and efforts to make this book possible, and we extend our gratitude also to the expert reviewers who were indispensable in ensuring the quality and accuracy of each chapter. It has been a true privilege to work with the rye research community on this book, which we hope serves as valuable resource for both experts and initiates.

Seeland, Germany Nils Stein
M. Timothy Rabanus-Wallace

Acknowledgements We are indebted to our peer reviewers, for their interest, time, and effort they invested to help improving the individual chapter contributions to this volume: Eva Bauer, Micha Bayer, Perry Gustafsson, Andreas Houben, Peter Langridge, Adam Lukaszewski, Klaus Mayer, Gary Muehlbauer, Klaus Pillen, Joanne Russell, Karl Schmid, Ingo Schubert, and Alan Schulman. Thank you very much!

Contents

Contributors

A. M. Banaei-Moghaddam Laboratory of Genomics and Epigenomics (LGE), Department of Biochemistry, Institute of Biochemistry and Biophysics (IBB), University of Tehran, Tehran, Iran

Maja Boczkowska The Plant Breeding and Acclimatization Institute - National Research Institute, Błonie, Poland

Hanna Bolibok-Brągoszewska Institute of Biology, Warsaw University of Life Sciences, Warsaw, Poland

Jaroslav Doležel Center of the Region Haná for Biotechnological and Agricultural Research, Institute of Experimental Botany of the Czech Academy of Sciences, Olomouc, Czech Republic

Magdalena Dzięgielewska Department of Plant Genetics, Breeding and Biotechnology, West Pomeranian University of Technology in Szczecin, Szczecin, Poland

Evgeny A. Elisafenko Institute of Molecular and Cellular Biology, SB RAS, Novosibirsk, Russia;
Institute of Cytology and Genetics, SB RAS, Novosibirsk, Russia

Takashi R. Endo Kyoto, Japan

Agnieszka Grądzielewska Institute of Plant Genetics, Breeding and Biotechnology, University of Life Sciences in Lublin, Lublin, Poland

Heidrun Gundlach Helmholtz Zentrum München - German Research Center for Environmental Health, Plant Genome and Systems Biology (PGSB), Neuherberg, Germany

Bernd Hackauf Institute for Breeding Research on Agricultural Crops, Julius Kühn-Institute, Groß-Lüsewitz, Germany

A. Houben Leibniz Institute of Plant Genetics and Crop Plant Research (IPK) Gatersleben, Seeland, Germany

Jens Keilwagen Institute for Biosafety in Plant Biotechnology, Julius Kühn-Institute, Quedlinburg, Germany

Takao Komatsuda Institute of Crop Science, National Agriculture and Food Research Organization (NARO), Tsukuba, Japan;
Crop Research Institute, Shandong Academy of Agricultural Sciences (SAAS), Jinan, Shandong, China

Viktor Korzun KWS SAAT SE & Co. KGaA, Einbeck, Germany

Guangwei Li College of Agronomy, State Key Laboratory of Wheat and Maize Crop Science, Center for Crop Genome Engineering, Henan Agricultural University, Zhengzhou, China

Thomas Lux Helmholtz Zentrum München - German Research Center for Environmental Health, Plant Genome and Systems Biology (PGSB), Neuherberg, Germany

W. Ma Science Center for Future Foods, Jiangnan University, Wuxi, Jiangsu, China

Martin Mascher Leibniz Institute of Plant Genetics and Crop Plant Research (IPK) Gatersleben, Seeland, Germany;
German Centre for Integrative Biodiversity Research (iDiv) Halle-Jena-Leipzig, Leipzig, Germany

Klaus F. X. Mayer Helmholtz Zentrum München - German Research Center for Environmental Health, Plant Genome and Systems Biology (PGSB), Neuherberg, Germany

Joanna Melonek ARC Centre of Excellence in Plant Energy Biology, School of Molecular Sciences, The University of Western Australia, Crawley, WA, Australia

Thomas Miedaner University of Hohenheim, State Plant Breeding Institute, Stuttgart, Germany

Elena Mikhailova Saint-Petersburg State University, Saint-Petersburg, Russia

Kinga Moskal The Plant Breeding and Acclimatization Institute - National Research Institute, Błonie, Poland

Beata Myśków Department of Plant Genetics, Breeding and Biotechnology, West Pomeranian University of Technology in Szczecin, Szczecin, Poland

Hakan Özkan Department of Field Crops, Faculty of Agriculture, Çukurova University, Adana, Turkey

Mira L. Ponomareva Tatar Research Institute of Agriculture Federal Research Center, "Kazan Scientific Center of Russian Academy of Sciences", Kazan, Russian Federation

M. Timothy Rabanus-Wallace Leibniz Institute of Plant Genetics and Crop Plant Research (IPK), Seeland, Germany

Monika Rakoczy-Trojanowska Institute of Biology, Warsaw University of Life Sciences, Warsaw, Poland

Mona Schreiber Leibniz Institute of Plant Genetics and Crop Plant Research (IPK) Gatersleben, Seeland, Germany;
Department of Biology, University of Marburg, Marburg, Germany

Mark E. Sorrells Plant Breeding and Genetics, Cornell University, Ithaca, USA

Manuel Spannagl Helmholtz Zentrum München - German Research Center for Environmental Health, Plant Genome and Systems Biology (PGSB), Neuherberg, Germany

Nils Stein Leibniz Institute of Plant Genetics and Crop Plant Research (IPK), Seeland, Germany;
Center of Integrated Breeding Research (CiBreed), Department of Crop Sciences, Georg-August-University, Göttingen, Germany

Stefan Stojałowski Department of Plant Genetics, Breeding and Biotechnology, West Pomeranian University of Technology in Szczecin, Szczecin, Poland

Alexander V. Vershinin Institute of Molecular and Cellular Biology, SB RAS, Novosibirsk, Russia;
Novosibirsk State University, Novosibirsk, Russia

Daowen Wang College of Agronomy, State Key Laboratory of Wheat and Maize Crop Science, Center for Crop Genome Engineering, Henan Agricultural University, Zhengzhou, China

Peer Wilde KWS LOCHOW GMBH, Bergen, Germany

Jianping Yang College of Agronomy, State Key Laboratory of Wheat and Maize Crop Science, Center for Crop Genome Engineering, Henan Agricultural University, Zhengzhou, China

Economic and Academic Importance of Rye

1

Viktor Korzun, Mira L. Ponomareva, and Mark E. Sorrells

Abstract

Rye has been playing an important agronomic, nutritional and social role throughout human civilization. In the last 50 years, rye grain yields have increased but not enough to offset the decrease in cropping area to maintain production. In this context, hybrid rye has great potential due to high yield performance and greater resilience to climate variability. The production area of hybrid rye has been increasing for several years and is expected to continue increasing. In the last decade, uses, such as biogas as well as greening, are potential new markets for rye biomass production. Although rye genomics has lagged behind other cereal crops, it has made significant contributions to understanding the evolution of the grass family through comparative genomics analyses. Rye genomics and breeding have made great strides in the past 50 years and led to exciting new areas of research, in particular, hybrid varieties of rye that out-yield conventional synthetic varieties by 20–30% for both biomass and grain. While rye may be considered a minor crop in terms of production, contributions to cereal genomics have been substantial.

1.1 Background

Rye (*Secale cereale* L.) has the remarkable capability to grow in a wide range of environments, and more specifically performs well in low input environments where other cereals fail. The most widely grown type of rye is winter rye, also called fall rye. Rye grain has considerable value for functional and healthy humans' food. Rye is climate resilient and able to survive cold temperatures, semi-arid and high-altitude zones, and marginal soils. Until now, most of the rye production uses population or synthetic rye varieties.

Nevertheless, rye was the first small grain cereal to be successfully bred to produce hybrid varieties with the first hybrids launched in 1984. The importance of rye has continued to increase due to high yield, resilient agronomic performance, and stable, good grain quality.

V. Korzun (✉)
KWS SAAT SE & Co. KGaA, Einbeck, Germany
e-mail: viktor.korzun@kws.com

M. L. Ponomareva
Tatar Research Institute of Agriculture Federal Research Center, "Kazan Scientific Center of Russian Academy of Sciences", Kazan, Russian Federation

M. E. Sorrells
Plant Breeding and Genetics, Cornell University, Ithaca 14853, USA

© The Author(s), under exclusive license to Springer Nature Switzerland AG 2021

M. T. Rabanus-Wallace and N. Stein (eds.), *The Rye Genome*, Compendium of Plant Genomes,
https://doi.org/10.1007/978-3-030-83383-1_1

1.2 World Rye Production

According to the FAOSTAT (statistical platform of the Food and Agriculture Organization of the United Nations) in 2018 around 11.27 million tonnes (Mt) of rye grain were harvested worldwide, of which 9.13 Mt were grown on the European continent (Table 1.1). Thus, the contribution of Europe to world rye production was 81.0% while other continents contributed much less with Asia—13.0%, Africa—0.9%, North America—4.0%, and Australia—0.9%.

Average world production of rye from 2009 to 2018 amounted to 14.2 million tonnes per year. Rye production in the last decade declined slightly at an average annual rate of 0.2% due to the increasing availability of high performing wheat varieties and low prices for rye grain. The maximum rate of decline was recorded in 2008 at 19%. The peak of global rye production was in 2009 at the level of 18.3 Mt. Although it is slowly declining, rye production has been stable relative to other cereals with moderate peaks and valleys.

Around the world, the area of cultivated land dedicated to growing rye (Bushuk 2001) has decreased substantially since the 1970s. In 1996, 17 Mha were harvested, but this dropped by 61% to 6.7 Mha by 2008. This negative trend has continued in the last decade (2009–2018). During this period, the harvested world area decreased by 2.2 Mha or 31%. In 2018, the cultivated area of rye in the world amounted to 4.12 Mha. The decrease in cultivated area has been largely offset by an increase in yield. This significant yield increase was achieved through improvement of agronomic practices, especially in the use of chemical fertilizers and crop rotation, decline in the use of less fertile land, and development of high-yielding varieties, especially hybrids.

Table 1.1 Rye production per continent and Top 10 countries and regions Data from FAOSTAT (2020)

Region/Country	Area (Million ha, Mha)	Yield level (Tonnes per ha)	
Production (Million tonnes, Mt)			
World	*4.12*	*2.74*	*11.27*
Europa	*3.32*	*2.75*	*9.13*
Belarus	0.25	2.00	0.50
Denmark	0.09	5.20	0.48
Germany	0.52	4.21	2.20
Poland	0.89	2.42	2.17
Russian Federation	0.96	2.00	1.92
Spain	0.14	2.85	0.39
Ukraine	0.15	2.65	0.39
Asia	*0.46*	*3.21*	*1.46*
China	0.27	3.90	1.04
Turkey	0.11	2.89	0.32
North America	*0.23*	*2.64*	*0.60*
United States of America	0.12	2.15	0.27
Canada	0.08	2.99	0.24
Australia	*0.04*	*0.70*	*0.03*
Africa	*0.05*	*1.88*	*0.10*
South America	*0.05*	*1.93*	*0.10*

Climate change is a global driver of farmers' interest in growing hybrid rye varieties. In Europe, the increasingly stringent regulations on the use of fertilizers and chemicals for plant protection are incentives for farmers to grow hybrids because of their tolerance to marginal soils and diseases.

Winter rye is of great importance in the world economy and food traditions of those seven countries, where the crop is grown on more than 90 thousand hectares (Belarus, Denmark, Germany, Poland, Russian Federation, Spain, and Ukraine) (Table 1.1). The Russian Federation leads in the area of cultivation of winter rye since this cereal crop has traditionally been grown in a country where the conditions for growing crops are tough. In recent years, China, Canada, and the United States have begun to cultivate increasing amounts of rye.

Retrospective analysis showed that on average for the period 1994–2018, the Russian Federation contributed most to the world production of rye followed by Poland and Germany. In recent years, production of rye has changed significantly. In total, the Russian Federation, Germany, and Poland produced between 6 and 8 Mt of rye that accounted for more than 70% of European rye and about 57% of the total world grain harvest of this crop. In 2018, Germany and Poland became the leaders in the production of rye. The top 10 countries include Belarus, Ukraine, China, Denmark, Canada, Turkey, and Spain.

Rye production in Germany shifted over the last 25 years from population to hybrid varieties. Despite a smaller cropping area of 0.52 Mha in Germany, winter rye grain production in Germany increased due to higher yield (4.2 t per hectare) resulting from the strong yield performance of hybrid rye varieties. At the same time, winter rye in Poland was harvested on 0.89 Mha at a yield of only 2.4 tonnes per hectare because of a lower level of agrotechnology use and limited use of hybrid rye varieties.

Traditionally, the cultivation of Russian rye grain was the most economically beneficial practice, especially since the agro-climatic conditions of Russia are optimal for rye. Rye is relatively undemanding crop, is resistant to severe winter conditions, and grows well even in less fertile soils (including sandy). Therefore, the expenditures for fertilizers and plant protection chemicals on rye cultivation are lower than for other cereals. In 2018, rye grain production dropped from 2.4 to 1.92 Mt. The large reduction can be explained by reduced demand for rye and severe damage to winter rye by snow mold disease caused by several types of fungal pathogens like *Microdochium nivale, M. majus, Typhula ishikariensis, T. incarnata, Myriosclerotinia borealis* and *Pythium iwayami, P. okanoganense* (Gorskov et al. 2020; Ponomareva et al. 2020) Significant producers of rye are also China (1.04 Mt), Denmark (0.48 Mt), Belarus (0.50 Mt), and Ukraine (0.39 Mt).

Rye yields over the past six decades have increased in European countries and worldwide (Fig. 1.1). This can be explained by the expansion of the hybrid rye growing area and increased hybrid rye breeding efforts. Especially in Europe (Germany and Denmark), this has led to increased yield (Table 1.1).

The average rye yield was about 2.8 t/ha between 2008 and 2018. The lowest rye yield was in 2010 at less than 2.4 t/ha due to extremely unfavorable weather conditions with severe frost in winter and drought in summer. The highest yield of rye in Europe and on a global scale was recorded in 2017 with more than 3.0 t/ha.

From 1961 to 1978, rye and wheat yield in the world were very similar with about 2 t/ha, but since 1978 wheat yields were usually higher than rye by about 0.6 t/ha, and in Germany even 2.0 t/ha. This difference can be explained by the much higher investment in wheat breeding and the fact that rye is usually cultivated on poor soils. Nevertheless, yield potential of winter rye was much higher. For example, in 2001 and 2014 yield in Germany was above 6.1 t/ha on a total harvested area of 0.84 and 0.63 Mha, respectively. Breeding progress in Germany, especially hybrid breeding, prevented an even greater negative trend of area reduction (Fig. 1.2) due to increasing the yield per hectare with modern hybrids now delivering grain yields more than 12 t/ha.

Since early 2000, hybrid rye breeding investments have been increased resulting in

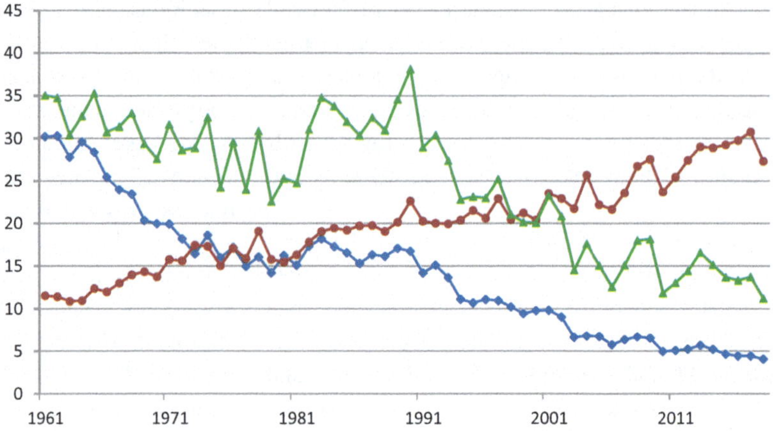

Fig. 1.1 Worldwide rye harvest area (blue, in Mha), production (green, in Mt), and yield (red, dt/ha) during 1961–2018. Data from FAOSTAT (2020)

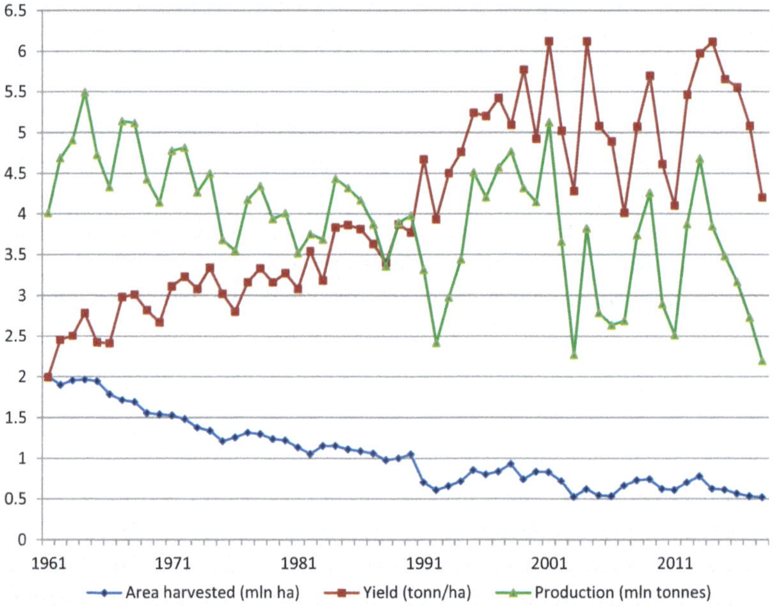

Fig. 1.2 Rye harvest area (blue, in mha), production (green, in million tonnes), and yield (red, in t/ha) in Germany during 1961–2018. Data from FAOSTAT (2020)

growing yield increases combined with strong agronomic performance and improved grain quality (*cf.* Wilde and Miedaner, Chap. 2 of this volume). Miedaner and Huebner (2011) reported that hybrids out-yielded population varieties by 20–25%.

In terms of global production, rye is a minor cereal (Table 1.2), since its production is less than 50% that of oats, ~8% that of barley, and 1.5% that of wheat.

1.3 Rye End Uses

Rye grain is used to bake bread and other products through the sour dough process that confers a unique taste with specific nutritional benefits and market opportunities. Rye-derived products benefit from reduced gluten compared to wheat. Rye is favored because its grain is rich in dietary fiber, carbohydrates, proteins, and several key

Table 1.2 Rye production compared to other cereals in the world in 2018. Data from FAOSTAT (2020)

Crop	Area (Mha)	Yield level (t per ha)	Production (Mt)
Maize	193.73	5.92	1147.62
Wheat	214.29	3.43	734.05
Rice	167.13	4.68	782.00
Barley	47.93	2.95	141.42
Sorghum	42.14	1.41	59.34
Oats	9.85	2.34	23.05
Triticale	3.81	3.36	12.80
Rye	4.12	2.74	11.27

minerals and nutrients. Trends in rye end use include an increase in rye being fed to animals and a decrease in human consumption. In recent years, rye has been well established as a feed component rich in energy for livestock, especially cattle and pigs, and for use in industry, distillery, and energy production.

According to Goncharenko (2014) potential consumption of winter rye is high. From 100 kg of winter rye grain, it is possible to make 160 kg of rye bread, 50 kg of pork meat, 230 l of milk, 36 l of ethanol, 60,000 l of biogas, or 450 kWh of energy.

The highest consumption on a country basis belongs to the Russian Federation. Russia consumed 2.55 Mt of rye in 2017, followed by Belarus (approximately 0.8 Mt), the USA (0.5 Mt), and Ukraine (0.38 Mt). Russian rye consumption is distributed as follows: 60% of grain is used in baking, about a third of the gross harvest is for animal feed, and 10% for other needs (Ponomareva and Ponomarev 2019).

According to the European Commission in 2020, the EU total supply of rye was 9.76 Mt of which 7.5 Mt were for domestic use. Most of domestic rye was used as food (2.96 Mt, 39%) and 2.70 Mt (36%) of domestic rye was used as animal feed (Fig. 1.3). For industrial processing, 0.60 Mt of rye (approximately 12%) were used for bioenergy (Cereals Supply and Demand 2020). The remaining part is represented by seed, stocks, and export.

Statistics show that in the European Union, rye produced for bread making has decreased or

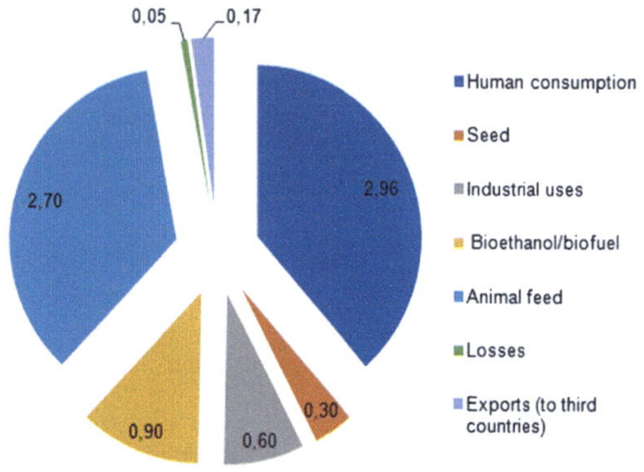

Fig. 1.3 Total supply of rye in EU (domestic use in Mt), 2020/2021 Projection (data from https://data.europa.eu/euodp/en/data/dataset/cereals-supply-and-demand)

stagnated, but other market segments such as feed are increasing. Since the 1990s, rye has been increasingly used to produce alcohol and plastics, as well as for the generation of renewable energy (Schlegel 2013).

Rye is mainly consumed in the countries of Northern and Eastern Europe, which exceed the European average of 5.4 kg/year per capita (2017; Table 1.3). Rye consumption is highest in Belarus consumption at an annual consumption per capita of 31.9 kg/year, although this value has decreased by 41% compared with 1995. In the EU, Poland ranks second in rye consumption at 25.7 kg/capita/year. Denmark ranks third in annual rye consumption per person for food production (23.5 kg/capita/year) with a strong increase in comparison with 1995. Rye has always been the national crop, most important

Table 1.3 Annual rye food consumption in the world and in selected countries (in kg/capita/year, data from FAOSTAT 2020)

Country/Year	1995	2005	2017
World	*1.4*	*0.9*	*0.6*
Europa	7.1	6.8	5.4
Belarus	78.0	33.4	31.9
Canada	0.5	0.6	0.6
China	0.5	0.1	0.2
Denmark	15.5	13.0	23.5
Estonia	21.5	18.9	20.8
Finland	15.8	15.4	18.4
Germany	12.0	9.9	9.3
Latvia	19.2	18.9	21.5
Lithuania	44.8	17.1	10.6
Norway	7.3	6.9	7.7
Poland	32.0	31.1	25.7
Russian Federation	11.9	8.9	5.3
Sweden	9.0	12.0	9.3
Turkey	2.4	3.4	2.4
Ukraine	12.8	8.2	8.1
USA	0.3	0.3	0.7

food product, and brand for Denmark. In Denmark, rye culture, eating and cultivation traditions have been passed down from generation to generation, with love and respect for rye and rye bread and increasing attention to health aspects of rye bread.

Among the Nordic and Baltic countries, Norway has the lowest consumption of rye at 7.7 kg/capita/year and the highest is Latvia and Estonia (21.5 and 20.8 kg/capita/year, respectively). The largest decline in 2017 compared to 1995 from the Baltic countries occurred in Lithuania. Rye consumption per capita has decreased over the last 20 years both in European countries and globally (Table 1.3). Worldwide, per capita consumption of rye decreased from 1.4 (1995) to 0.6 kg/capita/year (2017). In Europe consumption of rye has been quite stable and is about 9 times higher than worldwide consumption.

More than 3.2 Mt of rye produced in the EU is used for diversification. In Europe, rye is mostly used for feed, ethanol processing, and biogas. In Germany, 66% of rye was used for animal feeding and 15% for human nutrition, mainly for bread making (StatJ 2015).

Rye is an ideal crop for agricultural biogas production in regions with low fertility and sandy soils. Rye biomass is increasingly being used as a renewable raw material for biogas production (Geiger and Miedaner 2009). Bioethanol and biogas production may be a growing market for rye. Maximum methane yield per hectare is the main goal for the farmer. According to Huebner et al. (2011), mean rye methane yield was 4424 m^3/ha.

The European Biodiesel Board estimated that in 2017, Germany was the biggest producer of biofuel in Europe (more than 4 Mt) (http://www.ebb-eu.org/stats.php 2017). In 2007, 25% of Germany's rye harvest was used for bioenergy production. Rye as raw material for bioethanol can produce up to 5.4 t/ha fresh matter biomass yield, 420 l/t biomass, or 2268 l/ha of bioethanol yield, with 2.4 kg/l required biomass per liter of fuel (FNR, BDBe, harvest report of BMEL 2015).

In some countries, including Russia, rye is used as an early feed for green mass and as a cover crop. In the United States, rye is primarily grown as forage. The sowing of rye for green forage is expanding worldwide. This is the first culture to form a green conveyor, giving a high yield of biomass (up to 15 t/ha), suitable for all types of livestock and birds in spring and summer. Farms have additional opportunities to produce early fodder for silage, grass, flour, and hay (Sysuev et al. 2014). Forage is used in the form of green chop, pasture, haylage, or hay. Rye makes excellent forage, especially when combined with red clover and ryegrass. For best quality, rye should be cut between early heading and the milk stage of seed growth. Rye matures earlier and has higher crude protein levels than wheat and triticale. Although rye forage is less palatable than other forages, rye has greater cold tolerance, quicker growth at low temperatures, and more uniform seasonal forage production compared to wheat (*Triticum*), oats (*Avena*), barley (*Hordeum*), or triticale (*Triticosecale*) (Bruckner and Raymer 1990). Rye cultivars used for green fodder and hay in spring and summer grow fast and have thick foliage. They can regrow after being mown or grazed and their herbage is very nourishing (Schlegel 2013).

Rye is the most common and reliable cover crop in the Midwest and Northeast of the United States and in Canada, as it is one of the few cover crops that can be successfully established when planting in autumn after harvesting corn or soybeans. It is winter hardy throughout the region and accumulates significant amounts of biomass before spring planting of other crops (Snapp et al. 2005). As a cover crop, rye is multifunctional in no-till agroecosystems. This is the most effective way to reduce nitrogen and phosphorus losses. The deep root system of rye, especially rye hybrids, captures excess nitrogen and prevents it from entering groundwater or leaching. These nutrients are stored and then made available in the residual biomass for the next harvest. In general, fertilizer costs are decreased and labor is distributed evenly throughout the year.

1.4 Academic Importance of Rye

Our search in the Scopus international database, as per January 2021, revealed only 15,411 publications included rye as a keyword in contrast to over 150,000 publications on wheat. Most often the term rye is mentioned in publications with titles Agricultural and Biological Sciences, Biochemistry, Genetics, and Molecular Biology, Medicine, and Environmental Science and Chemistry (Fig. 1.4).

The aim of agricultural and biological research on rye has mainly focused on the study of economic traits such as grain yield, biomass, nutritional factors, early maturity, and grain quality (see Chap. 9 of this volume). The agronomic advantages and improved end use properties of rye grains achieved by research and development efforts make rye an attractive option for increasing global food production, particularly, for marginal land and stress-prone growing conditions.

As the only outcrossing Triticeae species, rye is of considerable interest both from an evolutionary and a genetic perspective, especially in comparative genomic studies with other grains species.

Genome sequences of Triticeae species are critical resources for understanding the biology and evolution of these species through comparative genomic approaches and for associating phenotypic traits with underlying genes. Rye was the last of the Triticeae species to be sequenced (see Chap. 7), thus closing a major gap in Triticeae genome research. Bauer et al. (2017), using comparative genome analyses, reported genomic diversity in ten rye inbred lines and one accession of the wild relative *Secale vavilovii*, revealing more than 90 million single nucleotide variants and insertions/deletions in the rye genome. These genomic resources have facilitated map-based cloning and functional characterization of genes underlying agronomic traits and advanced Triticeae genomics. Earlier, Hackauf et al. (2009) used marker sequences to construct a comparative map between rice and rye.

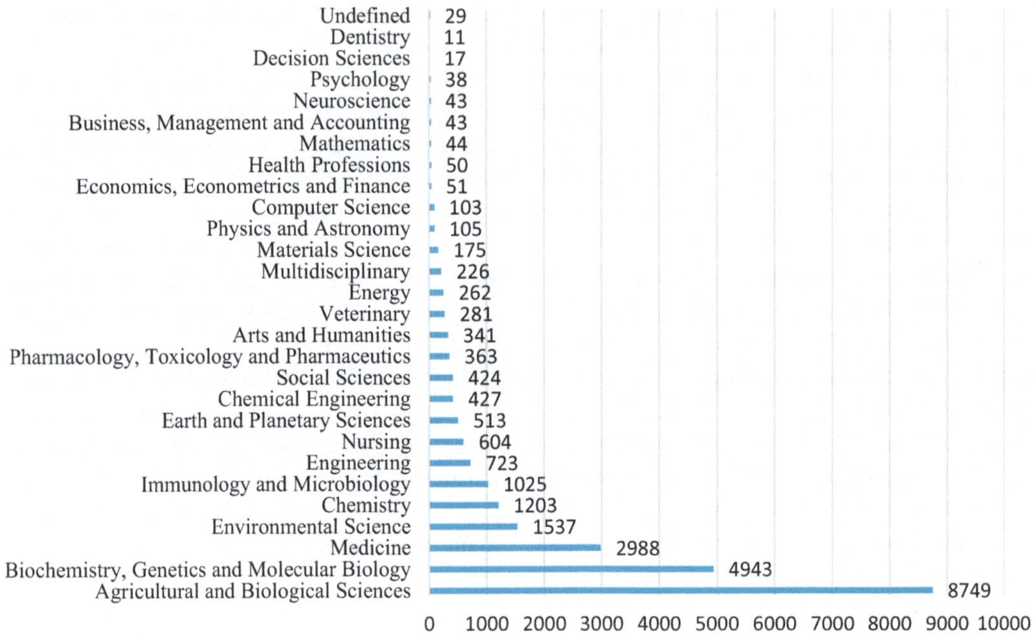

Fig. 1.4 Heading of publications using 'rye' as a string for database search. Scopus [Electronic resource]. Accessed: 09.01.2021

Genome wide sequence-based comparisons revealed many more chromosomal rearrangements between the grass genomes than previously reported based on RFLP analyses, thus exposing more complexity to the orthologous relationships between Triticeae genomes.

1.4.1 Rye as Genetic Resource for Wheat Improvement

Rye, a close relative of wheat (*Triticum aestivum* L.), which is the economically most important cereal (Feldman and Levy 2015), is providing a vast and largely untapped reservoir of genetic variation for traits such as stress tolerance, biomass, yield, and photosynthetic potential, not only for the commercial crop triticale (x *Triticosecale* Wittmack), but also for wheat (Lukaszewski 2015).

Rye chromosomes or their segments can be introgressed into wheat in the creation of substitution or translocation lines (Ren et al. 2017). There have been numerous reports on the introgression of rye for wheat improvement dating back to the 1800s (Franke 1991; Driscoll and Anderson 1967). Rye can be crossed with wheat and its agronomic traits can be transferred via classic pre-breeding from wheat/rye hybrids into the wheat genome. Therefore, this crop had a major impact on plant breeding strategies both through the production of the synthetic hybrid triticale as well as through the introgression of rye chromatin in wheat varieties, particularly by the short arm of chromosome 1R (1RS), as a source of genes for agronomic traits and disease resistance (Baum and Appels 1991).

Rye has proven to be a useful source of genes for improving important traits and diversity in wheat breeding (Saulescu et al. 2011; Johansson et al. 2020), especially for disease resistance genes. The short arm of rye chromosome 1R carries resistance genes for leaf rust (*Lr26*), stem rust (*Sr31*), stripe rust (*Yr9*), and powdery mildew (*Pm8*) (McIntosh et al. 2011; Crespo-Herrera et al. 2017), therefore it was incorporated into tetraploid and hexaploid wheats. The most prominent has been the widely used 1BL.1RS or 1AL.1RS translocations where the short arm of rye chromosome 1R from Petkus rye has replaced the short arm of wheat chromosome 1B or 1A. The 1BL.1RS wheat-rye translocation has contributed immensely to global wheat production as a source of resistance genes (*Sr31/Yr9/Lr26/Pm9*) to wheat fungal diseases (Schlegel 2020). This translocation has been used extensively in wheat breeding by CIMMYT and other breeding programs globally resulting in hundreds of wheat varieties with the rye chromosome arm (Crespo-Herrera et al. 2017) or segments of it (Lukaszewski 2000). It was reported to be present in about 1050 wheat cultivars (Schlegel and Korzun, 1997). This introgression was also found to increase root biomass leading to drought tolerance (Howell et al. 2019). Furthermore, new disease resistance genes from other rye chromosomes have been introgressed into wheat (Driscoll and Jensen 1965; Rabinovich 1998; An et al. 2019; http://www.rye-gene-map.de/rye-introgression/). Rye is one of the most winter hardy crops (Erath et al. 2017). While rye quality/utility does not compare to wheat, however, it can be reliably grown in harsher environments. As such, it has always been viewed with much envy by wheat breeders and many efforts have been made to utilize its gene pool for wheat improvement.

With a full reference genome sequence, inexpensive low-density high throughput sequencing (HTS) of a wheat panel proved sufficient to identify the positions of rye introgressions. Crop improvement in rye, as well as in wheat and triticale, will profit from investigations of rye gene families implicated in pathogen resistance, low temperature tolerance, and fertility control systems for hybrid breeding. Consortium scientists (Rabanus-Wallace et al. 2021) showed that rye introgressions in wheat breeding panels can be characterized at high throughput to predict the yield effects and trade-offs of rye chromatin (see Chap. 7 of this volume).

1.4.2 Rye Grain as a Source for Human Health Benefit

A healthy diet and lifestyle are currently in the spotlight, and the demand for healthy foods is growing. Rye was an essential part of the daily diet in northern and eastern parts of Europe because of its high energy value and beneficial agricultural properties (Liukkonen et al. 2007). Scientific evidence shows that rye contains a mixture of biologically active substances and possesses a wide range of protective properties in the prevention and treatment of metabolic syndrome, including cardiovascular diseases and type 2 diabetes as well as intestinal health and certain types of cancer (Jonsson et al. 2018). Their studies have shown that rye helps reduce development of childhood asthma, promotes weight loss, helps to prevent ulcers and stones in the gallbladder, and can improve the metabolic parameters of cells.

The main chemical constituents of the rye grain are the same as in other cereals: starch, dietary fiber (DF), protein, and mineral matter. Some of these key components include manganese, copper, magnesium, phosphorous, B-complex vitamins, and phenolic antioxidant compounds. Barley, oat, and rye grains are all rich sources of (1,3;1,4)-b-D-glucan, whereas wheat, rice, and maize have much lower concentrations. The benefits of DF in human nutrition, enhanced health, and lifestyle-related non-communicable disease prevention are well known and cereals play an important role. Rye flour mixed with wheat flour in various proportions up to 40%, has been shown to increase DF in whole meal rye bread without reducing the acceptability of certain types of bread or pastry products (Kołodziejczyk et al. 2020; Angioloni and Collar 2011; Ragaee and Abdel-Aal 2006). 73% of the diet art fiber in rye is insoluble and 27% soluble (Feng 2019). However, the health effects of rye can be associated not only with fiber content, but also with the so-called "rye fiber complex", which is a mixture of various biologically active compounds including arabinoxylans, oligosaccharides, lignans, phytates, and phenolic acids. Rye grains can be used in the production of foods enriched with bioactive components as these substances protect against many diet-related diseases (Meija and Krams 2019). Breeding rye varieties rich in DF can satisfy this new social need. In addition, rye flour and baking products contain more antioxidants than wheat products (Angioloni and Collar 2011; Michalska et al. 2008).

From a nutritional approach, rye proteins are recognized to be superior to those of wheat and other cereal grains because of their better composition of essential amino acids (Wrigley and Bushuk 2017). In rat feeding experiments, lysine was the first and threonine the second, most limiting amino acid. Rye protein has the highest content of lysine (up to 0.619 g 100 g^{-1}), valine, threonine, and methionine compared to wheat and barley (Sabirov et al. 2018). Continuous dialog between researchers, breeders, the food industry, and consumers are needed to enable the health properties of rye to benefit more people worldwide. Studies have demonstrated the benefits of including rye whole meal flour in food products (Jonsson et al. 2018).

1.4.3 Abiotic Stress Tolerance of Rye

Rye is known for its tolerance to abiotic stresses of various kinds. Rye generally tolerates marginal soils better than wheat or barley, especially acidic soils that release the phytotoxic Al^{3+} cation into the soil solution where it inhibits root growth, thus reducing the ability of plants to acquire water and nutrients (Ma et al. 2004). Among the Triticeae, rye has the highest tolerance to aluminum, and there has been considerable effort made to identify and clone the genes for aluminum tolerance (Alt) located on chromosomes 3RS, 4RL 6RS, and 7RS (Miftahudin et al. 2002, 2005; Matos et al. 2005).

The role of root architecture on water use efficiency (WUE) and agronomic performance in the field and greenhouse under well-watered and water stressed conditions was investigated by Ehdaie et al. (2008) using the well known translocations of the short arm of chromosome 1R from rye in bread wheat (1RS.1BL and 1RS.1AL). The 1RS translocations in 'Pavon 76' wheat from CIMMYT delayed maturity and increased root biomass resulting in increased grain yield and grain weight. The 1RS translocations, were more tolerant to field environmental stresses than Pavon 76, indicating one of the benefits for these translocations in wheat improvement. More recently, the increased root biomass and drought tolerance was isolated to a short segment of the rye 1RS:1BL translocation (Howell et al. 2019).

1.5 Conclusion

Rye is the minor cereal crop in the world based on growing area and production with major usage in human consumption and animal feed as well as in industrial processing and bioenergy production. Genomic tools in rye were developed only recently because of the large genome size, the low-international recognition of this crop, and the challenges in genetically analyzing a cross-pollinator with a genetic self-incompatibility mechanism.

Nevertheless, the recent advance of genomic resources in rye has now allowed researchers to more rapidly and precisely (1) uncover the genetic architecture of qualitative and quantitative traits by quantitative trait loci (QTL) mapping, (2) achieve balanced introgression of small genome segments from diverse genetic resources, and (3) introduce genome-based selection in the breeding process (Miedaner et. al. 2019) and enhance breeding progress towards the rapid and successful development of rye varieties with adaptation to targeted growing regions, stable and high yield, and human health benefit.

In contrast to declining rye growing area covered by conventional synthetic varieties, the growing area planted to hybrid varieties has increased substantially in recent decades (*cf.* Wilde and Miedaner, Chap. 2 of this volume). Rye hybrids have remarkable agronomic potential and new molecular and genomic tools and analyses will likely continue to increase in popularity in the future.

References

An D, Ma P, Zheng Q, Fu S, Li L, Han F, Han G, Wang J, Xu Y, Jin Y, Luo Q, Zhang X (2019) Development and molecular cytogenetic identification of a new wheat-rye 4R chromosome disomic addition line with resistances to powdery mildew, stripe rust and sharp eyespot. Theor Appl Genet 132:257–272

Angioloni A, Collar C (2011) Nutritional and functional added value of oat, Kamut®, spelt, rye and buckwheat versus common wheat in breadmaking. J Sci Food Agric 91:1283–1292

Bauer E, Schmutzer T, Barilar I, Mascher M, Gundlach H, Martis MM, Twardziok SO, Hackauf B, Gordillo A, Wilde P, Schmidt M, Korzun V, Mayer KFX, Schmid K, Schön CC, Scholz U (2017) Towards a whole-genome sequence for rye (*Secale cereale* L.). Plant J 89:853–869

Baum M, Appels R (1991) The cytogenetic and molecular architecture of chromosome 1R—one of the most widely utilized sources of alien chromatin in wheat varieties. Chromosoma 101:1–10

Bruckner PL, Raymer PL (1990) Factors influencing species and cultivar choice of small grains for winter forage. J Prod Agric 3:349–355

Bushuk W (2001) Rye production and uses worldwide. Cereal Foods World 46:70–73

Cereals Supply & Demand (2020) EU open data portal. https://data.europa.eu/euodp/en/data/dataset/cereals-supply-and-demand

Crespo-Herrera LA, Garkava-Gustavsson L, Åhman I (2017) A systematic review of rye (*Secale cereale* L.) as a source of resistance to pathogens and pests in wheat (*Triticum aestivum* L.). Hereditas 154:1–9

Driscoll CJ, Anderson LM (1967) Cytogenetic studies of transec—a wheat-rye translocation line. Can J Genet Cytol 9:375–380

Driscoll CJ, Jensen NF (1965) Release of a wheat-rye translocation stock involving leaf rust and powdery mildew resistances. Crop Sci 5(3):279–280

Ehdaie B, Alloush GA, Waines JG (2008) Genotypic variation in linear rate of grain growth and contribution of stem reserves to grain yield in wheat. Field Crops Res 106:34–43

Erath W, Bauer E, Fowler DB, Gordillo A, Korzun V, Ponomareva M, Schmidt M, Schmiedchen B, Wilde P, Schön CC (2017) Exploring new alleles for frost tolerance in winter rye. Theor Appl Genet 130:2151–2164

European Biodiesel Board http://www.ebb-eu.org/stats.php

FAOSTAT (2020) http://www.fao.org/faostat/en/#data/QC

Feldman M, Levy AA (2015) Original and evolution of wheat and related Triticeae species. In: Molnár-Láng M, Ceoloni C, Doležel J (eds) Alien introgression in wheat. Springer International Publishing, Switzerland, pp 21–76

Feng G (2019) Rye. Bioactive factors and processing technology for cereal foods. Springer, Singapore, pp 151–169

Franke R (1991) Crosses involving W. Rimpau's Triticale. Plant Breeding 107:303–308

Geiger HH, Miedaner T (2009) Rye breeding. In: Carena MJ (ed) Cereals (Handbook of Plant Breeding), 1st ed. Springer, New York, NY, USA, pp 157–181

Goncharenko AA (2014) Pressing questions of the winter rye breeding. Moscow FGBNU Rosinformagrotekh, pp 1–372 (in Russian)

Gorshkov V, Osipova E, Ponomareva M, Ponomarev S, Gogoleva N, Petrova O, Gogoleva O, Mescherov A, Balkin A, Vetchinkina E, Gogolev Y, Korzun V (2020) Rye snow mold-associated Microdochium nivale strains inhabiting a common area: variability in genetics, morphotype, extracellular enzymatic activities, and virulence. J Fungi 6(4):335

Hackauf B, Rudd S, Van Der Voort JR, Miedaner T, Wehling P (2009) Comparative mapping of DNA sequences in rye (*Secale cereale* L.) in relation to the rice genome. Theor Appl Genet 118:371–384

Howell T, Moriconi JI, Zhao X, Hegarty J, Fahima T, Santa-Maria GE, Dubcovsky J (2019) A wheat/rye polymorphism affects seminal root length and yield across different irrigation regimes. J Exp Bot 70:4027–4037

Huebner M, Oechsner H, Koch S, Seggl A, Hrenn H, Schmiedchen B, Wilde P, Miedaner T (2011) Impact of genotype, harvest time and chemical composition on the methane yield of winter rye for biogas production. Biomass Bioenerg 235:4316–4323

Johansson E, Henriksson T, Prieto-Linde ML, Andersson S, Ashraf R, Rahmatov M (2020) Diverse wheat-alien introgression lines as a basis for durable resistance and quality characteristics in bread wheat. Front Plant Sci 11:1067

Jonsson K, Andersson R, Knudsen KEB, Hallmans G, Hanhineva K, Katina K, Kolehmainen M, Kyrø C, Langton M, Nordlund E, Lærke HN, Olsen A, Poutanen K, Tjønneland A, Landberg R (2018) Rye and health-where do we stand and where do we go? Trends Food Sci Technol 79:78–87

Kołodziejczyk P, Michniewicz J, Buchowski MS, Paschke H (2020) Effects of fibre-rich rye milling fraction on the functional properties and nutritional quality of wholemeal rye bread. J Food Sci Technol 57:222–232

Liukkonen KH, Heinio RL, Salmenkallio-Marttila M, Autio K, Katina K, Poutanen K (2007) Rye. Bakery products: science and technology. Blackwell Publishing, Ames, Iowa, USA, pp 109–122

Lukaszewski AJ (2000) Manipulation of the 1RS.1BL translocation in wheat by induced homoeologous recombination. Crop Sci 40:216–225

Lukaszewski AJ (2015) Introgressions between wheat and rye. In: Molnár-Láng M, Ceoloni C, Doležel J (eds) Alien Introgression in Wheat, Springer, Cham, pp 163–189

Ma JF, Shen RF, Nagao S, Tanimoto E (2004) Aluminum targets elongating cells by reducing cell wall extensibility in wheat roots. Plant Cell Physiol 45:583–589

Matos M, Camacho MV, Perez-Flores V, Pernaute B, Pinto-Carnide O, Benito C (2005) A new aluminum tolerance gene located on rye chromosome arm 7RS. Theor Appl Genet 111:360–369

McIntosh RA, Zhang P, Cowger C, Parks R, Lagudah ES, Hoxha S (2011) Rye-derived powdery mildew resistance gene Pm8 in wheat is suppressed by the Pm3 locus. Theor Appl Genet 123:359–367

Meija L, Krams I (2019) Rye. In: Johnson J, Wallace TC (eds) Whole grains and their bioactives: composition and health, John Wiley & Sons Ltd, pp 169–208

Michalska A, Amigo-Benavent M, Zielinski H, del Castillo MD (2008) Effect of bread making on formation of Maillard reaction products contributing to the overall antioxidant activity of rye bread. J Cereal Sci 48:123–132

Miedaner T, Huebner M (2011) Quality demands for different uses of hybrid rye. 61. Tagung Der Vereinigung Der Pflanzenzuechter Und Saatgutkaufleute Oesterreichs 2010:45–49

Miedaner T, Korzun V, Bauer E (2019) Genomics-based hybrid rye breeding. In: Miedaner T, Korzun V (eds) Applications of genetic and genomic research in cereals, Woodhead Publishing, pp 329–348

Miftahudin R, Scoles GJ, Gustafson JP (2002) AFLP markers tightly linked to the aluminum -tolerance gene Alt3 in rye (*Secale cereale* L.). Theor Appl Genet 104:626–631

Miftahudin T, Chikmawati T, Ross K, Scoles GJ, Gustafson JP (2005) Targeting the aluminum tolerance gene Alt3 region in rye, using rice/rye microcolinearity. Theor Appl Genet 110:906–913

Ponomareva ML, Ponomarev SN (2019) Scientific bases of winter rye breeding. Kazan: FEN Publishing House. pp 1–352 (in Russian)

Ponomareva ML, Gorskov VY, Ponomarev SN, Korzun V, Miedaner T (2021) Snow mold of winter cereals: a complex disease and a challenge for resistance breeding. Theor Appl Genet 134:419–433

Rabanus-Wallace MT, Hackauf B, Mascher M, Lux T, Wicker T, Gundlach H, Báez M, Houben A, Mayer KFX, Guo L, Poland J, Pozniak CJ, Walkowiak S, Melonek J, Praz C, Schreiber M, Budak H, Heuberger M, Steuernagel B, Wulff B, Börner A, Byrns B, Čížková J, Fowler DB, Fritz A, Himmelbach A, Kaithakottil G, Keilwagen J, Keller B, Konkin D, Larsen J, Li Q, Myśków B, Padmarasu S, Rawat N, Sesiz U, Sezgi B, Sharpe A, Šimková H, Small I, Swarbreck D, Toegelová H, Tsvetkova N, Voylokov AV, Vrána J, Bauer E, Bolibok-Bragoszewska H, Doležel J, Hall A, Jia J, Korzun V, Laroche A, Ma X-F, Ordon F, Özkan H, Rakoczy-Trojanowska M, Scholz U, Schulman AH, Siekmann D, Stojałowski S, Tiwari V, Spannagl M, Stein N (2021) Chromosome-scale genome assembly provides insights into rye biology, evolution, and agronomic potential. Nature Genet 53:564–573

Rabinovich SV (1998) Importance of wheat-rye translocations for breeding modern cultivars of *Triticum aestivum* L. Euphytica 100:323–340

Ragaee S, Abdel-Aal E-SM (2006) Pasting properties of starch and protein in selected cereals and quality of their food products. Food Chem 95:9–18

Ren T, Tang Z, Fu S, Yan B, Tan F, Ren Z, Li Z (2017) Molecular cytogenetic characterization of novel wheat-rye T1RS. 1BL translocation lines with high resistance to diseases and great agronomic traits. Front Plant Sci 8: 799

Sabirov AA, Barakova NV, Samodelkin EA (2018) Effect of impact-activating-disintegration treatment on grain protein fraction of autumn rye. Agron Res 16 (2):1466–1474

Saulescu NN, Ittu G, Ciuca M, Ittu M, Serban G, Mustatea P (2011) Transferring useful rye genes to wheat, using triticale as a bridge. Czech J Genet Plant Breed 47:56–62

Schlegel RHJ (2013) Rye: genetics, breeding, and cultivation. FL, CRC Press, Boca Raton

Schlegel R (2020) Current list of wheats with rye and alien introgression 1R.1B - 1R/1B - 1R-1B - 1RS.1BL - 1R(1B) - alien. http://www.rye-gene-map.de/rye-introgression/index.html (Accessed 15 Dec 2020)

Schlegel R, Korzun V (1997) About the origin of 1RS.1BL wheat-rye chromosome translocations from Germany. Plant Breed 116:537–540

Snapp SS, Swinton M, Labarta R, Mutch D, Black JR, Leep R, Nyiraneza J, O'Neil K (2005) Evaluating cover crops for benefits, costs and performance within cropping system niches. Agron J 97:322–332

StatJ (2015) Statistisches Jahrbuch über Ernährung, Landwirtschaft und Forsten der Bundesrepublik Deutschland (2014). Landwirtschaftsverlag GmbH, Münster-Hiltrup

Sysuev VA, Kedrova LI, Utkina EI (2014) Priority directions of research in decision of a problem of multipurpose use of winter rye. Agrarnaya nauka Evro-Severo-Vostoka 6(43):4–8 (in Russian)

Wrigley C, Bushuk W (2017). Rye: grain-quality characteristics and management of quality requirements. In: Cereal Grains, Woodhead Publishing, pp 153–178

2 Hybrid Rye Breeding

Peer Wilde and Thomas Miedaner

Abstract

With the advent of hybrid breeding in the last decades of the twentieth century, the image of rye changed from being perceived as a 'forgotten crop' to a pioneer among the cereals. Meanwhile, the genetic basis of the hybrid system with relevant components such as self-fertility, cytoplasmatic male sterility and fertility restoration, heterotic groups and inbreeding tolerance are understood and a sustained reduction for the costs of the hybrid system has advanced. New enabling technologies, including DNA markers, high-throughput phenotyping and gene discovery became rapidly absorbed into the breeding process. Enhancing genetic diversity is a key success factor for long-term genetic gain. Broadening established heterotic patterns and introgressing new QTLs for simple and complex inherited traits are indispensable. With the advent of genomic selection-based breeding schemes, recurrent selection and commercial inbred line development are changing dramatically. The impact on the architecture of a breeding program is illustrated by several practice-orientated examples. Essential preconditions for the inscription of a variety are highlighted with examples of official testing systems and traits relevant for market acceptance. Progress from breeding has been substantial as revealed by the German official tests and can be regarded as a key driver to maintain competitiveness of the crop in Europe but also in other regions of the world. The economic effects of breeding research have been highly profitable from a societal and environmental protection perspective.

2.1 Introduction

Rye (*Secale cereale* L.) was cultivated worldwide on 4.1 million hectares in 2018 (FAOSTAT 2020) and is grown mainly in Northern and Eastern Europe, where about 70% of the world harvest is produced. The top-producing country is Germany, followed by Poland, the Russian Federation, China and Finno-Scandinavian countries. Here, rye is traditionally used for bread making, but can also be utilized as grain feed for animals, for distilling spirits, or for bioenergy production as a substrate for biomethane or ethanol production. Outside Europe, rye is often used as pasture, hay, or cover crop (Oelke et al. 1990). Rye is outstanding for its early, vigorous growth, high tolerance to abiotic

P. Wilde
KWS LOCHOW GMBH,
Ferdinand-von-Lochow-Straße 5, 29303 Bergen, Germany

T. Miedaner (✉)
University of Hohenheim, State Plant Breeding Institute, 70599 Stuttgart, Germany
e-mail: miedaner@uni-hohenheim.de

© The Author(s), under exclusive license to Springer Nature Switzerland AG 2021

M. T. Rabanus-Wallace and N. Stein (eds.), *The Rye Genome*, Compendium of Plant Genomes,
https://doi.org/10.1007/978-3-030-83383-1_2

and biotic stress factors and adaptation to acid or sandy soils with low water and nutrient availability making it a competitive alternative in regions where growing other crops would not be profitable (Geiger and Miedaner 2009).

The origin of rye is in Southwest Asia, particularly in Turkey, Lebanon, Syria, Iran, Iraq, and Afghanistan (Sencer and Hawkes 1980). Three species are generally recognized in the genus *Secale*: *Secale silvestre* (annual, autogamous), *S. strictum* (perennial, allogamous), and *S. cereale* (annual, allogamous). Within the latter species, cultivated, weedy and wild forms are existing (Fredericksen and Petersen 1998). Cultivated rye is a diploid ($2n = 2x = 14$) species and the only cross-pollinating small-grain cereal of the temperate zone.

2.2 Types of Varieties in Rye

In rye, populations and hybrids are well established as the main types of varieties. Recognizing their characteristics is relevant for diverse applications like breeding methodology, selection purposes, maintenance, seed multiplication, or plant variety protection.

The term 'population' comprises open-pollinated and synthetic varieties as sub-types. In both cases, only self-incompatible germplasm should be used to avoid inbreeding depression from any self-pollination and to enhance random mating. In a population based on many genotypes different from each other, random mating promotes high heterozygosity, which is beneficial for the per se performance of the population. At the same time, a genetic equilibrium (Hardy–Weinberg equilibrium) is maintained, which keeps the phenotypic appearance of the population stable over subsequent generations of seed multiplication. Random-mating or panmictic populations can be regrown by farm-saved seed without performance reduction caused by genetic factors.

To create a new variety, most population rye breeders use half or full sib families and select those showing high performance (Geiger 1982). Because per se performance of a family is correlated with its general combining ability (GCA), an improvement in population performance will be the result. In the case of open-pollinated varieties, there is no methodological difference between breeding for population improvement and for maintenance. In the case of synthetic varieties, families identified as well performing are kept separate and serve as parental components. By targeted intercrossing or random mating, they are regularly used to reconstruct and maintain the final population.

In the case of hybrid varieties, self-fertile inbred lines derived from two genetically distinct populations expressing a heterotic pattern ('heterotic groups') are used as parental components. Lines are selected for per se performance and for GCA effects to the respective opposite pool. Seed parent lines must be carriers of non-restorer genes allowing development of a male fertile and male sterile near-isogenic version. In contrast, pollen parent lines carry restorer genes allowing reconstitution of male fertility in the final hybrid. In practical hybrid breeding, a male sterile single cross of two seed parent lines is crossed with a synthetic built up from 2 to 4 pollen parent lines. Thus, the genetic basis of a typical rye hybrid is comparable to a four-way hybrid or a top cross hybrid in maize. Producing farm-saved seed from hybrids is not advisable due to their narrow genetic basis. Inevitably, resultant partially inbred plants will suffer from inbreeding depression and a correspondingly reduced yield.

2.3 Genetic Basis for Hybrid Breeding

2.3.1 Self-fertility and Self-incompatibility

Rye has a highly effective gametophytic self-incompatibility (SI) system that prevents self-fertilization and secures a high level of heterozygosity in open-pollinated populations (see also Chap. 10). SI is governed by multiple alleles at the two loci *S* and *Z* (Lundqvist 1956) that were first localized by isozymes on chromosomes 1R and 2R, respectively (Wricke and Wehling 1985; Gertz and

Wricke 1989). Both loci are gametophytically expressed and cause an SI response when both the *S* and *Z* alleles of a haploid pollen grain match the same alleles in the diploid stigma (Hackauf and Wehling 2005). Then, pollen tube growth is inhibited and the pollination is completely restricted. More recently, a third SI locus is known as *S5 (T)* locus on chromosome 5R (Voylokov et al. 1993). Trang et al. (1982) reported 6–7 alleles at the S and 12–13 alleles at the Z locus in the population cultivar Halo. Self-fertility may be caused by a dominant gene but could also follow a more complex pattern of inheritance (Melz et al. 1990). Lundqvist (1956) showed that self-fertility (*Sf*) mutations occur in both, *S* and *Z* genes. The self-fertility used in German hybrid rye material traces back to the early work of Ossent (1938) who selfed 10,000s of rye heads of different populations. Starting with a few self-fertile plants, he reached full self-fertility after a few cycles of recurrent selection due to dominant inheritance.

2.3.2 Cytoplasmic-Male Sterility

Several sources of cytoplasmic-male sterility (CMS) have been detected in rye (Table 2.1, see also Chap. 10). Of these, only the Pampa (P) cytoplasm from an Argentinean landrace of rye ("Waldstaudenroggen") gained importance for hybrid production. It is environmentally very stable and easy to maintain, but hard to restore. In an experiment across 10 European locations in two years, no pollen fertility occurred in maintainer lines and their crosses (Geiger et al. 1995). In European material, maintainer genotypes dominate and only 3–5% of the gametes are (partial) restorers. All other mentioned cytoplasms belong to the V (Vavilov) type that is, in contrast, hard to maintain and easy to restore due to a high frequency of gametes with partial or full restoration. Łapiński and Stojałowski (2001) reported that the majority of male sterility sources from 50 rye populations belonged to the Vavilov type. In a follow-up study, Stojałowski et al. (2008) detected P cytoplasm only in two Iranian sources ('IRAN I', 'IRAN IX') that have been reported earlier to contain CMS (Geiger and Morgenstern 1975) and in Argentinean populations ('Pico Massaux', 'San Jose', 'Trenelense') similar to those where the P cytoplasm was detected. Further, PCR-based markers can be used to identify CMS phenotypes, namely the differentiation of the normal cytoplasm from the P and V type cytoplasm is possible (Stojałowski et al. 2004).

2.3.3 Restoration of Male Fertility

Pollen-fertility restoration is of crucial importance for hybrid rye growing. Incomplete restoration might result in poor seed set, but definitely increases the infection by the ergot fungus (*Claviceps purpurea*) that produces toxic alkaloids in the purple-black sclerotia growing on rye heads after infection at flowering. The less pollen available, the higher is the incidence of ergot infection (Miedaner and Geiger 2015). Low tolerance limits for ergot sclerotia in commercial grain lots for food and feed (<0.05% and <0.1% per 500 g grain, respectively) exist in the European Union and are expected to tighten in the coming years. European restorer sources for the P cytoplasm are scarce, highly dependent on environment and the seed parent genotype, often leading to incomplete restoration (Geiger et al. 1995). Even worse, they show oligogenic inheritance, often with a major gene that requires several minor genes for full restoration (Miedaner et al. 2000). Restorer of fertility (*Rf*) genes are nuclear encoded and should display full dominance to be useful in hybrid breeding. In those germplasms where CMS has been found, effective *Rf* genes were detected, notably in Iranian primitive rye (e.g., 'IRAN IX', 'Altevogt 14161') and Argentinean landraces (e.g., 'Pico Gentario'). Male-fertility restoration from these sources is mainly monogenically inherited (Miedaner et al. 2000) and shows much lower environmental and seed parent effects than the European sources (Miedaner et al. 2005). The *Rf* gene of 'IRAN IX' was first introduced into the commercial hybrid cultivar 'Pollino' released in 2005. Hybrid cultivars carrying this gene have a much lower ergot incidence after inoculation by

Table 2.1 Different sources of CMS cytoplasm (Schlegel 2016, adjusted)

Type	Description	References
P	Pampa cytoplasm	Geiger and Schnell (1970)
C	Cytoplasm of wild rye, *S. montanum*	Łapiński (1972)
R	Russian cytoplasm	Kobyljanski and Katerova (1973)
S	Mutated cytoplasm of Kärtner rye	Warzecha and Salak-Warzecha (2003)
G	Mutated cytoplasm of Schlägler Alt, Norddeutscher Champagner	Melz et al. (2003)

Claviceps purpurea than those without this restorer gene (Miedaner et al. 2005; Miedaner and Geiger 2015). However, non-adapted *Rf* genes display negative side effects such as lower grain yield, lower 1000-grain weight and taller plant stature, even in testcross progenies (Miedaner et al. 2017). Due to a large range in yield penalty between lines derived from the same non-adapted source, the opportunity exists to select for better adaptation to the elite gene pool.

2.3.4 Heterotic Groups

Information about heterotic groups is fundamental for the maximum exploitation of heterosis (Melchinger and Gumber 1998). The seed and pollen parents should be derived from genetically unrelated germplasm in order to maximize heterosis. According to a 7 × 7 diallel of European rye populations, two germplasm groups from Germany showed the highest panmictic mid-parent heterosis relative to the mean with 18.7% for Petkus and 10.8% for Carsten (Hepting 1978). They were bred by two German rye breeders, Ferdinand von Lochow from the small village Petkus near Berlin and Dr. h.c. R. Carsten from Bad Schwartau, starting with different open-pollinated landraces. Since then, all hybrids released in Germany belong to the Petkus × Carsten type. This pattern was confirmed by analysis with about 180,000 SNPs clearly showing two separate clusters and strong differentiation between the seed and pollen parent pool (F_{ST} = 0.229, Bauer et al. 2017). The seed parent (Petkus) pool shows high yield performance, high kernel weight and lodging resistance, whereas the pollen parent (Carsten) pool exhibits large spikes and a good seed set, but a high lodging susceptibility, a high level of pre-harvest sprouting and a low stand density contributing to a considerably lower per se performance than the Petkus derived populations 'Nomaro' and 'Kustro' (Hepting 1978). This contributed, together with its wide adaptability and high yield performance, to a preponderance of the Petkus population in worldwide rye breeding.

2.3.5 Inbreeding Tolerance

In rye, as an outcrossing crop, performance greatly depends on the degree of heterozygosity. During subsequent selfing, the performance drops dramatically due to inbreeding depression. Crossing inbred lines, however, results in substantial heterosis (H) defined as H = F1-MP with F1 and MP indicating the F1- and the mid-parent performance of a pair of inbred lines, respectively. Inbreeding depression is highest for traits showing high heterosis. In maize, Schnell (1974) showed that with extended breeding cycles, the relative amount of heterosis (expressed as a percent of MP performance) drops because the inbred lines and subsequently, the hybrid performance, increase. This could also be shown for rye, where first-cycle lines yielded relative heterosis for grain yield of 192% (Geiger and Wahle 1978), whereas 22 years later the relative heterosis was estimated as 139% (Geiger et al. 2001) caused by strong selection for inbreeding tolerance that raised the performance of inbred lines considerably.

2.3.6 GCA/SCA Variance Relationship

The general (GCA) and specific combining ability (SCA) variances are important parameters for predicting hybrid performance (Melchinger et al. 1987) and for designing hybrid-breeding programs. The higher the GCA relative to the SCA variance, the better superior hybrids can be predicted from their GCA effects and the more effective early testing procedures become. Producing hybrids between heterotic groups generally decreases the relative importance of SCA as shown in maize (Melchinger and Gumber 1998). Estimations of GCA and SCA variance in rye showed a high preponderance of GCA variance for grain yield and plant height and, consequently high importance of additive gene action (Tomerius et al. 1997). However, in crosses with inbred lines pre-selected for high GCA effects, the importance of SCA variance might increase considerably (Geiger 1982). For practical breeding, the proportion of GCA/SCA variances greatly determines the number of testers needed for combining ability tests. Based on model calculations one tester was optimal for the first stage of GCA selection and three testers for the 2nd stage (Tomerius et al. 2008).

2.3.7 Correlation Inbred Lines: Testcrosses

The correlation between inbred lines and their testcrosses determines which traits can be reliably selected based on per se performance among inbred line populations or must be selected on a testcross basis. Selection on per se performance in early generations is highly advantageous for the breeder because it saves time, seed production costs and a larger proportion of additive genetic variance can be exploited in lines compared to hybrids. The genotypic correlation depends on the complexity of the trait under consideration as well as on the predominant type of gene action (Hallauer et al. 2010; Mihaljevic et al. 2005). In elite breeding populations, the genotypic correlations were highest ($r_g \geq 0.7$) for plant height, test weight, thousand-kernel weight, falling number and starch content (Miedaner et al. 2014). In contrast, for grain yield a much lower genotypic correlation (Table 2.2) has been found, providing the main reason for the need to perform testcrosses in early generations.

2.4 Enabling Technologies

2.4.1 Marker Technology

For a long time, rye lagged behind other cereals in developing new marker technologies because of its low international importance. With the advent of next generation sequencing (NGS) it became feasible for the first time to develop a public single nucleotide polymorphism (SNP) Illumina® Infinium iSelect HD Custom BeadChip comprising more than 5000 markers (Haseneyer et al. 2011). Based on this array, an extended custom 16 k Illumina BeadChip was produced (Auinger et al. 2016) and recently even a high-density Affymetrix Axiom® Rye600k genotyping array was designed (Bauer et al. 2017). Alternatively, the application of NGS technology for genotyping-by-sequencing (GBS) is well established and is offered as DArTseq by Diversity Arrays Technology Pty Ltd., Australia. A medium-density 16 k chip was successfully used for mapping quantitative traits (Miedaner et al. 2012; Hackauf et al. 2017a). However, although some quantitative trait loci (QTL) with high effects were detected in adapted populations, they have not been used in practical rye breeding, mainly caused by missing validation experiments. In contrast, molecular markers are a perfect tool for marker-assisted selection (MAS) of monogenic traits in breeding populations.

A good example is the introgression of non-adapted *Rf* genes by molecular markers. In inbred lines developed from the accessions ‘IRAN IX’ and ‘Pico Gentario’, pollen-fertility restoration of the P cytoplasm was assigned to two genes, *Rfp1* and *Rfp2*, respectively, that were both mapped on chromosome 4RL; one of these genes is sufficient to reach full pollen fertility (Stracke et al. 2003). Studies with newly developed co-

Table 2.2 Estimates of genotypic correlation between inbred lines and testcrosses

Reference	Grain yield	Plant height	1000-grain weight	Falling number
Köhler (1986)	0.51	0.87	0.81	–
Wilde (1987)	0.56	0.83	0.73	–
Hartmann (1997)	0.56	0.81	0.76	0.90
Miedaner et al. (2014)[a]	–	0.48/0.80	0.68/0.67	0.94/0.73

[a]Two inbred line populations and their testcrosses (N = 220 per population)

dominant markers permitted to delimit *Rfp1* within a 0.7 cM genetic interval and allowed prediction of *Rfp1* genotypes with high precision (Hackauf et al. 2012). A third restorer gene, *Rfp3*, was detected in the Iranian primitive rye 'Altevogt 14161' (Falke et al. 2009) that could also be mapped to the same interval as *Rfp1* and *Rfp2* (Hackauf et al. 2017b). This subgenomic region is syntenic to the *Rfm1* locus of barley on chromosome 6HS. Interestingly, *Rf* genes for the G (Börner et al. 1998) and C (Stojałowski et al. 2005) cytoplasms in rye, which both are functionally different from P (Geiger et al. 1995), have been mapped to the same segment of chromosome 4RL (Hackauf et al. 2009). Likewise, the rye gene *Rfc4*, which restores male fertility in hexaploid wheat with *Triticum timophevii* sterility-inducing cytoplasm was mapped on chromosome 4RL (Curtis and Lukaszewski 1993). For *Rfp1* from 'IRAN IX', the marker interval could be reduced to 0.2 cM, equivalent to about 120 kb, by establishing new markers and mapping a population of about 5,000 plants of a backcross population (Wilde et al. 2017). Two tightly linked and equivalent, but independently acting *Rf* genes were detected in the respective interval (*Rfp1a*, *Rfp1b*). Compared to earlier studies, the introgression segment from Iranian primitive rye was shortened and the yield penalty considerably lowered but, not totally suspended, yet. This example illustrates the enormous advantage of applying molecular markers when the breeder is targeting background selection. Moreover, markers tremendously shorten the selection procedure for *Rf* genes in foreground selection, because testcrossing (1st generation) of genotypes putatively carrying the target gene and phenotyping of the testcross progenies for pollen-fertility restoration (2nd generation), is no longer necessary or can be restricted to the final step confirming the success of the backcross procedure.

The potential of genomic selection (GS) for rye breeding has been investigated recently by the use of SNP arrays (Auinger et al. 2016; Bernal-Vasquez et al. 2017; Wang et al. 2014, 2015). In spite of the fast decline of linkage disequilibrium (LD) detected in rye candidate genes (Li et al. 2011), the use of medium-sized arrays (e.g., 16 k Infinium iSelect HD BeadChip, Auinger et al 2016) suffices. GS uses information from the whole genome, i.e., on all polymorphic markers, that are available and thus can be expected to take into account also small effect gene loci for a given trait that cannot be captured by QTL analyzes (Jannink et al. 2010). The ultimate goal for the rye breeder would be to predict the genetic value of non-phenotyped entries to reduce cycle length and costs. This is especially useful for hybrid rye breeding, where every line has to be testcrossed before phenotyping for grain yield, thus needing two years for one stage of yield selection. GS improved prediction accuracy compared to MAS in all tested quantitatively inherited traits (grain yield, plant height, starch and total pentosane content), especially in cases where only a low proportion of genotypic variation could be explained by MAS (Wang et al. 2014). Accuracy of prediction, which is defined as the correlation (r_{MG}) of the effect as estimated by *M*arker to the true *G*enetic effect, highly decreased when an estimation set from one biparental population was compared to a test set from another population, although both populations shared one parental line (Wang et al. 2014). Even for analyzing the very resource-demanding parameter phenotypic stability, a GS approach could detect stable QTLs for quality traits (test weight, soluble pentosane content and falling number), but not for yield-related traits (Wang et al. 2015), indicating that quality traits have a simpler genetic architecture. Another application for exploiting the potential of GS in breeding programs is the prediction of breeding values across selection cycles. Pedigree, genomic and phenotypic data of four consecutive breeding cycles from a commercial rye breeding program were used for a detailed cross-validation analysis. It could be demonstrated that GS for grain yield, plant height and TKW yielded improved prediction accuracies when data across cycles were accumulated (Auinger et al. 2016). For the three traits mentioned, prediction accuracies obtained from cross-validation with large calibration sets ($N = 832$) derived from all four available cycles were around 0.70, which looks rather promising. However, prediction accuracies usually yielded much lower means and showed considerable variance of estimates when lines from a given cycle had to be predicted across-cycle from calibration sets where the given cycle is missing. The authors concluded that uncertainty of prediction is an important factor, which should not be neglected when discussing the potential of GS. Among other factors, connectivity over breeding cycles via common ancestors is of particular importance to ensure persistency of prediction accuracy.

The authors suggest a number of opportunities of GS specifically for hybrid rye: (1) reduction of cycle length, (2) applying indices combining genomic and phenotypic information and (3) better exploitation of segregation variance within families. Some more applied examples on how to exploit these opportunities are illustrated in Sect. 2.6 of this chapter. Implementing GS for practical breeders is challenging. For example, a sophisticated management and design of crosses and knowledge of the respective familial structures is necessary when the breeder strives to maintain genetic variance and to maximize or balance short- but also long-term gain from selection.

2.4.2 High-Throughput Phenotyping

Phenotyping is a bottleneck for practical selection in plant breeding (Würschum 2019). To identify superior genotypes 10,000s of candidates have to be tested in field trials for mainly complex inherited quantitative traits. Although technical achievements for improved machinery equipment enable today for high power field-work, it is still a labor- and time-intensive procedure. Given the decreasing costs for marker assays, phenotyping is an even higher restriction for exploiting the genetic architecture of important traits. Using sensor technology to assess multiple traits might be the first step to high-throughput phenotyping.

For quality traits, like water (dry matter) and protein content, the application of near-infrared spectroscopy (NIRS) provided a breakthrough in the 1990s. More recently, water content can be measured very precisely on-field by specially equipped harvesters. This made obvious the advantages of a fast, precise, non-destructive measurement for plant breeding and boosted research into high-throughput phenotyping platforms (review by Würschum 2019). Multiple sensors are available with either morphological or spectral measurements and they can be applied either as field-based mobile platforms or as unmanned aerial vehicles, like drones (Haghighattalab et al. 2016). Hyperspectral imaging is able to measure water content, abiotic stress and disease severity, composition of plants and ingredients of plant parts. Vegetation Indices (VIs) summarize canopy reflectance information in simple algorithms used for qualitative and quantitative assessment of a wide range of plant parameters, like water, chlorophyll, or carotenoid content (Xue and Su 2017). Prediction of biomass yield via hyperspectral data is an interesting application when candidate lines are routinely phenotyped for grain yield. Thus, factorial experiments with two harvest dates, at milk ripening for biomass yield and at full ripening for grain yield, can be made superfluous.

The VIs alone reached a prediction ability (PA) of maximal 0.42 when both flights were combined (Table 2.3).

In a second attempt, the most informative bands from all 400 assessed hyperspectral wavelengths (410–993 nm) were chosen by variable selection methods, a procedure that increased PA (Galán et al. 2020). Even higher values were achieved by combining the hyperspectral information with the routinely assessed plant height. An additional increase could be achieved by combining hyperspectral and genomic data with plant height. The high information value of plant height is due to a positive correlation between plant height and biomass ($r = 0.57$, $P < 0.001$). For most parameters, the late flight had higher PA values. Other applications for high-throughput phenotyping are the identification of drought-tolerant genotypes by growing them on sandy soils under rainfed conditions or the quantification of disease severities by either hyperspectral or thermal imaging (Mahlein et al. 2012).

2.4.3 Gene Discovery

In practice, breeding populations do not contain the full genetic diversity needed to fast-forward genetic gain or to supply specific traits of interest, which are mandatory to enter new market areas. To compensate for these shortcomings, the breeder aims to make genes or major QTL available that are explaining a major part of the genotypic variance of the respective trait. Germplasm collections from gene banks or indigenous populations adapted to unfavorable climatic or edaphic conditions, for example, can be valuable sources of genetic variation. A high amount of time and resources are required to (1) phenotype accessions for the trait of interest, (2) identify the most relevant genes/QTL controlling the trait and their mode of inheritance and finally (3) transfer them into adapted germplasm. Current technologies help to increase the chance to mine new and valuable QTL by narrowing the number of accessions to be analyzed in-depth down to the most promising ones, for example by combining genomic (Yu et al. 2016) or passport data (Crossa et al. 2017) with environmental data, which indicate the desired tolerance to biotic or abiotic stresses. Current experimental approaches targeting the identification of QTL and candidate genes in crops can be divided into natural or experimental populations (Cavanagh et al. 2008).

Gene discovery in rye is a specifically difficult task due to self-incompatibility, low agronomic performance, low inbreeding tolerance and the unknown heterotic group of genetic resources (Haussmann et al. 2004). Frost tolerance of high yielding European breeding populations is, for example, insufficient for the Canadian and Russian cropping areas. In a paradigmatic and successful case study, new alleles for frost tolerance (Erath et al. 2017) could be explored in a biparental QTL mapping population from a cross between a European inbred line and a gamete from the Canadian cultivar Puma. Phenotypic variance has been estimated for several traits recorded in field experiments and controlled test environments at −20 and −23 °C. A QTL at the Frost resistance locus 2 *(Fr-R2)* on chromosome 5R, explained a high proportion of the phenotypic variance in recombinant inbred lines and

Table 2.3 Prediction ability for biomass yield assessed by hyperspectral information collected on an early (after flowering) and late (yellow ripening) drone flight and by plant height for 274 three-way hybrids across 4 locations and 2 years. Data from Galán et al. (2019, 2020)

Parameter	Early flight	Late flight	Both flights
All vegetation indices (VI_{all}, $N = 20$)	0.35	0.42	0.42
Hyperspectral bands (HYP, $N = 32$ selected wavelengths)	0.54	0.52	0.59
VI_{all} + Plant height (PH)	0.55	0.62	0.58
HYP + PH	0.55	0.50	0.62
HYP + PH + genomic data	0.72	0.75	0.75

their testcrosses. This QTL was mapped close to the well-known Vernalization-response (*Vrn-R1*) locus. Other consistent QTLs were found on chromosomes 4R and 7R. As a selection strategy, the results suggest a three-stage procedure with (1) MAS based on markers from the *Fr-R2* locus, (2) applying a genomic prediction model capturing also effects of smaller QTLs and (3) a final verification in a multi-environmental field and lab phenotyping platform. From an enabling technology point of view the study demonstrates some crucial lessons to be considered in further gene discovery studies: (1) Gaining insight into the architecture of frost tolerance in winter rye should make use of pre-existing knowledge (Pasquariello et al. 2014) from homoeologous groups in other related species, i.e., the *Triticeae*, (2) development of genomic tools in the target species itself (Bauer et al. 2017) is of utmost importance, (3) mapping populations tracing back to parents that are highly diverse for the trait of interest should be built up and last but not least, (4) it pays also for all subsequent breeding steps to invest in a solid phenotyping platform under controlled and field environments.

2.5 Enhancing Genetic Diversity

2.5.1 Broadening of Central European Pools

Continuous selection within heterotic pools is expected to reduce genetic diversity as pointed out by Duvick et al. (2004). Due to selection on GCA to the opposite pool, the decrease in genetic diversity within pools will correspond with an increase in difference between them. In rye, genome-wide selection signals (Bauer et al. 2017) could be identified and the genetic differentiation between breeding pools revealed. Candidate genes identified to be under selection affect morphological traits such as plant height and traits connected with the hybrid system, such as restoration of male fertility.

Further, genetic drift also leads to a loss of genetic diversity that is hard to avoid when generating inbred lines by second cycle breeding, because only a limited number of elite lines is used for intercrossing. Actually, as revealed from the long-term experiment in corn, investigating reciprocal recurrent selection over 18 cycles between the Iowa Stiff Stalk Synthetic (BSSS) and the Iowa Corn Borer Synthetic No1 (BSCB1) (Gerke et al. 2015), most of the observed reduction in genetic diversity could be attributed to genetic drift.

The established Central European heterotic pools trace back to the Petkus (Pool P) and Carsten pools (Pool C). Benefits from broadening the genetic diversity of these pools could potentially arise from; (1) enhancing genetic gain from selection, which is proportional to the genetic standard deviation and (2) from expanding the traditional Central European target environment to new markets, such as Eastern Europe, Canada or Asia, which require specific traits such as overwintering capability.

The potential of Eastern European populations to broaden established heterotic pools has been investigated (Wilde et al. 2006; Fischer et al. 2010) by exploiting both phenotypic and genomic data. In the latter study (Table 2.4) around 30 S_0-plants randomly sampled from the candidate populations had been cloned in each case, outcrossed to pool P and C and genotyped with 30 SSR markers evenly distributed on the seven rye chromosomes. In total, P and C were represented with 121 and 142 S_0-plants respectively. Testcrosses with P and C were grown in separate yield trials at 4 locations and 2 replications in Germany in 2007.

For grain yield, parameters, such as means, genetic variances and usefulness, were estimated. The usefulness criterion (Schnell 1983) with U (α) = mean + $i\ h\ \sigma_G$ combines population mean and the expected genetic gain, which is proportional to the genetic standard deviation (σ_G) of a candidate population, the selection intensity (i) and the square root of the heritability (h). Thus, the parameter reflects the performance of the selected fraction from the candidate population. The lower selection intensity ($i \sim 1$) might consider that after discarding for non-yield traits only a limited number of S_0-plants are available for selection on grain yield. The higher selection

Table 2.4 Parameter estimates for grain yield (dt ha^{-1}) of testcrosses from S_0-clones sampled from 5 Eastern European populations (POP 1–5) and the Central European Pool P and C. Means, genetic variances (σ^2_G) and usefulness values (U, explanation see text) given for a low and a high selection intensity (i)

Parameter	Population						
	POP 1	POP 2	POP 3	POP 4	POP 5	Pool P	Pool C
	Tester: Pool P						
Mean	55.1	56.0	54.0	53.8	57.8		58.2
σ^2_G	3	7	9	21	9		3
U($i_{0.05=2.063}$)	58.1	60.8	59.7	62.5	63.8		62.6
U($i_{0.40=0.966}$)	56.5	58.3	56.6	57.9	60.6		60.3
	Tester: Pool C						
Mean	58.0	59.6	58.2	55.8	58.6	61.1	
σ^2_G	14	13	17	35	11	13	
U($i_{0.05} = 2.063$)	64.9	66.6	66.0	67.4	64.8	70.8	
U($i_{0.40} = 0.966$)	61.3	62.9	61.9	61.2	61.5	65.6	

intensity ($i \sim 2$) might be appropriate if the breeder can reduce the number of candidate populations beforehand, which will then allow more investment of capacity into the most promising populations. Within the testcross series with Pool P, S_0-plants sampled from Pool C showed the highest mean for grain yield as expected which, was closely followed by the mean of POP 5. Genetic variances were smallest for Pool C underlying the need for broadening this pool. Usefulness was highest for POP 5 benefitting from both an above average mean and genetic variance. Thus, this population can be regarded as the first choice to enhance genetic diversity in Pool C. Within the testcross series with C as a tester, the mean of P ranked first followed by POP 2. Except for POP 4, genetic variances were similar to those of P. Usefulness of the POP 1–5 did not surpass the Pool P value.

Gene diversity found in the Eastern European populations (Table 2.5) support the high genetic variances estimated from phenotyping. In addition, molecular analyzes suggested that there is "new" variance because a number of SSR alleles present in POP 1–5 were found to be absent in Pool P and C. As expected, the highest genetic distance between populations was found between Pool P and C. Conversely, this means that including Eastern European populations into one of the pools will reduce diversity among pools and possibly also lead to a reduction of heterosis.

Phenotyping candidate populations and estimating the relevant parameters is cost and time consuming. Therefore, a multi-stage approach to select between candidates and to focus on the most promising individuals within selected populations is highly meaningful:

1. Populations can be screened by analyzes of genotypic data for criteria such as gene diversity and genetic distance to established pools and to identify other candidate populations with little cost.
2. An evaluation of population per se performance will provide valuable information on non-yield traits such as disease resistance, plant height, quality.
3. A sample of 30–50 S_0-plants or S_0-clones from the pre-selected populations should be outcrossed to the established heterotic pools for testcross seed production. Due to self-incompatibility, the gametic array of the S_0-plants cannot be maintained by producing S_1-lines but their genetic content can be conserved by crossing them to elite lines derived from the established pools.

Table 2.5 Parameter estimates from 30 genome-wide distributed SSR loci for gene diversity (Nei 1987) and Modified Rogers distances (MRD, Wright 1978)

Parameter	Population						
	POP1	Pop2	POP3	POP4	POP5	Pool P	Pool C
Gene diversity	0.56	0.55	0.56	0.56	0.55	0.53	0.43
Pool P	0.21	0.25	0.2	0.22	0.2		0.29
Pool C	0.28	0.29	0.26	0.25	0.27	0.29	

4. Testcrosses grown in multi-environmental trials provide estimates for most relevant parameters such as mean, genetic variance and usefulness.
5. Superior S_0-plants from finally selected populations may enter with their progeny from elite line crosses into a synthetic.
6. The synthetic has to undergo several cycles of recurrent selection by using a selection index based on phenotypic and genomic-estimated GCA effects.

Occasionally and quite rightly, genetic diversity in landraces or exotic populations is regarded as "gold reserve" (Böhm et al. 2017) for plant breeding. Converting the "gold" into productive "working capital" needs to close the performance gap between resources and elite material. This is a demanding task because not only performance for individual traits, but a minimum threshold for all relevant traits should be reached when used, for example, in intercrossing a genotype derived from exotic resources.

Apart from rare lucky punches and from the authors' experience, usually two to three breeding cycles must be completed before benefitting from exotic resources in commercial lines. In consequence and in order to prevent raising false or unrealistic expectations, exotic resources should be integrated into a breeding program as a mid- to long-term approach.

Breeding designs should carefully be analyzed to reduce loss of genetic diversity due to genetic drift. Deficiencies made in this field cannot be compensated by using exotic genetic resources which can be a costly and cumbersome process and can negatively affect the competitiveness of a breeding program.

2.5.2 Use of Non-adapted Material for Introgressing Single Genes

In contrast to the tedious work of broadening gene pools for quantitative traits, the introgression of monogenic traits into elite germplasm is more straightforward. An example is the improvement of leaf rust (*Puccinia recondita*) resistance. In the 1990s, all first-cycle inbred lines were highly susceptible to this important disease. Due to the introduction of Eastern European resistance sources, this has changed. Similarly, the introgression of non-adapted pollen-fertility genes (see Sect. 2.4.1) or stem rust (*Puccinia graminis* f.sp. *secalis*) resistance genes are future challenges. The necessary steps are: (1) detection of the desired trait expression in donor genotypes, (2) establishing segregating populations by backcross and selfing steps, (3) discovery of linked molecular markers by either linkage or association mapping, (4) introgression into elite pools by marker-assisted backcrossing. In rye, hundreds of self-incompatible plant genetic resources (PGR) are stored in gene banks. They comprise wild species, weedy rye populations, old European landraces and elite populations from different rye-growing countries. PGR is usually a rich source of monogenically inherited resistance (*R*) genes. Musa et al. (1984), for example, detected ten *R* genes for leaf-rust resistance in only six inbred lines. Further, four *R* genes were found in the Russian rye cultivars 'Sanim' (from 'Sangaste' × 'Immunaya1'), 'Immunaya 1' (origin *Secale strictum),* 'Chulpan', and 'Novozybkovskaya 4', respectively (Solodukhina, 1994; Kobylanski and Solodukhina, 1996). In

Germany, Wehling et al. (2003) and Roux et al. (2004) identified in total five dominant *R* genes for leaf-rust resistance, either derived from local inbred lines (*Pr1, Pr2*) or from rye genetic resources: 'Jaroslavna'/Russia (*Pr3*), 'Turkey'/Canada (*Pr4*), 'WSR'/Germany (*Pr5*). Molecular markers for these sources are available, so the target genes can be easily introgressed into elite material (Wehling et al. 2003; Roux et al. 2004). Similarly, for stem rust, resistance sources were found in Austrian landraces, Russian populations and the US forage rye (Miedaner et al. 2016). The proportions of fully resistant plants within the populations varied from 2 to 70%. Indeed, a gene for qualitative resistance was detected in two Russian populations on chromosome 7RL and a quantitative resistance with at least three QTL in another population (Gruner et al. 2020). Association mapping revealed additional *R* genes on chromosomes 1R, 2R and 6R in two Austrian landraces, 'Oberkärntner' and 'Tiroler'. Other candidates for MAS of monogenic traits are dwarfing genes, like the dominant *Ddw1* on chromosome 5R (Kalih et al. 2014), that was recently used in practical breeding of rye and triticale (X. *Triticosecale* Wittmack).

Depending on their origin, target genes might be linked with negative agronomic traits. An example is the pollen-fertility restorer genes on chromosome 4RL, derived from Iranian and Argentinean sources (Fig. 2.1). The *Rfp3* gene significantly enhanced the restorer index by 80% on average, as expected, but also led to taller plants and decreased grain yield. The yield penalty ranged from 0.95 to 10.0 dt ha^{-1} among seven introgression lines and illustrates a potential for selecting lines with shorter introgression intervals with markers. Similar yield penalties were observed for *Rfp* genes from the primitive ryes 'IRAN III', 'IRAN IX' and the Argentinean landraces 'Pico Gentario' and 'Trenelense' (Wilde et al. 2017). In the latter study, the yield penalty ranged from 3.05 to 7.00 dt ha^{-1} depending on the marker haplotype.

2.5.3 Enhancing Quantitative FHB Resistance

Genetic diversity for quantitative traits is a key objective for high selection gain in breeding. Because of the well-known challenges for introgressing quantitative traits from exotic plant genetic resources (Haussmann et al. 2004), the breeders firstly search within their elite gene pools. When the observed trait level is too low, recurrent selection (RS) is a method for improving the population mean without decreasing genetic variance. In hybrid breeding, S_1 line selection is commonly used (Hallauer and Carena 2009). Resistance to Fusarium head blight (FHB) is an example of a quantitative trait

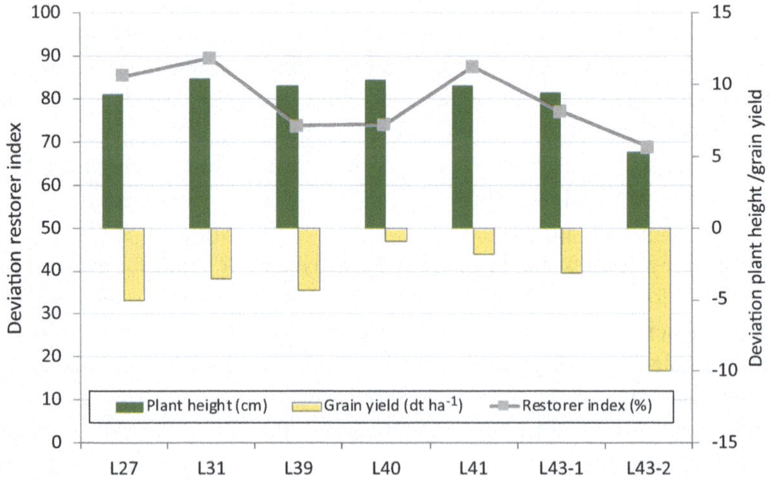

Fig. 2.1 Deviation of restorer index, plant height and grain yield for seven introgression lines with the pollen-fertility restorer gene *Rfp3* compared to near-isogenic lines without this gene. Data from Miedaner et al. (2017)

that is not easy to handle. FHB is caused by an array of *Fusarium* species, reducing grain yield and quality and contaminating the harvest with mycotoxins, of which deoxynivalenol (DON) is one of the most frequently found. For FHB resistance in rye, large genotypic variation in breeding populations has been reported. However, resistant genotypes are scarce (Miedaner et al. 2003b). Making the situation even more difficult, the correlation of line per se to testcross performance has been regarded to be low for this trait (Miedaner et al. 2003b). A re-evaluation of a RS program showed that indeed within the materials selected in previous cycles, the correlation was absent ($r = 0.09$), while after adding a sample of 111 unselected S1-L tested in the same experiment, the correlation changed to moderate ($r = 0.68$). The S_1 lines have been evaluated after the fifth cycle of an RS where each cycle selection was based on an index of line and testcross performance. The effect of recurrent selection can be tremendous (Fig. 2.2).

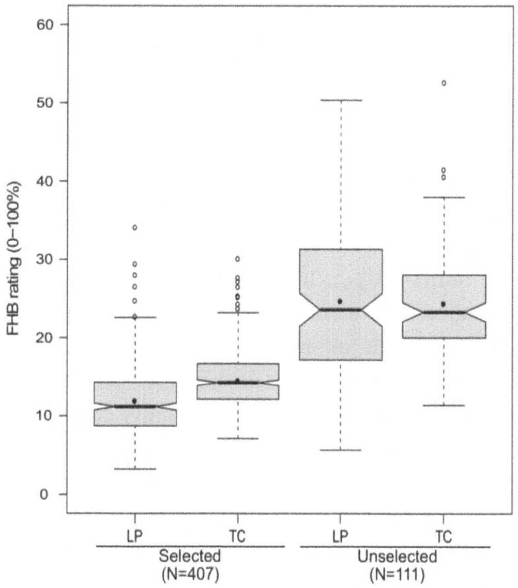

Fig. 2.2 Boxplots of Fusarium head blight (FHB) ratings (0–100%) of 407 S_1 lines from the fifth cycle of a recurrent selection program and 111 unselected S_1 lines for line performance (LP) and their corresponding testcross performance (TC) after inoculation by *Fusarium culmorum* in two locations. Data from Miedaner and Wilde (2019)

Low FHB infection generally leads to a lower DON content in grain (Miedaner et al. 2003a, b). High genotypic correlation coefficients between both traits (0.8–0.9) promote indirect selection for reduced DON content by assessing FHB resistance. Genotype × environment interaction played a major role for both traits (Miedaner et al. 2003b) illustrating the necessity of selecting in several environments (location × year combinations). Resistance QTL on rye chromosomes 1R, 4R, 5R and 7R were reported in a mapping study with four triticale populations (Kalih et al. 2015). The dominant rye-dwarfing gene *Ddw1* had a significantly negative impact on FHB resistance in triticale (Kalih et al. 2014). Additionally, it reduced plant height as expected and delayed heading.

In a recent genome-wide association study (GWAS), the lines from the RS program mentioned above were genotyped by a 15 k SNP assay (Gaikpa et al. 2020). Data were corrected for population structure by the genomic kinship matrix (K) and the first principal component. In total, 15 QTLs for FHB resistance on all rye chromosomes, except chromosome 7, were identified that jointly explained 74% of the genotypic variance. Among them, two major QTLs were detected on chromosomes 1R and 5R explaining 33% and 14% of the genotypic variance, respectively. Genome-wide prediction resulted in 44% higher cross-validated prediction abilities than marker-assisted selection revealing the quantitative nature of FHB resistance also in rye. Genomic approaches, thus, may accelerate breeding for complex traits.

2.6 Breeding Schemes

2.6.1 Recurrent Selection to Improve Base Populations

Recurrent selection (RS) is a cyclic process of selection and recombination targeted at improving genetically broad-based populations by (1) increasing the frequency of all favorable alleles and (2) maintaining genetic variability (Hallauer and Carena 2009). From heterozygous genotypes identified as superior in the RS procedure, inbred progeny could be advanced to form parental lines for Product Development (PD). This integrated breeding approach was first formulated in the 1970s (Sprague and Eberhart 1977). It gained priority with the advent of genomic selection (GS, Gaynor et al. 2017; Rembe et al. 2019). A key feature of GS applications in RS and PD, i.e., commercial line development, is the chance to shorten cycle times (Fig. 2.3). Mass selection (fast and slow) in a population of heterozygous genotypes aims to improve GCA with the opposite pool by using genomic-estimated GCA effects (GEGCA) as a selection criterion. Adopting a simple genetic model, additive and GCA effects can be linearly converted into each other. By applying advanced glasshouse technology (Hickey et al. 2017) cycle length can be reduced from traditionally one year to six or even less months. As pointed out by Rembe et al. (2019), persistency of prediction accuracy over several cycles has been analyzed only based on computer simulations (Müller et al. 2017) and experimental proof-of-concept, in terms of a high realized long-term gain, has still to be provided.

An RS scheme based on genomic and phenotypic selection (PEGCA + GEGCA in Fig. 2.3) and well adapted to hybrid rye breeding (for details see Fig. 2.4) involves intercrossing of superior S1-genotypes selected in the previous selection cycle, dissecting the population into S1 candidate lines and assessing their GCA to the opposite heterotic pool. Phenotypically estimated GCA (PEGCA) effects of candidate lines are based on using CMS testers to produce testcross seed. Then testcrosses are grown in multi-environmental trials, which allow high accuracy assessment of grain dry matter yield (GDY) as the most important trait. At the same time, phenotypic data assessed from testcrosses, can be used regularly to update the genome-wide prediction model.

Specifically, genomic prediction of GCA effects can address different sources of genetic variabilities, such as the variance within crosses (V_{wc}) and within S_0-plants (V_{wS_0}). This is less costly compared to the phenotypic approach or is

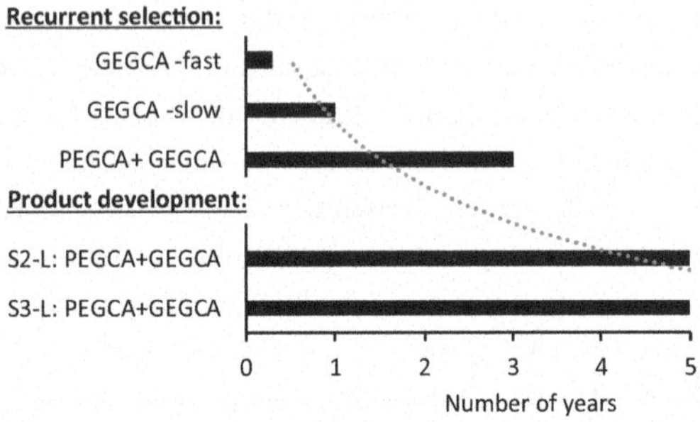

Fig. 2.3 Cycle length in Recurrent Selection and product development schemes; GE = *G*enomic *E*stimated, PE = *P*henotypic *E*stimated; GCA = *G*eneral *C*ombining *A*bility

even not feasible with the latter (V_{wS_0}) in the given RS procedure. GS models are trained with historical data from the previous cycles. Well performing S1-SP and their derived S2-L identified in the RS scheme can directly be fueled into product development.

2.6.2 Product Development

The RS procedure presented above might be integrated into concrete breeding schemes for line development. In the scheme, purely based on phenotypic selection (PS, Fig. 2.5), we start with intercrossing S2-L, self the progeny in two subsequent generations down to receive S2-L seed. With S2-L, we can exploit both V_{bc} and V_{wc}, and also half of V_{wS_0}, to select for line per se performance. The main selection pressure is exerted on traits such as resistance to lodging, plant height, heading date or disease resistances, inbreeding tolerance and grain quality characters, e.g., 1000-grain and test weight and falling number. Once the lines have overcome this first hurdle, candidate lines are outcrossed to CMS testers in year 4. In the same year, S2-SP are selfed to S3-L to speed up the further inbreeding process in case the respective S2-L is selected. In the following year 5, the testcross seed is used for phenotyping in multi-environmental trials. At least for so-called 'second cycle' lines, their phenotypically assessed GCA estimates (PEGCA) for GDY will be the main selection

Year	Generation	Technical procedure	Selection criterion	Exploitable additive variance (V_A) V_{bc}	V_{wc}	V_{wSo}	Selection stage	Comments
1	1	< S1-SP × S1-SP' >	GEGCA $_{S1\text{-}SP\ within\ S1\text{-}L}$			0.50	3	Intercross GEGCA selected S1-SP and generate pair cross seed
	2	< S0-SP >	GEGCA $_{S0\text{-}SP}$		0.25		1	Self S0-SP, which have been GEGCA selected
2	3	[T-CMS × S1-L]						Produce testcross seed of selected S1-L by outcrossing them to CMS
3	4	T-CMS • S1-L	I=b_g GEGCA + b_p PEGCA	0.75	0.25		2	Combine PEGCA (from testcross mean) and GEGCA (from parental S0-SP) into index I for final selection of S1-L to be intercrossed with their selected S1-SP

Legend:	
S0-, S1-SP	S0 and S1 single plants, resp.
L, T-CMS	Line, cytolasmatic male sterile tester from the opposite breeding pool
GEGCA, PEGCA	Genomically or Phenotypically Estimated General Combining Ability effects, resp.
<>, [ ]	intercrossing by pair crosses, selfing by paper bags or testcross seed production in plastic walls, resp.
▭	phenotypic or genomic assessment of the selection criterion
V_{bc}, V_{wc}, V_{wSo}	Variance between and within crosses,within S0-plants, resp.
b_g, b_p	weights for GEGCA and PEGCA in an index

Fig. 2.4 Technical description of an RS scheme involving phenotypic (PEGCA) and genomic (GEGCA) selection of GCA (General Combining Ability) effects

criterion. Lines with a positive PEGCA will continue to a second selection stage. This second selection stage—not displayed in the schemes below—will usually be the last step before using candidates as parental lines for commercial hybrids. Synchronously, excellent S2-lines are intercrossed to build up an improved population for the next breeding cycle. The steps described so far in scheme PS follow a classical and well-known phenotypic selection procedure (Geiger and Miedaner 2009), thus being an obvious object of comparison for alternate breeding schemes.

Alternatively, combined genomic and phenotypic selection (COM S2-L, Fig. 2.5) could be designed and an additional selection stage implemented with selection on GEGCA for GDY after the line per se test with historical data from previous cycles. Thus, assuming a sufficiently high prediction accuracy, candidate lines with an inferior GEGCA are discarded. The breeding program will benefit insofar as the large effort in producing and phenotyping testcrosses can be reserved to the putatively better part of the candidate population. In year 5, PEGCA effects from phenotyping testcrosses and GEGCA effects from genomic prediction enter into a selection index combining both sources of information, which can be weighted, e.g., by a classical Smith-Hazel approach (Dekkers 2007) or by empirical weighting factors. The index should be more accurate compared to PEGCA alone for two main reasons: (1) The latter might be impeded by loss of testing environments caused by technical hazards or high genotype x environment interaction. (2) If model training comes from aggregating data across multiple breeding cycles, GEGCA can deliver highly valuable complementary information to PEGCA as it mitigates biases from genotype x year interaction. As described for PS, S2-L with a positive index value will continue to a second selection stage and will be intercrossed for setting up a new breeding population.

An alternative combined genomic and phenotypic selection (COM S3-L, Fig. 2.5) affords selfing of a number of S2-single plants to S3-L within all those S2-L tested for per se performance in year 3. After having identified the S2-L with the best GEGCA, the breeder selects the best S3-L via GEGCA within those S2-L. Thus, compared to the other two schemes, we can exploit a larger part of the segregation variance (V_{wSo}). Only those S_3-L excelling by a superior

Scheme	PS		COM S2-L		COM S3-L	
Year	Technical procedure	Selection criterion	Technical procedure	Selection criterion	Technical procedure	Selection criterion
1	< S2-L xS2-L' >		< S2-L xS2-L' >		< S3-L xS3-L' >	
	< So-SP >		< So-SP >		< So-SP >	
2	< S1-SP >		< S1-SP >		< S1-SP >	
3	*a1=No/N1*	L *per se* $_{S2\text{-}L}$	*a1=No/N1*	L *per se* $_{S2\text{-}L}$	<S2-SP: *a1=No/N1*	L *per se* $_{S2\text{-}L}$
			a2= N2/N1	GEGCA $_{S2\text{-}L}$	*a2= N2/N1*	GEGCA $_{S2\text{-}L}$
					a3= N3/N2	GEGCA $_{S2\text{-}SP:\,S2\text{-}L}$
4	<S2-SP> [T x S2-L]		<S2-SP> [T x S2-L]		[T x S3-L]	
5	*a2= Nf/N1* T-CMS x S2-L	PEGCA$_{S2\text{-}L}$	*a3= Nf/N2* T-CMS x S2-L	I= b_p PEGCA$_{S2\text{-}L}$ +b_g GEGCA$_{S2\text{-}L}$	*a4= Nf/N3* T-CMS x S3-L	I= b_p PEGCA$_{S3\text{-}L}$ +b_g GEGCA$_{S3\text{-}L}$
6	< S2-L xS2-L' >		< S2-L xS2-L' >		< S3-L xS3-L' >	T xS2-L

L *per se* $_{S2\text{-}L}$= Line per se performance of S2-L

Fig. 2.5 Technical description of breeding schemes based on phenotypic selection (PS) and combined genomic and phenotypic selection in two generations (COM S2-L and COM S3-L) *Nx* = number of lines per stage x, *Nf* = final number of lines); other abbreviations see Fig. 2.4

GEGCA are outcrossed to testers. As previously described for COM S2-L, an index combining PEGCA and GEGCA from S3-L is used as a selection criterion. S3-L with a positive index value continues to a second selection stage and is intercrossed for setting up a new breeding population.

For comparing the efficiency of different breeding schemes, we estimate the annual gain from selection as the evaluation criterion. If model calculations are used as analytical tools restrictions, for e.g., the budget, effective population size should be incorporated to enable a fair and realistic comparison of alternative approaches and to allow for optimum calculations. In the following, GDY is taken as the only trait under selection, which is realistic, particularly for second cycle breeding populations that have already reached a good performance level for the other traits of interest. The respective variance components for GCA effects, their interactions with locations, years, years × locations, and single plot error are assumed to adopt a ratio of 1.0: 0.25: 0.25: 1.0: 2.0, regarded as typical for Central European conditions (taken from calculations of Laidig et al. 2017).

For all breeding schemes described above a fixed budget is assumed. Without presenting further details here, all activities necessary to process the breeding schemes above are assigned for glasshouse, nursery, genotyping, testcross seed production and phenotyping in yield plots, valued in money terms and summed up until the predefined budget maximum is reached. For line test, a selection rate of $\alpha = 0.1$ is predefined according to practical experience. This allows easy calculation of start-up costs per selected line, including costs for intercrossing, selfing and lines discarded in the per se test.

Further, technical risk prevention and high accuracy strategy (Wilde 1996) is assumed. When assessing PEGCA, all breeding schemes use a fixed number of testers ($T = 2$), locations ($P = 4$) and replications ($R = 2$). For most variance component ratios, these assumptions lead to lower gains from selection but also to lower variance in gains compared to other scenarios with lower testing but higher selection intensity. Assessing the GCA effect of individual candidate lines accurately produces some further positive side effects: (1) the false-positive rate of inferior candidates included into the second and final selection stage, intercrossing and other cost-driving activities can be reduced and (2) updating the genomic prediction model will benefit if an accurate phenotypic test is used for recalibration.

For all breeding schemes $N_f = 60$ has been fixed as the number of candidate lines entering into a final phenotypic evaluation. With this test, the putative best lines are identified as hybrid parents and hybrid performance can be predicted based on parental GCA effects.

For the calculation of genetic gains (Fig. 2.6), well-known approximations (Utz 1984) to obtain the selection intensity and the gain in multi-stage selection are applied. For selection indices combining information from phenotyping testcrosses (PEGCA) and from genomic prediction (GEGCA), weights were calculated by use of the Smith-Hazel index approach.

No detailed discussion on how to maximize genetic gain can be presented here and only selected topics are highlighted in the following. Instead of an optimum allocation of breeding resources, only an allocation fitting to the above-mentioned risk prevention strategy is considered. Under this restriction the effect of the GS selection rate (Fig. 2.6) on total genetic gain is investigated for a range of prediction accuracies ($0.1 \leq r_{GEGCA,\ GCA} < 0.5$) which seems to be realistic according to empirical results. As can be derived from the graphs, $r_{GEGCA,\ GCA}$ should exceed a minimum threshold value of around 0.3 to make COM S2-L and COM S3-L becoming more attractive than PS. Further, a moderate GS selection rate of around 0.4–0.5 is advisable to reduce the risk of achieving a genetic gain lower than that of PS. Comparing COM S2-L and COM S3-L, the latter benefits from a larger proportion of segregation variance and thus can achieve the highest genetic gains.

Assuming a GS selection rate of 0.4, the allocation of breeding resources for the three breeding schemes is displayed in Fig. 2.7. Compared to PS, budget demands for glasshouse, nursery and genotyping increase substantially from about 30% to 54% (COM S2-L) and 66% (COM S3-L), respectively. Correspondingly, expenses for testcross seed production and yield plots decline from 62% (PS) to 38% and 26% for the combined schemes, respectively. Numbers of entries to be processed at subsequent selection stages, as shown in Table 2.6, reflect the budget demands for each selection stage. These figures highlight the necessary re-organization of a breeding program when integrating GS as a tool.

Summarizing the results of the model calculation, implementation of GS into the breeding process is highly attractive in terms of a higher

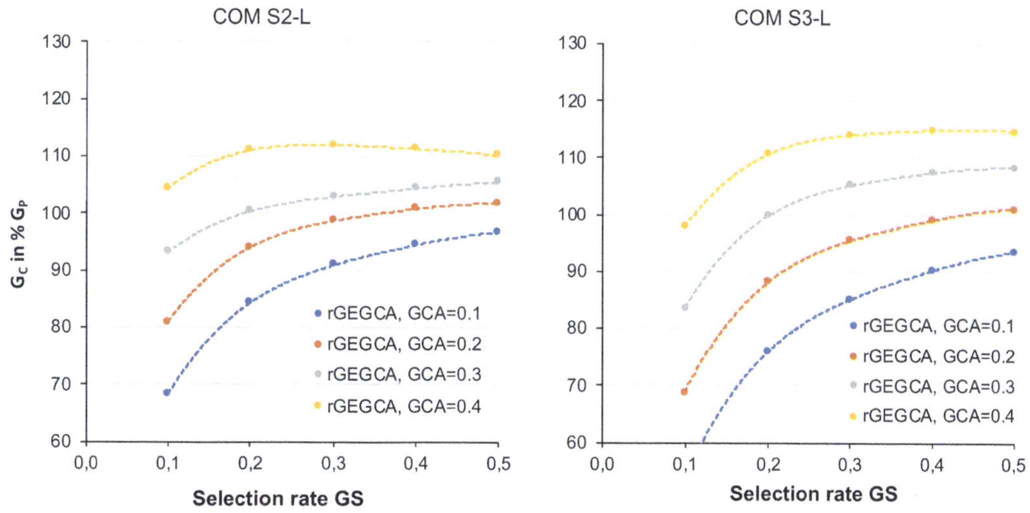

Fig. 2.6 Total gain relative to gain from phenotypic selection (G_C in % G_P) from the combined genomic and phenotypic selection schemes COM S2-L and COM S3-L for varying prediction accuracies of GCA ($r_{GEGCA,\ GCA}$) and varying selection rates from genomic selection (GS)

expected genetic gain. This finding is in line with the results of Marulanda et al. (2016) which are based on somewhat different assumptions. By model calculations, pitfalls and zones of risks can be identified. For example, taking care of high and stable prediction accuracy and an adapted selection intensity for all GS steps involved appears to be of high relevance.

2.7 Preconditions for Inscription of a Variety

Developing a new rye cultivar is cost- and time-intensive and particularly challenging because several traits have to be combined and most show complex inheritance. Before the registration of a variety to their respective National lists, official authorities test varieties for (1) Value for Cultivation and Use (VCU) and (2) Distinctness, Uniformity and Stability (DUS). Further, a suitable variety denomination is needed to start the registration process.

Taking Germany as an example for other EU member states: "the legal basis of national listing is the Seed Act (SaatG). It serves to protect consumers and ensure that agricultural and horticultural industries are provided with high-quality seed and plant material derived from healthy, high-quality and productive varieties" (Bundessortenamt 2019).

The German Federal Plant Variety Office ("Bundessortenamt") considers a variety to have VCU "if its qualities taken as a whole offer a clear improvement for cultivation, for use of the harvest or use of products derived from the harvest compared to comparable listed varieties" (Bundessortenamt 2019).

In Germany, VCU testing for rye requires a three-year test period with 10 to 20 locations per year. VCU testing in rye is usually carried out with two cultivation variants, i.e., an extensive variant without any use of fungicides and growth regulators and a more intensive variant mirroring the actual common practice of farmers with full use of growth regulators and fungicides whenever needed.

DUS tests are embedded into a system providing protection with an intellectual property right for new plant varieties at the National or European Community levels. Plant breeders' rights serve the plant breeding industry and breeding advancements. Anyone who breeds or discovers a new rye variety can apply for national plant breeders' rights at the Federal Plant Variety Office under the Plant Breeders' Rights Act (SortG). Plant breeders' rights can be obtained if a plant variety is new, distinct, uniform, stable and designated with a suitable denomination.

The definition of DUS criteria in the variety definition process is traditionally based on the

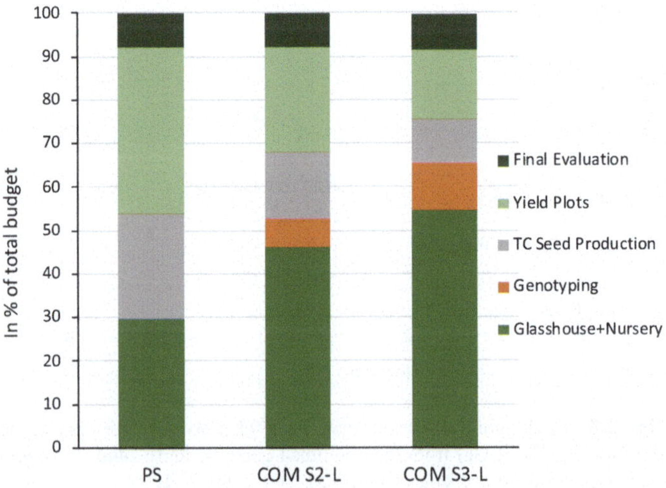

Fig. 2.7 Allocation of breeding resources to phenotypic (PS) and combined (COM) genomic and phenotypic selection in two generations (COM S2-L and COM S3-L); TC = testcross

Table 2.6 Number of entries (N_0-N_f) processed in the breeding schemes at subsequent selection stages; N_f = final number

Number of entries	Breeding scheme		
	PS	COM S2-L	COM S3-L
$N0$	7090	10,960	7520
$N1$	**709**	1096	752
$N2$	–	**439**	902[a]
$N3$	–	–	**301**
N_f	**60**	**60**	**60**

[a]Assuming 3 S3-L/selected S2-L: N1 × 0.4 × 3 = 902
Bold = Number of entries phenotyped by testcrosses

analysis of phenotypic data in standardized field trials. In agreement with a general guideline, a rye-specific procedure has been defined in document TG/58/6 of the UPOV (https://www.upov.int/tgp/de/). This guideline applies to open-pollinated varieties, but also to hybrids and their parental components. Based on characteristics such as ploidy level (2n *versus* 4n) and seasonal type (winter *versus* spring), varieties are subdivided into major groups. Then, each of the groups' 20 characteristics is described in consecutive phenological states known to be optimal for expression. These characteristics reflect the variety-specific morphology and physiology of plant organs such as grain, stem, leaf, or ear. For each trait, the method of assessment is defined either by actual measurements (M), by visual assessments based on a single observation of a group of plants or parts of plants (VG), or visual assessments from observations of a number of individual plants or plant parts (VS). The statistical analysis of the data assessed in the field trials is detailed in a Technical Working Party paper (TWC/18/10) resulting in the so-called combined-over years distinctness (COYD) and uniformity criteria (COYU), (https://www.upov.int/tgp/de/).

An actual research project investigates the use of DNA markers for DUS testing in rye (Schmid, pers. commun.). Application of DNA marker technology might accelerate the DUS procedure and thus could lead to advantages for plant breeding companies, Plant Variety Protection granting authorities and the agricultural sector. The potential use of markers in DUS testing is based on the premise that the difference between varieties, the uniformity within varieties and the stability of varieties can be determined with appropriate measures of marker diversity and, therefore, complements currently used phenotypic traits used in DUS testing.

Only if both VCU and DUS tests are successfully passed, breeding companies are allowed to sell certified seed. When buying certified seeds, farmers pay royalties to breeders, which allows them to invest in further research activities and to develop new and superior varieties. Thus, the whole value chain including farmers, food and feed producers and finally consumers benefit from better varieties.

2.8 Breeding Goals

In practical breeding programs, the assessment of candidates always involves consideration of multiple traits under selection. A breeding goal could then be formalized by an index incorporating all traits of relevance. For several decades there is a well-developed theory available to construct such indices (Sölkner et al. 2008). For example, in the optimal Smith-Hazel selection index the weight given to a specific trait can be derived from estimating its respective economic value, heritability and genetic correlations between traits (Falconer and Mackay 1996).

Because often the latter parameters are difficult to estimate with sufficient accuracy, plant breeders or variety offices use multi-trait indices, which are constructed in a more simple and intuitive way and nevertheless aim to reflect the overall breeding goal. The trait performance of a candidate line is then expressed as the difference to well-known standard entries or to the trial mean. These differences are then weighted and summed up as partial indices such as resistance or quality indices or as overall varietal indices. A candidate hybrid will be discarded by the breeder or the inscription will be refused by the variety office if a predefined threshold value is not achieved.

Taking Germany as an example for the Central European target environment, a generic overview on traits relevant for rye breeding (Table 2.7) is provided by the so-called German "descriptive variety list" (BSL 2019). The traits can be grouped into those associated with phenology, biotic and abiotic stress, yield, or quality. The high number of traits comes from the highly versatile use of rye for human food, animal feed stuff and different industrial uses (bioethanol, biogas). A more detailed description of the traits and their heritabilities can be found in Miedaner and Laidig (2019). Compared to other cereals, such as wheat and barley, rye has been regarded as feed stuff of lower value in the past. Based on new nutritional findings, rye may be experiencing a renaissance as a high value feed stuff for fattening pigs. Although crude fiber content is low, rye excels by high contents of fructans and arabinoxylans. These are metabolized in the colon to butyrate, which has positive effects on mucosal membrane health and animal welfare. Further, boar taint and infection by salmonella can be prevented. Phosphorous content in rye is low, but the grains own high phytase activity, which helps the pig to better utilize phosphate and reduce phosphate excretion (Kamphues et al. 2019).

Obviously, global climate change will also have an impact on re-defining actual and future breeding targets. Taking Central Europe and the years following 2000 as an example, there was massive drought stress in rye-growing areas in 2003, 2007, 2011, 2018 and 2019. Because drought stress occurs episodically and thus is unpredictable, farmers might adopt risk prevention strategies and replace more drought prone cereals such as wheat with the more stress tolerant rye (Schittenhelm et al. 2014). Clearly, this strategy only provides a benefit if rye hybrids show superior performance under drought but also under normal or even optimal conditions. To ensure broad adaptability of germplasm, managed drought stress environments should be included in test environments (Haffke et al. 2015).

Over the last two decades, the traditional market areas for hybrid rye in Central Europe could be expanded to other regions of the world including Russia and Canada (Fig. 2.8). With the diversification of the target areas, the breeder has to consider new breeding objectives such as frost tolerance or snow mold resistance (Miedaner and Wilde 2019). Further, genotype × environment interactions and specifically cross-over interactions have to be considered as a challenge which makes it difficult or even impossible to predict the performance of candidate genotypes across changing environments.

Inevitably, there is an impact on the correlated genetic gain for the individual traits when an index mirroring a complex breeding goal is taken as a selection criterion. The higher the number of traits and the higher their economic weights and the more they are correlated in an undesired direction, the lower will be the gain for a given trait. Thus, the breeder faces a dilemma arising from this multi-trait situation. Briefly, some ways to escape from this dilemma can be sketched out.

1. Rye breeders can apply cost-efficient multistage selection procedures as outlined in Sect. 2.6 (this chapter). Because the genetic correlation of line per se performance of a candidate to its GCA effect is high for many traits, they can be easily selected at the beginning of line development.
2. High-throughput technologies such as GS or HYP are available, allowing for testing of a large number of candidate lines.
3. The breeder could establish satellite programs besides the main program with regard to increasing the diversity of target environments. QTL for traits that are mandatory for specific environments ("must-have QTL") can then be introgressed.
4. Finally, a breeding program could be restricted to just one segment × environment-

Table 2.7 Traits for different usages of rye for human food, animal feed stuff, and industrial purposes

Type of trait	Grain			Whole-plant mass	
	Baking	Feed	Ethanol		
Early		Late			
Phenology-associated traits:					
Ear emergence	+	+	+	++	+
Ripening	+	+	+	0	+
Plant height	0	0	0	+	++
Susceptibility to stress-associated traits					
Winter killing	––	––	––	––	––
Lodging	––	––	––	0	––
Culm buckling	0	0	0	0	0
Powdery Mildew	––	––	––	––	––
Rhynchosporium	–	–	–	–	–
Leaf rust	––	––	––	0	0
Ergot	––	––	––	0	0
Yield-associated traits					
Ear density	+	+	+	+	+
No. of grains per ear	+	+	+	0	0
Thousand-grain weight	++	++	++	0	0
Grain yield (extensive)	++	++	++	0	0
Grain yield (intensive)	++	++	++	0	0
Dry matter yield	0	0	0	++	++
Quality-associated traits					
Falling number	++	+	+	0	0
Protein content	+	++	––	0	0
Amylogram viscosity	++	––	––	0	0

++, +, 0, –, –– = Very high, high, neutral, negative, and very negative importance for the trait

combination or a few of them rather than diluting the capacity on all of them.

2.9 Breeding Progress

In Germany, hybrid and population varieties have been tested in the same VCU trials since the beginning of the 1980s and recently the respective data have been analyzed for the period 1985–2016 (Miedaner and Laidig 2019). Further, from the annual special harvest survey (Besondere Ernteermittlung, BEE 2017) estimates for the average national-wide on-farm performance are available. VCU trials offer an ideal tool to dissect the genetic and non-genetic part of the overall progress made over time. Separately for each type of variety, populations and hybrids, the genetic trend can be formalized as a regression coefficient of genetic effects of new candidate lines on years of testing (Laidig et al. 2017). Analogously, the non-genetic trend can be estimated for the innovation coming from agronomy. For grain yield and related characters (Table 2.8), a high increase at the trait level from 1985 to 2016 can be observed. Whereas at the beginning hybrids out-yielded populations by

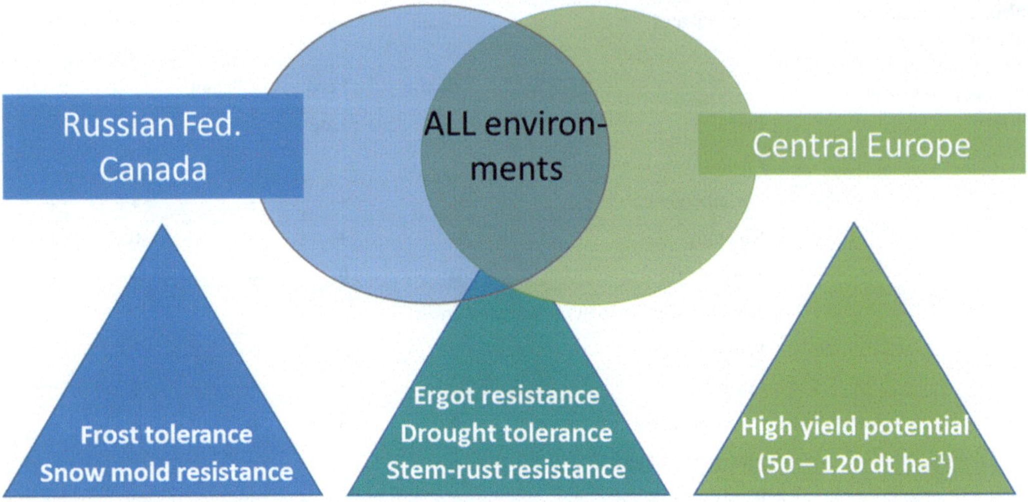

Fig. 2.8 Specific breeding goals for hybrid rye cultivars in different regions of the world

about 10% in 1985, their superiority increased to about 20% in 2016. The gap observed between types of varieties is confirmed by corresponding annual genetic trends differing by a factor of about 3 for hybrids [0.773 dt (ha year)$^{-1}$] and populations [0.237 dt (ha year)$^{-1}$]. Assuming a simple genetic model and *ceteris paribus* conditions, annual genetic progress should be similar for both types of variety (Sprague and Eberhart 1977). Breeding budget will be one of the factors probably contributing most to the actual observed large difference between them. In the case of hybrids, royalties paid by farmers each year in a foreseeable amount are re-invested into large research and development budgets benefitting from economy of scale effects and availability of modern technology. In contrast, in case of population cultivars, lower royalties and wide use of farm-saved seed substantially restrict financial returns to breeding companies. Compared to the genetic trends, the non-genetic annual trends are often lower reflecting the fact that effects from innovation in agronomy can be exploited only to a limited extent in VCU trials.

A comparison of grain yield achieved in the year 2016 on VCU (100.7 and 84.1 dt ha^{-1} for hybrids and populations, resp.) and on-farm level (55.7 dt ha^{-1}) reveals a considerable performance gap (Table 2.8). There is little impact from such performance gap under the assumption that the ranking of cultivars in the VCU and the on-farm environment is identical. However, as pointed out by Falconer and Mackay (1996, p. 322), a character measured in two different environments should not be regarded as one character, but as two. Potentially, genes required for high yield level in VCU trials represent a set of genes different from those required for the more stress-prone on-farm environment and thus the genetic correlation between the two traits might be lower than 1. Lowering the yield level in VCU trials might appear to be an obvious solution. However, trials on low-yielding or stress-prone environments often are impeded by higher error variances. Thus, there might be a trade-off between increasing the genetic correlation between the selection (VCU trial) and the target (on-farm) environment on one side and the accuracy of yield assessment on the other side.

A large difference is also found for the overall trends between VCU trials [0.866 and 0.551 dt (ha × year)$^{-1}$ for hybrids and populations, respectively] and on-farm [0.383 dt (ha × year)$^{-1}$]. Probably, a major part of this difference is due to the fact that prices for rye substantially decreased due to agro-political decisions made in the period of review which lead to shifting rye cultivation from better to less fertile soils. When

Table 2.8 Trait levels and regression coefficients of yield-related traits. Percent trends (%) given rel. to 1985 (Miedaner and Laidig 2019)

Trait	Unit	Source	Type	Trait level		Linear annual trends						r_g to
				1985	2016	Genetic		Non-genetic		Overall/on-farm		Grain
						abs[a]	%	abs	%	abs	%	Yield[b]
Grain yield	dt ha^{-1}	VCU trial	HYB	73.9	100.7	**0.773**	1.05	0.112	0.15	**0.866**	1.17	
			POP	67.0	84.1	**0.237**	0.35	**0.242**	0.36	**0.551**	0.82	
			HYB%	110.3	119.7							
		on-Farm		43.8	55.7					**0.383**	0.87	
Ear density	ears m^{-2}	VCU trial	HYB	475.9	593.1	**2.494**	0.52	1.510	0.32	**3.783**	0.80	0.55
			POP	487.9	539.8	**0.871**	0.18	0.793	0.16	1.675	0.34	
			HYB%	97.5	109.9							
Single-ear yield	g ear^{-1}	VCU trial	HYB	1.64	1.79	**0.007**	0.40	−0.002	−0.12	0.005	0.31	
			POP	1.46	1.64	**0.002**	0.17	0.002	0.15	**0.006**	0.41	—
			HYB%	112.3	109.1							
Number of kernels per ear	kernels ear^{-1}	VCU trial	HYB	51.7	49.0	0.046	0.10	0.002	−0.26	0.038	−0.17	−0.28
			POP	45.7	43.4	0.037	0.22	−0.067	−0.5	0.033	−0.17	
			HYB%	113.1	112.9							
Thousand-grain mass	g	VCU trial	HYB	34.1	37.2	**0.130**	0.30	−0.041	−0.03	0.090	0.29	−0.22
			POP	34.2	38.3	0.018	−0.09	**0.111**	0.43	**0.109**	0.38	
			HYB%	99.7	97.1							

[a]Regression coefficients, **bold letters**: significant at 5% level or higher; HYB = hybrid cultivar, POP = population cultivar, HYB%: hybrid relative to population performance
[b]Genetic correlations according to Laidig et al. (2017) (1989–2014)

farmers had to decide between genetic (population *vs.* hybrids) or agronomic (extensive *vs.* intensive use of fungicides, fertilizers and growth regulators) intensification, they often chose the first alternative as can be seen from a market share of hybrids that is now at more than 80% in Germany (Miedaner and Laidig 2019).

Driving forces for yield progress can be identified when analyzing yield components. Significant annual genetic trends can be found particularly for ear density, but also for single-ear yield. Whereas at the beginning of the reviewed period, ear density in hybrids was slightly lower (97.5%) compared to population varieties, it was substantially higher at the end (109.1%). This finding is fully in line with trends summarized in a pivotal paper on corn breeding (Duvick 2005). According to the author "newer hybrids yield more than older hybrids because of continuing improvement in ability of the hybrids to withstand the stress of higher plant density, which in turn is owed to their greater tolerance to locally important abiotic and biotic stresses." Obviously, actual rye hybrids are able to keep a higher number of tillers alive during pre-flowering reduction phases and they can maintain a sufficient sink in the later grain filling phase. The number of kernels per ear was already high in hybrid cultivars compared to population cultivars at the beginning (51.7 vs. 45.7) and did not improve further.

2.10 Conclusions and Future Developments

Rye breeding, as with breeding programs for many species, managed in the last decade to integrate molecular markers into commercial breeding schemes and to step forward to genomics-based breeding (Miedaner et al. 2019). For monogenic traits, like pollen-fertility restoration or rust resistances, KASP markers can be easily applied in large populations for the selection of the favorite plants at the seedling stage. The advent of medium-density SNP chips opened the way for assigning unknown material to heterotic groups and for developing genomic selection procedures that are now a must-have in each breeding program (see Sect. 2.6). These molecular techniques will make it easier to develop specific rye materials for new target areas and new traits.

For cross-pollinating rye, hybrid breeding allowed the systematic development of cultivars with specific performance traits, like biomass yield and disease resistance. This will foster the possible expansion of rye production to new target areas, like Russia, Canada, or the USA. While in Russia, bread and ethanol production is the main goal, in North America, stakeholders must be convinced that rye is an excellent foodstuff for livestock, because the use for bread making in these areas is limited. Another main advantage of rye, especially in areas with poor soils and regular drought stress, is the high stress-resilience of the crop. This will become even more important with the restriction of nitrogen use in industrial countries. In Denmark, for example, the rye acreage grew by 3.5 times since 2009 when the "Green Growth Agreement" released nitrogen application restrictions (https://www.statbank.dk/AFG07). Global climate change is predicted to result in less predictable weather and more episodic extreme weather events which will also contribute to the necessity for abiotic stress tolerance as a major breeding target in the future. Rye provides an excellent starting base. For this purpose, phenotyping in controlled environments is of key relevance. Techniques such as the use of robotics or automatic and remote sensing of plant stands will allow more reproducible results than classical field experiments in terms of abiotic stress. Key genomic regions identified by these techniques can be directly incorporated into elite germplasm. Further, the search for more effective alleles in the same genomic region will be facilitated and gene bank accessions (Varshney et al. 2018) can be effectively exploited. In any case, breeders should be aware that results from advanced phenotyping are indirect assessments (Fischer et al. 2014, Chap. 9). In consequence, their correlation to field performance under stress has to be verified.

Although the relative contribution of phenotyping will be reduced by the implementation of genomic selection (see Fig. 2.7), it will become even more important to produce high-quality phenotypic data. Breeders will need these data for manifold purposes including updating calibrations, identifying top performing commercial candidates, assessing their stability for yield and disease resistance across multiple locations. Organizing high-quality phenotypic data will be a challenge because advancement in this field is much slower compared to the ability to genotype large arrays of germplasm in the laboratory.

For the sake of commercial plant breeding as well as for the well-being of food production for a growing population, a steady increase of grain yield will be pivotal. This requires continuing investment into breeding programs and breeding research, but also a targeted broadening of gene pools (see Sect. 2.5 of this chapter). Because the hybrid breeder cannot benefit from breeding programs of competitors in the same effective way as the breeder of self-pollinating crops, it will get more and more important to widen the genetic variation by effective reciprocal recurrent selection programs supported by genomics and by the exploitation of new genetic materials. The main driver for effective genomic selection will be the prediction accuracy that is routinely achieved in commercial programs; this parameter will greatly benefit from the use of multiple cycles of training populations (Auinger et al. 2016). However, the use of new genetic materials will decrease the degree of relatedness between training and breeding populations. Therefore, the recalibration of genomic selection models by phenotyping training populations adjusted to the actual breeding progress will be a continuing task for the breeder.

Breeding rye as an internationally underutilized crop faces a particular challenge when it comes to training students with skills in quantitative genetics, plant molecular biology and classical breeding. To efficiently apply the new genomic tools, they should be given the opportunity to study in a multi-disciplinary and interactive learning environment.

The economic effects of plant breeding research investments are found to be highly profitable from a societal point of view (Witzke et al. 2004) and at the same time beneficial for reducing CO_2 emissions (Lotze-Campen et al. 2015).

References

Auinger H-J, Schönleben M, Lehermeier C et al (2016) Model training across multiple breeding cycles significantly improves genomic prediction accuracy in rye (*Secale cereale* L.). Theor Appl Genet 129:2043–2053

Bauer E, Schmutzer T, Barilar I et al (2017) Towards a whole-genome sequence for rye (*Secale cereale* L.) Plant J 89:853–869

Bernal-Vasquez AM, Gordillo A, Schmidt M, Piepho HP (2017) Genomic prediction in early selection stages using multi-year data in a hybrid rye breeding program. BMC Genet 18(1):51

Besondere Ernte-und Qualitätsermittlung (BEE) (2017) Reihe: Daten–analyse. (Special harvest and quality survey. series: data analysis) (In German). https://www.bmel-statistik.de/. Accessed on 29 June 2018

Böhm J, Schipprack W, Utz HF, Melchinger AE (2017) Tapping the genetic diversity of landraces in allogamous crops with doubled haploid lines: a case study from European flint maize. Theor Appl Genet 130:861–873

Börner A, Korzun V, Polley A, Malyshev S, Melz G (1998) Genetics and molecular mapping of a male fertility restoration locus (*Rfg1*) in rye (*Secale cereale* L.). Theor Appl Genet 97(1–2):99–102

BSL (2019) Beschreibende Sortenliste für Getreide, Mais, Öl- und Faserpflanzen, Leguminosen, Rüben, Zwischenfrüchte [Descriptive variety list for cereals, maize, oil and fibre plants, pulse crops, beets, catch crops, in German]. Bundessortenamt, Hannover

Bundessortenamt (2019) Admission to the National List. Internet: https://www.bundessortenamt.de/bsa/en/variety-testing/national-listing/. Accessed on 11 Sept 2019

Cavanagh C, Morell M, Mackay I, Powell W (2008) From mutations to MAGIC: resources for gene discovery, validation and delivery in crop plants. Curr Opinion Plant Biol 11:215–221

Crossa J, Pérez-Rodríguez P, Cuevas J et al (2017) Genomic selection in plant breeding: methods, models, and perspectives. Trends Pl Sci 22:961–975

Curtis CA, Lukaszewski AJ (1993) Localization of genes in rye that restore male fertility to hexaploid wheat with timopheevi cytoplasm. Plant Breed 111:106–112

Dekkers JCM (2007) Prediction of response to marker-assisted and genomic selection using selection index theory. J Anim Breed Genet 124(6):331–341

Duvick DN, Smith JSC, Cooper M (2004) Long-term selection in a commercial hybrid maize breeding program. In: Janick J (ed) Plant Breed Rev 24:109–151

Duvick DN (2005) Genetic progress in yield of United States maize (*Zea mays* L.). Maydica 50:193–202

Erath W, Bauer E, Fowler DB, Gordillo A, Korzun V, Ponomareva M, Schmidt M, Schmiedchen B, Wilde P, Schön CC (2017) Exploring new alleles for frost tolerance in winter rye. Theor Appl Genet 130:2151–2164

Falconer DS, Mackay TFC (1996) Introduction to quantitative genetics, 4th edn. Longman Science and Technology, Harlow, UK

Falke KC, Wilde P, Miedaner T (2009) Rye introgression lines as source of alleles for pollen-fertility restoration in Pampa CMS. Plant Breed 128:528–531. https://doi.org/10.1111/j.1439-0523.2008.01589.x

FAOSTAT (2020) Production. Crops. "Rye", "World", FAO, Rome, Italy. http://www.fao.org/faostat/en/#data/QC. Accessed on 09 June 2020

Fischer S, Melchinger AE, Korzun V et al (2010) Molecular marker assisted broadening of the Central European heterotic groups in rye with Eastern European germplasm. Theor Appl Genet 120:291–299

Fischer RA, Byerlee D, Edmeades GD (2014) Crop yields and global food Security. ACIAR: Canberra, ACT

Frederiksen SE, Petersen G (1998) A taxonomic revision of Secale (*Triticeae, Poaceae*). Nord J Bot 18:399–420

Gaikpa DS, Koch S, Fromme FJ, Siekmann D, Würschum T, Miedaner T (2020) Genome-wide association mapping and genomic prediction of Fusarium head blight resistance, heading stage and plant height in winter rye (*Secale cereale*). Plant Breed 139:508–520. https://doi.org/10.1111/pbr.12810

Galán RJ, Bernal-Vasquez AM, Jebsen C, Piepho HP, Thorwarth P, Steffan P, Gordillo A, Miedaner T (2019) Hyperspectral reflectance data and agronomic traits can predict biomass yield in winter rye hybrids. BioEnergy Res 13:168–182. https://doi.org/10.1007/s12155-019-10080-z

Galán RJ, Bernal-Vasquez AM, Jebsen C, Piepho HP, Thorwarth P, Steffan P, Gordillo A, Miedaner T (2020) Integration of genotypic, hyperspectral, and phenotypic data to improve biomass yield prediction in hybrid rye. Theor Appl Genet 133:3001–3015. https://doi.org/10.1007/s00122-020-03651-8

Gaynor RC, Gorjanc G, Bentley AR, Ober ES, Howell P, Jackson R, Mackay IJ, Hickey JM (2017) A two-part strategy for using genomic selection to develop inbred lines. Crop Sci 57:1–15

Geiger HH (1982) Breeding methods in diploid rye (*Secale cereale* L.). Tag-Ber. Akad. Landwirtsch.-Wiss. DDR 198:305–332

Geiger HH, Miedaner T (2009) Rye Breeding. In: Carena MJ (ed) Cereals (handbook of plant breeding), 1st edn. Springer, New York, pp 157–181

Geiger HH, Morgenstern K (1975) Angewandt-genetische Studien zur cytoplasmatischen Pollensterilität bei Winterroggen [Applied genetic studies on cytoplasmic pollen sterility in winter rye, in German]. Theor Appl Genet 46(6):269–276. https://doi.org/10.1007/BF00281148

Geiger HH, Schnell FW (1970) Cytoplasmic male sterility in rye (*Secale cereale* L.). Crop Sci 10:590–593

Geiger HH, Wahle G (1978) Struktur der Heterosis von Komplexmerkmalen bei Winterroggen-Einfachhybriden [Structure of heterosis regarding complex traits of single cross hybrids in winter rye, in German]. Z. Pflanzenzüchtg 80:198–210

Geiger HH, Yuan Y, Miedaner T et al (1995) Environmental sensitivity of cytoplasmic genic male sterility (CMS) in *Secale cereale* L. In: Kück U, Wricke G (eds) Genetic mechanisms for hybrid breeding. Adv Plant Breed 18:7–17

Geiger HH, Wilde P, Erfurt M, Pakac I (2001) Heterosis of factorial inter-pool single crosses among elite winter rye inbred lines. In: Proceedings of EUCARPIA Rye Meeting July 4–7, 2001, Radzików, Poland, p 19

Gerke JP, Edwards JW, Guill KE, Ross-Ibarra J, McMullen MD (2015) The genomic impacts of drift and selection for hybrid performance in maize. Genetics 201:1201–1211

Gertz A, Wricke G (1989) Linkage between the incompatibility locus Z and a *β-glucosidase isozyme locus (β-Glu)* in rye. Plant Breed 102:255–259

Gruner P, Schmitt A-K, Flath K, Schmiedchen B, Eifler J, Gordillo A, Schmidt M, Korzun V, Fromme F-J, Siekmann D, Tratwal A, Danielewicz J, Korbas M, Marciniak K, Kyrsztofik R, Niewińska M, Koch S, Piepho H-P, Miedaner T (2020) Mapping stem rust (Puccinia graminis f. sp. secalis) resistance in self-fertile winter rye populations. Front Plant Sci 11:667. https://www.frontiersin.org/articles/10.3389/fpls.2020.00667/full

Hackauf B, Wehling P (2005) Approaching the self-incompatibility locus Z in rye (*Secale cereale* L.) via comparative genetics. Theor Appl Genet 110:832–845

Hackauf B, Stojałowski S, Wortmann H, Wilde P, Fromme FJ, Menzel J, Korzun V, Wehling P (2009) Minimierung des Mutterkornbefalls im Hybridroggen durch Ansätze der Präzisionszüchtung [Minimizing ergot severity in hybrid rye by precision breeding approaches, in German]. J Kulturpfl 61:15–20. https://doi.org/10.5073/JfK.2009.01.03

Hackauf B, Korzun V, Wortmann H, Wilde P, Wehling P (2012) Development of conserved ortholog set markers linked to the restorer gene *Rfp1* in rye. Mol Breeding 30:1507–1518

Haffke S, Wilde P, Schmiedchen B et al (2015) Toward a selection of broadly adapted germplasm for yield stability of hybrid rye under normal and managed drought stress conditions. Crop Sci 55:1026–1034

Hackauf B, Bauer E, Korzun V et al (2017a) Fine mapping of the restorer gene *Rfp3* from an Iranian primitive rye (*Secale cereale* L.). Theor Appl Genet 130:1179–1189

Hackauf B, Haffke S, Fromme FJ et al (2017b) QTL mapping and comparative genome analysis of agronomic traits including grain yield in winter rye. Theor Appl Genet 130:1801–1817

Haghighattalab A, González Pérez L, Mondal S et al (2016) Application of unmanned aerial systems for high throughput phenotyping of large wheat breeding nurseries. Plant Methods 12:35. https://doi.org/10.1186/s13007-016-0134-6

Hallauer AR, Carena MJ (2009) Maize. In: Carena MJ (ed) Cereals (handbook of plant breeding), 1st edn. Springer, New York, pp 3–98

Hallauer AR, Carena MJ, Miranda Filho JD (2010) Quantitative genetics in maize breeding, vol 6. Springer Science & Business Media

Hartmann A (1997) Untersuchungen zur genetischen Variation von Winterroggen unter ortsüblicher und verringerter Stickstoffversorgung [Studies on genetic variation of winter rye under usual and reduced nitrogen supply, in German]. PhD thesis, University of Hohenheim, Stuttgart, Germany

Haseneyer G, Schmutzer T, Seidel M et al (2011) From RNA-seq to large-scale genotyping-genomics resources for rye (*Secale cereale* L.). BMC Plant Biol 11:131

Haussmann BIG, Parzies HK, Presterl T et al (2004) Plant genetic resources in crop improvement. Plant Genet Resour 2:3–21

Hepting L (1978) Analyse eines 7 × 7–Sortendiallels zur Ermittlung geeigneten Ausgangsmaterials für die Hybridzüchtung bei Roggen [Analysis of a 7 × 7–variety diallel for determination of suitable base materials for hybrid breeding in rye, in German]. Z Pflanzenzüchtg 80:188–197

Hickey LT, German SE, Pereyra SA, Diaz JE, Ziems LA, Fowler RA, Platz GJ, Franckowiak JD, Mark DMJ (2017) Speed breeding for multiple disease resistance in barley. Euphytica 213:64

Jannink JL, Lorenz AJ, Iwata H (2010) Genomic selection in plant breeding: from theory to practice. Briefings Funct Genom 9:166–177

Kalih R, Maurer HP, Hackauf B et al (2014) Effect of rye dwarfing gene on plant height, heading stage, and Fusarium head blight in triticale (x*Triticosecale* Wittmack). Theor Appl Genet 127:1527–1536

Kalih R, Maurer HP, Miedaner T (2015) Genetic architecture of Fusarium head blight resistance in four winter triticale populations. Phytopathol 105:334–341. https://doi.org/10.1094/Phyto-04-14-0124-R

Kamphues J, Grone R, Bunte S, Visscher C, Hartung C, Wilke V, Felde VA (2019) The nutritional value of rye—arguments from the veterinary point of view. Oral

presentation at the Danish crown meeting on rye feeding-Vejle
Kobyljanski VD, Katerova AG (1973) Inheritance of cytoplasmatic and nuclear male sterility in Oshima diploid rye. Genetika 9:5–11
Kobylanski VD, Solodukhina OV (1996) Genetic bases and practical breeding utilization of heterogenous resistance of rye to brown rust. Vortr Pflanzenzüchtg 35:155–163
Köhler KL (1986) Experimentelle Untersuchungen zur Aufspaltungsvariation und -kovariation der Linieneigenleistung und Kombinationsfähigkeit bei Roggen [Experimental studies on segregation variance and covariance of line per se performance and combining ability in rye. In German]. PhD thesis, University of Hohenheim, Stuttgart, Germany
Laidig F, Piepho HP, Rentel D, Drobek T, Meyer U, Huesken A (2017) Breeding progress, variation, and correlation of grain and quality traits in winter rye hybrid and population varieties and national on-farm progress in Germany over 26 years. Theor Appl Genet 130:981–998. https://doi.org/10.1007/s00122-017-2865-9
Łapiński M (1972) Cytoplasmic-genic type of male sterility in *Secale montanum* Guss. Wheat Inf Serv 35:25–28
Łapiński M, Stojałowski S (2001) Occurrence of male sterility-inducing cytoplasm in rye (*Secale* spp.) populations. Pant Breed Seed Sci 48:7–23
Li Y, Haseneyer G, Schön CC, Ankerst D, Korzun V, Wilde P, Bauer E (2011) High levels of nucleotide diversity and fast decline of linkage disequilibrium in rye (*Secale cereale* L.) genes involved in frost response. BMC Plant Biol 11:6
Lotze-Campen H, Witzke VH, Noleppa S, Schwarz G (2015) Science for food, climate protection and welfare: an economic analysis of plant breeding research in Germany. Agric Syst 136:79–84
Lundqvist A (1956) Self-incompatibility in rye. I. Genetic control in the diploid. Hereditas 42:293–348
Mahlein AK, Oerke EC, Steiner U, Dehne HW (2012) Recent advances in sensing plant diseases for precision crop protection. Europ J Plant Pathol 133:197–209
Marulanda JJ, Mi X, Melchinger AE et al (2016) Optimum breeding strategies using genomic selection for hybrid breeding in wheat, maize, rye, barley, rice and triticale. Theor Appl Genet 129:1901–1913
Melchinger AE, Gumber RK (1998) Overview of heterosis and heterotic groups in agronomic crops. In: Lamkey KR, Staub JE (eds) Concepts and breeding of heterosis in crop plants. Crop Science Society of America, Madison, WI, pp 29–44
Melchinger AE, Geiger HH, Seitz G, Schmidt GA (1987) Optimum prediction of three-way crosses from single crosses in forage maize (*Zea mays* L.). Theor Appl Genet 74:339–345
Melz G, Kaczmarek J, Szigat G (1990) Genetical analysis of rye (*Secale cereale* L.). Location of self-fertility genes in different inbred lines. Genet Polon 31:1–7
Melz G, Melz G, Hartmann F (2003) Genetics of a male-sterile rye of "G-type" with results of the first F1-hybrids, *Plant Breed.* Seed Sci 47:47–55
Miedaner T, Geiger HH (2015) Biology, genetics, and management of ergot (*Claviceps* spp.) in rye, sorghum, and pearl millet. Toxins 7:659–678. https://doi.org/10.3390/toxins7030659
Miedaner T, Laidig F (2019) Hybrid breeding in rye (*Secale cereale* L.). In: Al-Khayri JM, Jain SM, Johnson DV (eds) Advances in plant breeding strategies: Cereals, vol 5. Springer, Cham, pp 343–372
Miedaner T, Wilde P (2019) Selection strategies for disease-resistant hybrid rye. In: Ordon F, Friedt W (eds) Advances in breeding techniques for cereals. Burleigh Dodds Science Publishing, Cambridge, UK, pp 223–246. ISBN: 978-1786762443
Miedaner T, Glass C, Dreyer F et al (2000) Mapping of genes for male–fertility restoration in 'Pampa' CMS winter rye (*Secale cereale* L.). Theor Appl Genet 101:1226–1233
Miedaner T, Schneider B, Geiger HH (2003a) Deoxynivalenol (DON) content and Fusarium head blight resistance in segregating populations of winter rye and winter wheat. Crop Sci 43:519–526
Miedaner T, Wortmann H, Geiger HH (2003b) Genetics of deoxynivalenol (DON) contamination caused by Fusarium head blight in hybrid rye. Plant Breed Seed Sci 48:69–78
Miedaner T, Wilde P, Wortmann H (2005) Combining ability of non-adapted sources for male-fertility restoration in Pampa CMS of hybrid rye. Plant Breed 124:39–43
Miedaner T, Dänicke S, Schmiedchen B, Wilde P, Wortmann H, Dhillon BS, Geiger HH, Mirdita V (2010) Genetic variation for ergot (*Claviceps purpurea*) resistance and alkaloid concentrations in cytoplasmic-male sterile winter rye under pollen isolation. Euphytica 173:299–306. https://doi.org/10.1007/s10681-009-0083-5
Miedaner T, Hübner M, Korzun V et al (2012) Genetic architecture of complex agronomic traits examined in two testcross populations of rye (*Secale cereale* L.). BMC Genom 13:706. 10.1186/1471-2164-13-706
Miedaner T, Schwegler DD, Wilde P et al (2014) Association between line per se and testcross performance for eight agronomic and quality traits in winter rye. Theor Appl Genet 127:33–41
Miedaner T, Schmitt AK, Klocke B et al (2016) Analyzing genetic diversity for virulence and resistance phenotypes in populations of stem rust (*Puccinia graminis* f. sp. *secalis*) and winter rye (*Secale cereale* L.). Phytopathol 106:1335–1343
Miedaner T, Herter CP, Goßlau H et al (2017) Correlated effects of exotic pollen-fertility restorer genes on agronomic and quality traits of hybrid rye. Plant Breed 136:224–229
Miedaner T, Korzun V, Bauer E (2019) Genomics-based hybrid rye breeding. In: Miedaner T, Korzun V (eds) Applications of genetic and genomic research

in cereals. Elsevier, Amsterdam, NL, pp 329–348. ISBN: 9780081021637

Mihaljevic R, Schön CC, Utz HF, Melchinger AE (2005) Correlations and QTL correspondence between line *per se* and testcross performance for agronomic traits in four populations of European maize. Crop Sci 45:114–122

Musa GLC, Dyck PL, Samborski DJ (1984) The inheritance of resistance in rye to *Puccinia recondita* f. sp. *secalis* and f. sp. *tritici*. Can J Plant Sci 64:511–519

Müller D, Schopp P, Melchinger AE (2017) Persistency of prediction accuracy and genetic gain in synthetic populations under recurrent genomic selection. G3 (Bethesda) 7:801–811

Nei M (1987) Molecular evolutionary genetics. Columbia University Press, New York, USA

Oelke EA, Oplinger ES, Bahri H et al (1990) Rye. In: Alternative field crops manual. University of Wisconsin Extension, Madison and University of Minnesota, St. Paul. http://www.hort.purdue.edu/newcrop/afcm/rye.html. Accessed on 23 Aug 2019

Ossent HP (1938) Zehn Jahre Roggenzüchtung in Müncheberg [Ten years of rye breeding in Müncheberg, in German]. Züchter 10:255–261

Pasquariello M, Barabaschi D, Himmelbach A, Steuernagel B, Ariyadasa R, Stein N, Gandolfi F, Tenedini E, Bernardis I, Tagliafico E, Pecchioni N, Francia E (2014) The barley *frost resistance-H2* locus. Funct Integr Genom 14:85–100

Rembe M, ZhaoY JY, Reif JC (2019) Reciprocal recurrent genomic selection: an attractive tool to leverage hybrid wheat breeding. Theor Appl Genet 132:687–698

Roux SR, Hackauf B, Linz A et al (2004) Leaf-rust resistance in rye (*Secale cereale* L.). 2. Genetic analysis and mapping of resistance genes *Pr3, Pr4*, and *Pr5*. Theor Appl Genet 110:192–201

Schittenhelm S, Kraft M, Wittich KP (2014) Performance of winter cereals grown on field-stored soil moisture only. Eur J Agron 52:247–258

Schlegel R (2016) Hybrid breeding boosted molecular genetics in rye. Russ J Genet-Appl Research 6 (5):569–583

Schnell FW (1974) Trends and problems in breeding methods for hybrid corn. In: Proceedings of 16th British Poultry Breeders Round Table. Birmingham, UK, pp 86–98

Schnell FW (1983) Probleme der Elternwahl - ein Überblick [Problems of parental choice - an overview, in German] In: Ber. Arb.tag. Arb.gem. Saatzuchtleiter Gumpenstein, Austria, pp 1–11

Sencer HA, Hawkes G (1980) On the origin of cultivated rye. Biol J Linn Soc 13:299–313

Solodukhina OV (1994) Rye resistance to brown rust and powdery mildew: the potential of hereditary variability [in Russian]. Genetika 30:616–618

Sölkner J, Grausgruber H, Okeyo AM, Ruckenbauer P, Wurzinger M (2008) Breeding objectives and the relative importance of traits in plant and animal breeding: a comparative review. Euphytica 161:273–282

Sprague GF, Eberhart SA (1977) Corn breeding. In: Sprague GF (ed) Corn and corn improvement, 2nd ed. American Society of Agronomy, Madison, pp 305–363

Stojałowski S, Kociuba M, Stochmal B, Kondzioła M, Jaciubek M (2008) Determining the plasmotypic structure of rye populations by SCAR markers. J Appl Genet 49:229–232

Stojałowski S, Jaciubek M, Masojć P (2005) Rye SCAR markers for male fertility restoration in the P cytoplasm are also applicable to marker-assisted selection in the C cytoplasm. J Appl Genet 46(4):371–373

Stojałowski S, Łapiński M, Masojć P (2004) RAPD markers linked with restorer genes for the C-source of cytoplasmic male sterility in rye (*Secale cereale* L.) Plant Breed 123:428–433

Stracke S, Schilling AG, Förster J et al (2003) Development of PCR-based markers linked to dominant genes for male-fertility restoration in Pampa CMS of rye (*Secale cereale* L.). Theor Appl Genet 106:1184–1190

Tomerius A, Miedaner T, Knopf E, Wilde P, Wortmann H, Fromme FJ, Geiger HH (1997) Schätzung zuchtmethodischer Kenngrößen für die Entwicklung von Saatelterlinien in der Hybridroggenzüchtung [Estimation of breeding parameters for the development of seed parent lines in hybrid rye breeding, in German]. Ber. Arb.tag. Arb.gem. Saatzuchtleiter Gumpenstein 48:57–60

Tomerius AM, Miedaner T, Geiger HH (2008) A model calculation approach towards the optimization of a standard scheme of seed-parent line development in hybrid rye breeding. Plant Breeding 127:433–440

Trang QS, Wricke G, Weber WE (1982) Number of alleles at the incompatibility locus in *Secale cereale* L. Theor Appl Genet 63:245–248

Utz HF (1984) Calculating and maximizing the gain from selection. Vortr Pflanzenzüchtg 7:30–40

Varshney RK, Singh VK, Kumar A, Powell W, Sorrells ME (2018) Can genomics deliver climate-change ready crops? Curr Opinion Plant Biol 45:205–211

Voylokov AV, Fuong FT, Smirnov VG (1993) Genetic studies of self-fertility in rye (*Secale cereale* L.). 1. The identification of genotypes of self-fertile lines for the *Sf* alleles of self-incompatibility genes. Theor Appl Genet 87:616–618

Wang Y, Mette MF, Miedaner T et al (2015) First insights into the genotype-phenotype map of phenotypic stability in rye. J Exp Bot 66:3275–3284

Wang Y, Mette MF, Miedaner T et al (2014) The accuracy of prediction of genomic selection in elite hybrid rye populations surpasses the accuracy of marker-assisted selection and is equally augmented by multiple field evaluation locations and test years. BMC Genom 15:556

Warzecha R, Salak-Warzecha K (2003) A new source of male sterility in rye (*Secale cereale* L.). Plant Breed Seed Sci. 48:61–65

Wehling P, Linz A, Hackauf B et al (2003) Leaf-rust resistance in rye (*Secale cereale* L.). 1. Genetic analysis and mapping of resistance genes *Pr1* and *Pr2*. Theor Appl Genet 107:432–438

Wilde P (1987) Schätzung von Populationsparametern und Untersuchungen zur Effizienz verschiedener Verfahren der rekurrenten Selektion bei Winterroggen [Estimation of population parameters and studies on the efficacy of different recurrent selection procedures in winter rye, in German]. PhD thesis, University of Hohenheim, Stuttgart, Germany

Wilde P (1996) Multi-stage selection for combining ability among pollen parent lines in hybrid rye breeding. Vortr Pflanzenzüchtg 35:15–26

Wilde P, Korzun V, Menzel J et al (2017) Restorer-Pflanze. Patentschrift. DE102015016445A1, International: WO 2017/109012 AI

Wilde P, Korzun V, Schmiedchen B, Goncharenko AA, Torop AA (2006) Genetic diversity within and among Middle and Eastern European rye populations determined by phenotypic and SSR marker data. Vortr Pflanzenzüchtung 71:6–17

Witzke VH, Jechlitschka K, Kirschke D, Lotze-Campen H, Noleppa S (2004) Social rate of return to plant breeding research in Germany. Agrarwirt 53:206–210

Wricke G, Wehling P (1985) Linkage between an incompatibility locus and a peroxidase isozyme locus (*PrxT*) in rye. Theor Appl Genet 71:289–292

Wright S (1978) Evolution and genetics of populations, vol 4. University of Chicago Press, Chicago, IL

Würschum T (2019) Modern field phenotyping opens new avenues for selection. In: Miedaner T, Korzun V (eds) Applications of genetic and genomic research in cereals. Elsevier, Amsterdam, pp 233–250. ISBN: 9780081021637

Xue J, Su B (2017) Significant remote sensing vegetation indices: a review of developments and applications. J Sens. Article ID 1353691.https://doi.org/10.1155/2017/1353691

Yu X, Li X, Guo T, Zhu C, Wu Y, Mitchell S, Roozeboom K, Wang D, Wang M, Pederson G, Tesso T, Schnable P, Bernardo R, Yu J (2016) Genomic prediction contributing to a promising global strategy to turbocharge gene banks. Nat Plants 2:16150

Rye Cytogenetics and Chromosome Genomics

3

Elena Mikhailova and Jaroslav Doležel

Abstract

Rye (*Secale cereale* L., $2n = 2x = 14$) has a large genome of about 8 Gbp distributed across seven large chromosomes. Although they are easily observable by microscopy, their identification is difficult due to similar morphology. Thus, wheat-rye chromosome addition and substitution lines were originally employed to accomplish this and establish the homology of rye chromosomes with those of Triticinae species. The introduction of differential staining and fluorescence in situ hybridization (FISH) provided an important advance, but chromosome identification was still hampered by polymorphism of chromosome banding patterns. A different approach to identify chromosomes involves crosses of a sample to a tester set of wheat-rye chromosome addition lines and cytological analysis of chromosome pairing during meiosis in F_1 hybrids. While FISH enabled the analysis of long-range molecular organization of the chromosomes, genomic in situ hybridization (GISH) using rye genomic DNA as probe allowed identification of rye chromosomes introgressed to wheat, including interspecific chromosome translocations. The analysis of isolated mitotic metaphase chromosomes by flow cytometry enabled identification of chromosome 1R, and, if present, the accessory B chromosome. The two chromosomes could be purified by flow sorting for downstream analyses. Chromosomes 2R–7R could not be discriminated from each other and thus were flow-sorted individually from respective wheat-rye chromosomes addition lines, as was the short arm of 1R (1RS). Flow sorting of rye chromosomes facilitated the development of chromosome-specific molecular markers. Next generation sequencing of flow-sorted B chromosomes provided insights into their molecular organization and origin (see Chap. 4). Sequencing each of the seven rye chromosomes resulted in the first draft genome sequence, informing about rye gene complement and evolution, and recently facilitated the assembly of a rye reference genome.

E. Mikhailova (✉)
Saint-Petersburg State University, 7/9 Universitetskaya Emb., Saint-Petersburg, Russia199034
e-mail: emikhailova2010@gmail.com

J. Doležel
Center of the Region Haná for Biotechnological and Agricultural Research, Institute of Experimental Botany of the Czech Academy of Sciences, Šlechtitelů 31, 779 00 Olomouc, Czech Republic

3.1 Nuclear Genome Size

The first estimate of the amount of nuclear DNA in rye was published by Evans et al. (1972). Using Feulgen microspectrophotometry, he estimated a 1C amount to be 9.5 pg DNA (9291 Mbp), while

© The Author(s), under exclusive license to Springer Nature Switzerland AG 2021

M. T. Rabanus-Wallace and N. Stein (eds.), *The Rye Genome*, Compendium of Plant Genomes,
https://doi.org/10.1007/978-3-030-83383-1_3

in a subsequent study, Bennett and Smith (1976) estimated a 1C amount to be 8.8 pg DNA (8606 Mbp) in cv. Petkus Spring. The analyses using flow cytometry gave similar, but somewhat lower estimates. Thus, Doležel et al. (1998) estimated 1C equal to 8.095 pg DNA (7917 Mbp) for cv. Dankovske, Zonneveld et al. (2005) estimated 1C = 7.8 pg DNA (7623 Mbp) in cv. Petkus Spring, a mean 1C value of 8.65 pg DNA (8460 Mbp) was obtained by Eilam et al. (2007) after analyzing four different lines of rye, and Doležel et al. (2018) estimated a 1C DNA amount of 7.975 pg DNA (7800 Mbp) in inbred line Lo7. Most recent estimates of 1C DNA amounts by Rabanus-Wallace et al. (2021) in ten diploid accessions of rye ranged from 8.035 pg (7858 Mbp) in cv. Weining to 8.215 pg (8034 Mbp) in cv. Puma. Different estimates by various studies may be due to different methods and reference standards used, and may also reflect variation in DNA repeat content among the accessions. As the plants of cv. Puma analyzed by Rabanus-Wallace et al. (2021) contained either no, two or four B chromosomes, it was even possible to estimate the molecular size of the rye B chromosome to be 541 Mbp (Doležel and Čížková, own unpublished data). This value is in line with the estimate of ~580 Mbp given by Martis et al. (2012) (see also Chap. 4).

3.2 Rye Chromosome Morphology —Comparative Studies

Identification of individual rye chromosomes by microscopic observation of mitotic metaphase plates after simple staining methods was hampered by their morphological similarity. The first attempts of Avdulow (1931), Lewitsky (1931) and Aase (1935) focused on obtaining morphological characteristics of chromosomes (Hasegawa 1934; Oinuma 1953; Riley and Chapman 1958; Bhattacharyya and Jenkins 1960; Heneen 1962; Tarkowski and Stefanovska 1972; Merker 1973; Vosa 1974; Tikhonovich 1975; Moshkovich, Chebotar 1976). However, the use of morphometric methods led to discrepancies between the results of individual studies caused by chromosome polymorphisms due to structural rearrangements in different populations of wild rye species and cultivars (Müntzing and Prakken 1941; Thompson and Rees 1956; Sybenga 1959; Hrishi and Muntzing 1960; Rees 1961; Ahloowalia 1963; Candela et al. 1979; Smirnov and Sosnikhina 1984). To overcome this obstacle, chromosomes in inbred lines of rye (Heneen 1962; Tikhonovich 1975), primary trisomics (Kamanoi and Jenkins 1962, Sybenga 1965) and a translocation tester set (Sybenga and Wolters 1972) were analyzed. Chromomere analysis (Lima-de-Faria 1953) as well as densitometry of DNA along rye chromosomes (Heneen and Caspersson 1973) in inbred stocks paved the way to more reliable chromosome identification.

3.3 Functional Genetic Criteria for Chromosome Identification within Subtribe Triticinae

Wheat-rye chromosome addition and substitution lines (Sears 1952, 1966) were used for a genetically-based rye chromosome identification and to establish homoeology relationships within the subtribe Triticinae (reviewed by Gupta 1971; Miller 1984). As the result, homoeologous groups of chromosomes were introduced: the letter R standing for rye (Table 3.1). However, the functional criterion had certain limitations, namely the peculiar way the rye traits manifested in wheat-rye addition or substitution lines. This complicated the classification of rye chromosomes relying on this approach.

Crosses of wheat-rye addition lines of different origin were followed by the analyses of chromosome pairing in meiosis of F_1 hybrids in parallel with genetic data (Koller and Zeller 1976). The attempt was made to establish the identity of rye chromosomes present in wheat-rye chromosome addition lines Chinese Spring/Imperial CR, Holdfast/King II, and Kharkov/Dakold RAV. The homology of chromosomes C, IV, and V (see Table 3.1, lines 2, 3, 4) was established and some genes were localized on this chromosome (Riley and Chapman 1958; Evans and Jenkins 1960; Darvey 1973; Rao 1975).

Table 3.1 Possible conformance of chromosome designation in tester translocation lines and addition lines as proposed by different authors

N								Method/Plant material used for chromosome designation	References
1	1R	2R	3R	4R/7R*	5R	6R	7R/4R*	Functional test/'Imperial' rye addition lines	Gupta (1971)
2	E	B	G	C	A	F	D	Functional test, plant morphology/ 'Imperial' rye addition lines	Sears (1966)
3	V	III	VI	IV	I	II	VII	Functional test, plant morphology/ 'Imperial' rye addition lines	Riley (1965)
4	VII	II	I	V	VI	IV	III	Functional test, plant morphology/ 'Imperial' rye addition lines	Evans and Jenkins (1960)
5	7	2	3	1	6	4	5	Giemsa C-banding, morphometry/ 'Peterhof' rye genetic collection	Tikhonovich (1975)
6	VII	II	I	V	VI	IV	III	Giemsa C-banding, morphometry/Translocation tester set	de Vries and Sybenga (1976)
7	VII	I	II	IV	VI	V	III	Giemsa C-banding/'Imperial' rye addition lines, Sybenga's translocation tester lines, six types of trisomics	Zeller et al. (1977)
8	1R VII	2R II	3R I	4RV	5RVI	6R IV	7R III	Giemsa C-banding/'Imperial' rye addition lines, Sybenga's translocation tester lines	Schlegel and Mettin (1982)
9	VII	III	II	IV	VI	V	I	Giemsa C-banding/Translocation tester lines are intercrossed to 'Imperial' rye addition lines Meiosis pairing criterion is used for rye chromosome identification	Sybenga et al. (1985)

Comments to Table 3.1. Line 1 is based on the analyses of rye-wheat addition and substitution lines
*Partially functional substitution of wheat chromosomes by those of rye led to the prediction of 7R/4R and 4R/7R translocations (Gupta 1971; Darvey and Gustafson 1975; Zeller et al. 1977)

3.4 C-bands as Markers for Cytogenetic Analysis

Differential staining revolutionized cytogenetics in humans as well as in some animals and plants. It allowed identification of individual chromosomes and contributed to a better understanding of their longitudinal organization and characterization of structural changes (Vosa 1977). This raised hopes that it would facilitate chromosome identification also in rye. The hopes were fulfilled to some extent after the application of two different methods: fluorescent staining with Hoechst 33,258 (Sarma and Natarajan 1973; Vosa 1974) and chromosome denaturation-renaturation followed by staining with Giemsa (Merker 1973; Gill and Kimber 1974; Verma and Rees 1974; Shchapova 1974; Darvey and Gustafson 1975; Schlegel and Fridrich 1975; Singh and Röbbelen 1975; Tikhonovich 1975; Weimarck 1975; Gustafson et al. 1976; de Vries and Sybenga 1976; Kranz 1976; Zeller et al. 1977; Vosa 1977; Jones 1978; Pilch 1978; Giraldes et al. 1979).

Successful applications of Giemsa differential staining in rye included identification of all chromosomes of the complement (Gill and Kimber 1974; Verma and Rees 1974; de Vries and Sybenga 1976), in case of translocations (de Vries and Sybenga 1976, Singh and Röbbelen 1977), additional chromosomes in trisomics (Zeller et al. 1977; Pilch 1978), and addition lines of wheat (Darvey and Gustafson 1975; de

Vries and Sybenga 1976). A recognition of individual chromosomes became possible in interspecific hybrids and amphidiploids, mainly in Triticale (Merker 1973, 1976; Weimarck 1974; Darvey and Gustafson 1975, Iordansky et al. 1978; Lukaszewski and Gustafson 1987). Furthermore, the comparison of karyotypes of wild rye species (Fig. 3.1) with those of varieties fostered studies of the evolution of the genus *Secale* (Kranz 1976; Schapova and Kobylyansky 1976, Singh and Röbbelen 1977; Vosa 1977).

Differential staining revealed intra- and inter-species variability of heterochromatin in rye (Fig. 3.2) (Tikhonovich and Fadeyeva 1976; Giraldes et al. 1979; Atayeva et al. 1982; Semenov and Semenova 1982). Heterochromatin blocks of contrasting size were used as cytological markers to follow chromosome behavior during meiosis and at pre-meiotic stages (Kranz 1976; Thomas and Kaltsikes 1976; Bowman and Rajhathy 1977; Jones 1978; Giraldes and Orellana 1979; Tikhonovich et al. 1987). A series of studies of chromosome behavior in meiosis of inbred lines with respect to the amount of heterochromatin present at the telomeric ends of chromosome arms demonstrated that peculiar types of segregation were not dependent on the amount of heterochromatin, but were rather determined by the genotype of particular inbred lines (Tikhonovich et al. 1987). Polymorphisms for six C-bands on chromosome 1R were used to study the frequency and distribution of recombination along the chromosome in diploid rye (*S. cereale* L.) and in a hexaploid triticale (X *Triticosecale* Wittmack) derived from it (Lukaszewski 1992). Recombination was concentrated in the distal regions of both chromosome arms and was infrequent in the proximal regions. In hexaploid triticale, the total recombination frequency in the same chromosome was reduced to 51.7%. In both backgrounds, the distal half of the long arm showed similar recombination frequencies, 51.4% and 45.7% for rye and triticale, respectively. The remaining about two-thirds of the chromosome length showed 42.3% recombination in rye but only 6% recombination in triticale. The results demonstrated that the genetic background not only affected the total amount of recombination, but also its distribution along the chromosome length. Cytological chromosome markers namely heterochromatin bands were used to relate particular linkage groups with a chromosome. The collection of lines with differently sized C-blocks in three regions of 1R satellite chromosome (on two telomeric ends and abreast the nucleolar organizer) gave the opportunity to study the frequency of recombination of these cytological markers as well as the joint inheritance with genes encoding for morphological traits and isozyme loci (Mikhailova et al. 1994).

3.5 Rye Chromosome Polymorphism and a Standard Karyotype

Despite obvious advances, problems with the identification and classification of rye chromosomes persisted. More than twenty different ways of chromosome designation (Table 3.1) complicated the identification of each chromosome in the complement (Schlegel and Mettin 1982) due to heterozygosity of the used material and because of the presence of chromosome polymorphisms. The C-banding pattern is often not uniform on homologous chromosomes between varieties, nor within a variety (Verma and Rees 1974; Gill and Kimber 1974; Merker 1973; Singh and Röbbelen 1975; Giraldes et al. 1979). Moreover, even in rye inbred lines, Giemsa C-banding polymorphisms were detected (Weimarck 1975; Lelley et al. 1978; Fujigaki and Tsuchiya 1990).

To standardize rye chromosome nomenclature, two International Workshops on Rye Chromosome Nomenclature and Homoeology Relationships were held and adopted rye cv. Imperial chromosome additions to wheat cv. Chinese Spring as the standard rye chromosome set, despite the fact that these chromosomes were not identical to those found in the population of the variety Imperial (Sybenga 1983). The following conclusions were proposed during the

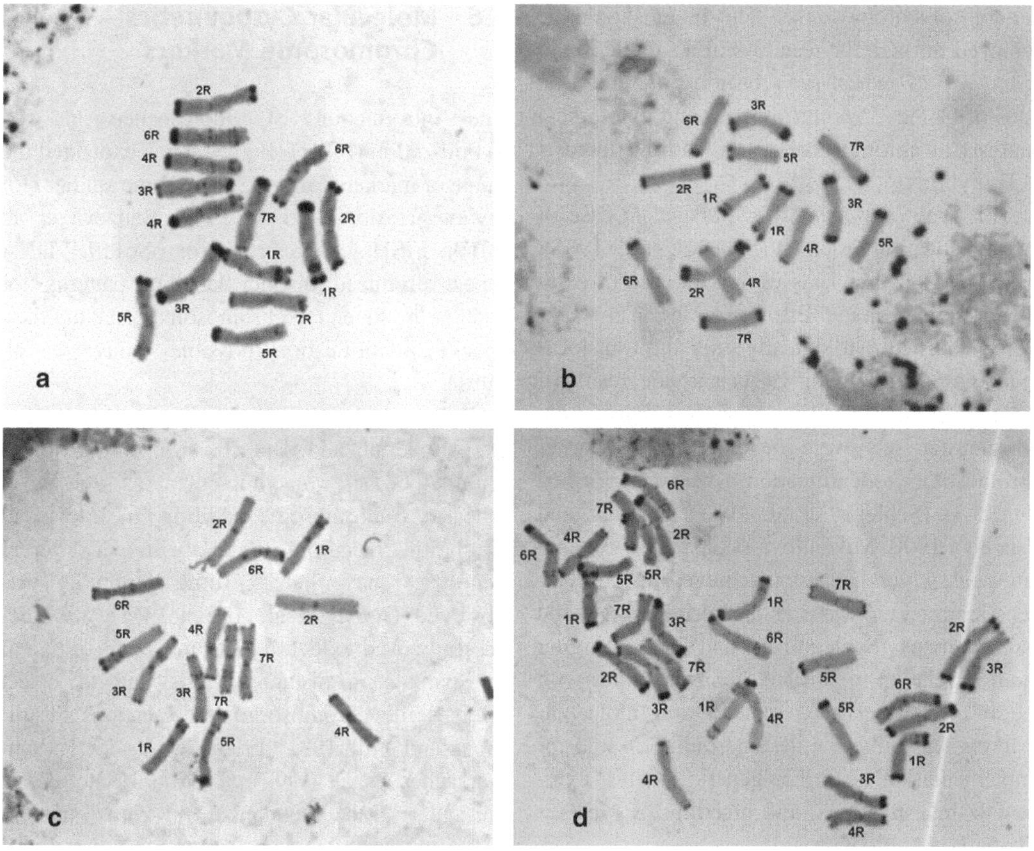

Fig. 3.1 C-banding of mitotic metaphase chromosomes in different species of genus *Secale*: (**a**, **d**)—*Secale cereale* L., (**b**)—*S. montanum* Guss., (**c**)—*S. silvestre* Host. (Courtesy of E. Badaeva)

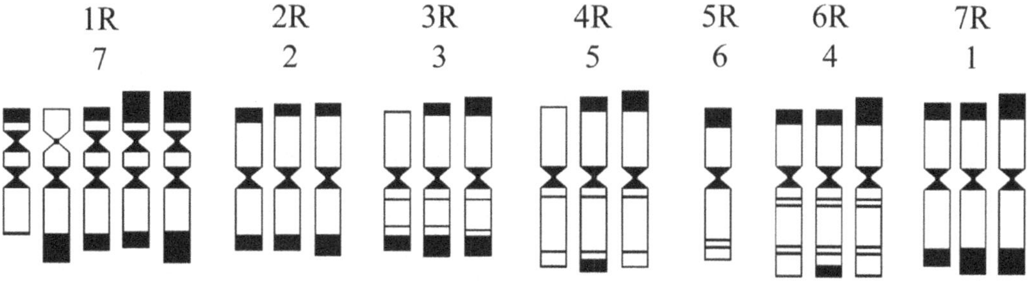

Fig. 3.2 C-banding variants of rye chromosomes revealed in four populations and inbred lines represented in the "Peterhof" genetic collection (1–7 initial designation of rye chromosomes). The idiogram was adapted from Tikhonovich (1975) and Smirnov and Sosnikhina (1984) with modifications (see also Table 3.1). Chromosome designation (1R–7R) was confirmed (Mikhailova et al. 1993) in testcrosses with the translocation tester set (Sybenga et al. (1985)) and by analyzing marker chromosome pairing in meiosis of F1 hybrids (for illustration see Fig. 3.3)

workshops: (1) a standard karyotype with the generalized C-banding pattern; (2) a designation of added rye chromosomes 1R–7R as homoeologous to wheat chromosomes which is based on functional substitution criterion (see Gupta 1971), and (3) a way how to identify new variants of chromosomes. The latter should be achieved through the testcross of a sample to the tester set of wheat-rye chromosome addition lines of Chinese Spring/Imperial and cytological analysis of chromosome pairing during meiosis in F_1 hybrids.

This type of analysis was pursued in case of the translocation tester set (Sybenga et al. 1985), see Table 3.1, line 9, as well as in case of extra chromosomes of cv. Esto trisomics. The latter were identified cytologically with the translocation lines (Melz et al. 1988) and the result of Sybenga et al. (1985) was confirmed. Further on, other tester sets were developed and marker chromosome identification was performed (Fig. 3.3) (Schlegel et al. 1987; Fujigaki and Tsuchiya 1990; Mikhailova et al. 1993). A separate goal, which had to be achieved in rye cytogenetic research was the anchoring of genetic linkage maps (Schlegel et al. 1986). Bringing them in relation to particular chromosomes with the use of tester sets, genetic and molecular markers was quite a milestone indispensable for implementation of further genetic studies, cytogenetic investigations, and breeding of rye (see Chap. 7).

Gradually it became obvious that C-banding alone was not sufficient to identify rye chromosomes. The identification of telomeric regions of a number of chromosomes was then achieved by the T-method, which was a C-staining derivative (Gustafson et al. 1983). Three out of the seven rye chromosomes could be identified using C and N staining techniques. The latter identified nucleolus organizing regions (NORs) and certain types of heterochromatin after specific processing of preparations and staining with Giemsa stain. The interstitial blocks of heterochromatin (C+ N+) detected by means of both methods corresponded to $(GAA)_m(GAG)_n$ sequences of satellite DNA (stDNA) (Schlegel and Gill 1984). The discovery of G-disks, which were originally identified in animals has also become a reality in plants. A G banding type on chromosomes of rye and barley (Xiaofeng and Zili 1988) was achieved.

3.6 Molecular Cytogenetics – New Chromosome Markers

The introduction of fluorescence in situ hybridization (FISH) (Fig. 3.4) has expanded the range of markers for rye chromosome studies (for review: Heslop-Harrison 1992; Badaeva et al. 2017). FISH with a probe for pSc119.1 DNA repeat produced distinct labeling patterns for each of the seven rye chromosomes, making their specific identification possible (Jouve et al. 1991).

When comparing the labeling pattern of rye cvs. Imperial and Blanco, rye chromosomes within the *Triticale* complement as well as in wheat-rye chromosome addition lines revealed similarities in chromosome labeling in different genotypes, nevertheless, some variability was observed (Jouve et al. 1991). FISH was successfully used to identify chromosomes of one of the parents in distant hybrids and to detect interspecific chromosome translocations (Tsujimoto and Gill 1991; Friebe et al. 1991; Schwarzacher et al. 1992). The ability to detect particular DNA sequences in situ enabled:

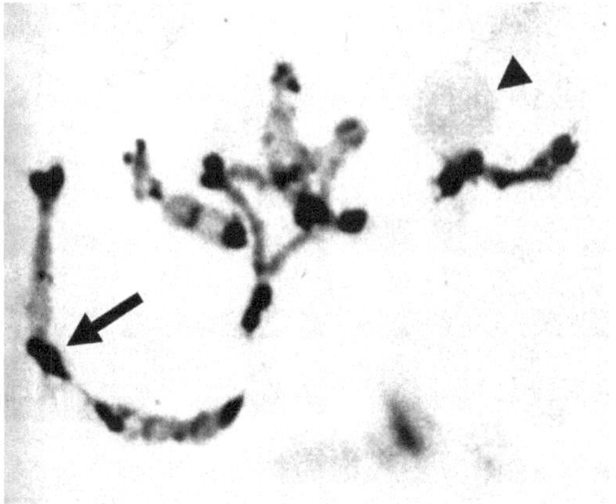

Fig. 3.3 Diakinesis in the F1 hybrid between the inbred line with the marker chromosome 1 (7R) (Tikhonovich 1975) and the line TsTs 2R–5R from the translocation tester set (Sybenga et al. 1985). Arrow points to chromosome 1(7R) with enlarged terminal heterochromatin band. The nucleolus at the NOR of the satellite chromosome 1R is marked with an arrowhead

(i) construction of physical maps of chromosomes (see Chap. 7); (ii) analysis of chromosome structure and aberrations; (iii) investigation of the structure, function, and evolution of chromosomes and genomes; (iv) determination of the spatial and temporal expression of genes; (v) identification and characterization of viruses, viral sequences and bacteria in tissues and other applications (Leitch et al. 1994).

Mikhailova et al. (2006) used FISH with probes for pSc200, pSc250, CCS1, 25S rDNA, and 5S rDNA, to define the positions of centromeres, subtelomeric domains, and rDNA sites on chromosomes of the Sy10 inbred population of the 'Peterhof' rye genetic collection (Sosnikhina et al. 2005). The probe pSc200 was a 521-bp insert in pUC18 comprising a 380-bp tandem repeat unit of subtelomeric DNA from rye (Vershinin et al. 1995, see also Chap. 8). The probe hybridized to thirteen major subtelomeric sites and ten minor sites in a haploid chromosome set. The probe pSc250 was a 476-bp insert in pUC18 and was a representative of a family of tandemly organized subtelomeric rye DNA sequences with an unusually extended monomer length of 500 bp (Vershinin et al. 1995). The probe hybridized to 13–14 major subtelomeric sites and six minor sites and was proximal to pSc200 loci. The probe CCS1 was a 260-bp motif (Aragon-Alcaide et al. 1996) of a centromere-specific clone (Hi-10) originally isolated from *Brachypodium sylvaticum* (Abbo et al. 1995). It hybridized exclusively to pericentromeric regions of all rye chromosomes, accurately marking centromeres and delimiting chromosome arms. The probe for 25S rDNA was a 2.3-kb subclone of the 25S rDNA coding region of *Arabidopsis thaliana* (Unfried and Gruendler 1990). The probe hybridized to a single locus and was a diagnostic feature of chromosome 1R. The 5S rDNA probe was derived from wheat clone pTa794 (Gerlach and Dyer 1980) and hybridized to two or three loci and was found suitable to identify chromosomes 1R, 3R, and 5R (Cuadrado et al. 1995; Cuadrado and Jouve 2002). FISH on chromosomes at early meiotic prophase using the telomeric probe HT100.3 and the centromeric probe INTR2 was performed according to

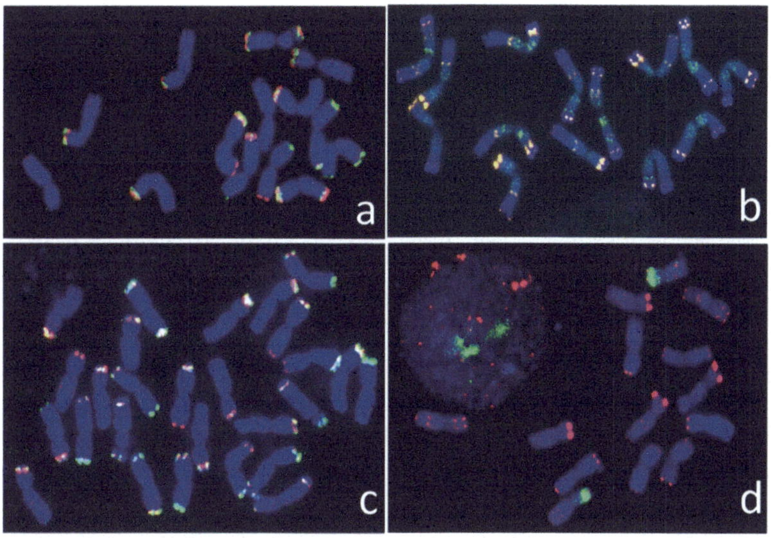

Fig. 3.4 FISH with various probes to mitotic chromosomes of rye belonging to different species. (**a**) *Secale cereale* L. 2n = 14: pSc200 (red) + pSc250 (green); (**b**) *S. montanum* Guss. pSc119 (red) + FAT (green); (**c**) *S. cereale* L. 2n = 4x = 28: pSc250 (red) + pSc119 (green); (**d**) *S. silvestre* Host.: pSc119 (red) + pTa71 (green). The Fat element was isolated from bacterial artificial chromosome (BAC) end sequences of wheat chromosome 3B. The Fat element represents a new family of Triticeae-specific, highly repeated DNA elements with a clustered-dispersed distribution pattern. Probe pTa71 is 45S rDNA amplified from wheat. (courtesy of E. Badaeva)

Hajdera et al. (2003), and Langdon et al. (2000), respectively. Multicolor FISH allowed to discriminate almost all chromosome arms. Using the FISH data together with information from Tikhonovich and Fadeyeva (1976), Tikhonovich et al. (1987), Mikhailova et al. (1993), and Alkhimova et al. (1999), the seven rye chromosomes could be identified and designated 1R to 7R in accordance with the Standard International Nomenclature (Sybenga 1983). Despite the fact that it was a population of high rate of inbreeding, chromosomes 2R and 7R showed heteromorphism with respect to the presence or abundance of the pSc200 and pSc250 repeat sequences; chromosome variation of this nature had been reported also in other lines of rye (Alkhimova et al. 1999). The structural differences between homologs made identification of chromosomes 4R, 6R, and 7R ambiguous in 18% of cases, but chromosomes 1R, 2R, 3R, and 5R could be identified with 99% confidence (Fig. 3.5). Identification of each chromosome allowed unambiguous analysis of the chromosomes involved in rare chiasmata revealed at MI in meiosis of the sy10 desynaptic mutant.

Genomic in situ hybridization (GISH) has been widely used to detect rye chromosomes or their parts in wheat introgression lines (Silkova et al. 2006) (Fig. 3.6). Koláčková et al. (2019) used GISH to reveal the position of rye chromosomes or chromosome arms in three-dimensional nuclear space in wheat-rye substitution and addition lines. In a standard version of GISH, labeled rye genomic DNA has been used as a probe. Recently, Fu et al. (2015) developed the oligonucleotide probes Oligo-1162, Oligo-pSc200 and Oligo-pSc250. The probes can be used for non-denaturing fluorescence in situ hybridization (ND-FISH) to label rye chromosomes in wheat background. According to the authors, oligonucleotide probes Oligo-pSc119.2-1, Oligo-pSc119.2-2, OligopTa535-1, Oligo-pTa535-2, Oligo-pTa71-2, Oligo-pAWRC.1 and Oligo-CCS1 can also be used for ND-FISH in wheat and rye.

ND-FISH with oligonucleotide probes provided a convenient and efficient way to identify individual rye chromosomes in wheat background. However, probes suitable for identification of specific segments of rye chromosomes were lacking. To fill this gap, Xi et al. (2020) developed five new probes (Oligo-5BL.46, Oligo-5A8080, Oligo-5A8080.1, Oligo-1AL.73, and Oligo-0R3) and used them with previously developed probes Oligo-44 and Oligo-45. The probes for Oligo-5BL.46, Oligo-5A8080, and Oligo-44 hybridized to intercalary regions of 1RS, 5RS, and 5RL chromosome arms, respectively. The probe Oligo-5A8080.1 combined with Oligo-45 identified intercalary regions of 1RS, 5RS, and 6RS arms simultaneously. Oligo-5A8080 and Oligo-5A8080.1 revealed variation in the distribution of 5S rDNA sequences and polymorphism among 5R chromosomes. Oligo-1AL.73 produced signals only on chromosomes 4R and 7R and contributed to the construction of an improved FISH map of chromosome 4R and to the confirmation of 4RL breakpoints in wheat-rye 4RL translocation chromosomes. Oligo-0R3 produced signals in the telomeric and subtelomeric regions of the seven rye chromosomes. These oligo probes also revealed five new tandem repeats in rye. Using the oligo probes, Xi et al. (2020) could discriminate short arms of 1R, 5R, and 6R and the long arms of 4R and 7R.

As mentioned above, the understanding of the Triticeae genome structure can be greatly facilitated by comparative cytogenetic analysis, which can determine chromosomal collinearity and structure alteration with single-gene probes of wheat or barley (Clark et al. 1989; Danilova et al. 2014; Said et al. 2018; Xie et al. 2020).

3.7 Flow Cytogenetics

Unlike classical and molecular cytogenetics, which analyze chromosomes using microscopy, flow cytogenetics is analyzing them using flow cytometry (Zwyrtková et al. 2021). During the analysis, metaphase chromosomes are classified according to light scatter and fluorescence properties when moving at high speed (10^3–10^4/s) in a narrow stream of liquid. Chromosomes in the flow interact individually with a light beam and,

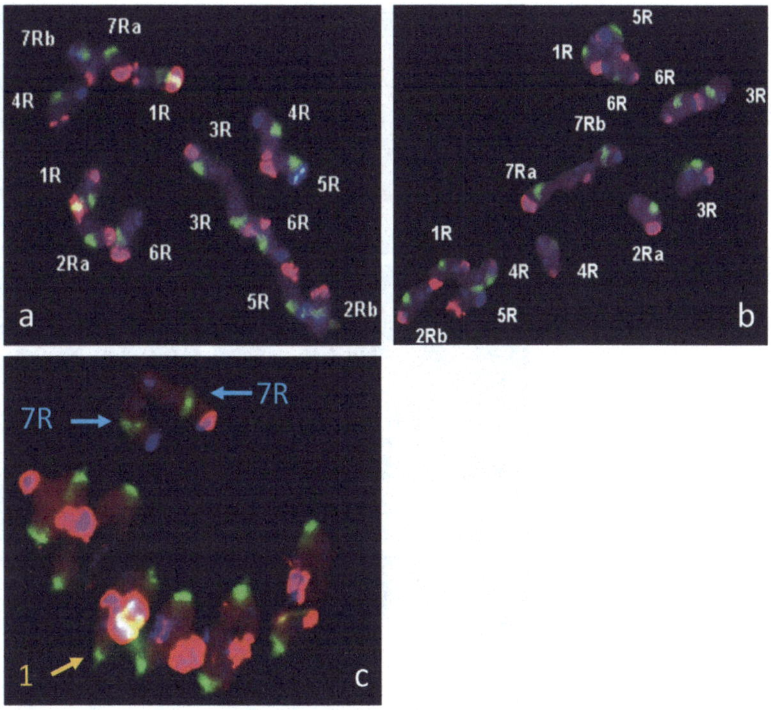

Fig. 3.5 Multicolor FISH to meiotic chromosomes of rye in inbred population Sy10. (**a**, **b**) metaphase I (MI) of the desynaptic sy10 mutant that forms mainly univalents; (**c**) MI in Sy10- wild type control. pSc200—red, pSc250—blue, CCS1 for centromeres—green, 25S rDNA—yellow on 1R chromosome, 5S rDNA—white-blue on 5R chromosome. A heteromorphic 7R bivalent can be seen in (**b**); seven bivalents can be seen in (**c**) and one of them, formed by the pair of 7R chromosomes, shows one chiasma and is heteromorphic (see text for probe description; for details of chromosome identification and the specificity of chiasma formation see Jenkins et al. 2005 and Mikhailova et al. 2006)

as large chromosome populations are analyzed, identification of minor subpopulations is possible. During the analysis, high resolution images cannot be captured, and thus flow-cytometric chromosome analysis (flow karyotyping) does not provide details on their longitudinal organization obtainable by microscopic observations. On the other hand, flow cytometry offers a unique opportunity to physically separate (sort) particular chromosomes or groups of chromosomes.

3.7.1 Flow Karyotyping and Chromosome Sorting

Flow karyotyping requires a liquid suspension of intact mitotic metaphase chromosomes. Kubaláková et al. (2003) developed a protocol for preparation of rye chromosome suspensions from root tips of hydroponically grown seedlings. The procedure includes cell cycle synchronization using hydroxyurea, short treatment with oryzalin to accumulate over 50% cells in metaphase, and a mild fixation of roots with formaldehyde. Root tip meristems dissected from the fixed roots are mechanically homogenized in LB01 buffer (Doležel et al. 1989) to release chromosomes. Using this protocol, 1 ml suspension containing $\sim 2 \times 10^5$ intact chromosomes can be prepared from 25 root tips. The analysis of chromosome suspensions stained by DNA fluorochrome DAPI results in histograms of relative fluorescence intensity (flow karyotypes) with a composite peak representing chromosomes 2R–7R and a peak of chromosome 1R (Fig. 3.7).

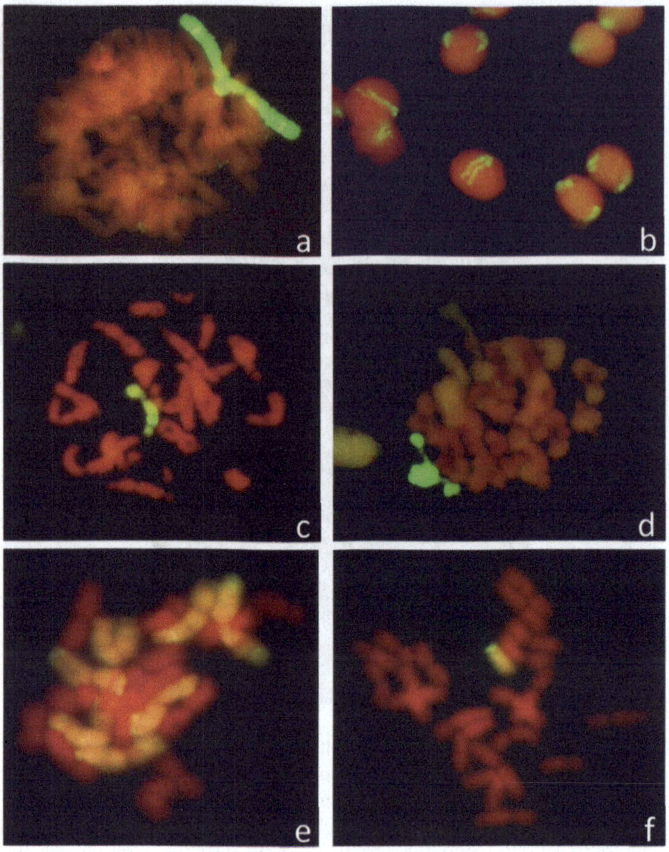

Fig. 3.6 GISH to mitotic (**a**, **e**, **f**) and meiotic (**c**, **d**) chromosomes in root tip (**e**, **f**) and tapetal (**a**) cells as well as to interphase nuclei in tapetum (**b**) of wheat-rye hybrid (**e**), mono- (**f**) and disomic (**a**, **b**, **c**, **d**) wheat-rye addition lines (for description of plant material see Mikhailova et al. 1998)

The degree at which the 1R peak can be discriminated depends on the genotype, indicating polymorphisms in DNA content. Chromosome 1R could be sorted with purities ranging from 85 to 95%, contaminated by a random mixture of doublets of chromosome arms and chromatids of various chromosomes. In order to characterize flow-sorted chromosome populations, Kubaláková et al. (2003) sorted 1000 chromosomes onto microscope slides and observed them after FISH with combinations of pSc119.2 and pSc250 DNA repeats, GAA multimers, and 5S rDNA, which in combination identify individual rye chromosomes.

Flow karyotypes obtained after analyzing a line of rye harboring the reciprocal chromosome translocations T2RS·2RL–5RL and T5RS·5RL–2RL, and a line with chromosomes T4RS·6RS and T4RL·6RL contained additional peaks representing the translocation chromosomes (Kubaláková et al. 2003). In case of the line with chromosomes T2RS·2RL–5RL and T5RS·5RL–2RL, the ratio of translocation chromosome peak areas to peak areas of wild type chromosomes agreed with a stable (homozygous) translocation. However, chromosome peak areas in flow karyotypes obtained from the line with chromosomes T4RS·6RS and T4RL·6RL suggested the presence of the translocation in only about 30% of seeds (Fig. 3.8). This result agreed with the observed low recovery rate of the translocation and confirmed the suitability of flow karyotyping to detect structural as well as numerical chromosome changes.

The analysis of samples prepared from a population of rye carrying B chromosomes (Bs) and from cv. Adams confirmed the suitability of flow karyotyping to detect the presence of particular chromosomes in rye (Fig. 3.9). Flow karyotypes obtained for both lines

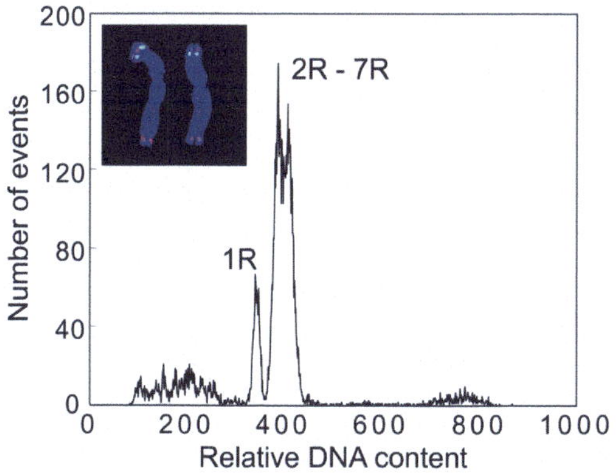

Fig. 3.7 Histogram of relative fluorescence intensity ('flow karyotype') obtained after flow-cytometric analysis of DAPI-stained chromosome suspension of rye cv. Imperial. The flow karyotype consists of a composite peak representing chromosomes 2R–7R, and a peak representing chromosome 1R. Inset: Images of flow-sorted chromosome 1R after FISH with 5S rDNA (green color) and pSc119 (red color) DNA sequences. The chromosomes were counterstained with DAPI (blue color). From Šimková et al. (2008a), with permission

comprised an additional and well-resolved peak (Kubaláková et al. 2003), however, the presence of Bs in cv. Adams was an unexpected finding. FISH on flow-sorted Bs revealed a cluster of the Afa family DNA repeat (Nagaki et al. 1995) on the long arm, making it a suitable cytogenetic marker for rye Bs.

The observation of large numbers of Bs on a single slide allowed the authors to detect rare A–B translocations, which occurred at the frequency of only about 0.5%. The most frequently observed translocation involved the short arm of 1R with a breakpoint on B located between the Afa repeat cluster and the centromere. Less frequently, a segment of an A chromosome was found translocated onto a B chromosome. However, the authors were not able to ascertain from which of the rye chromosomes 1R, 3R, or 5R the segment originated.

3.7.2 Chromosome Genomics

The ability to purify particular chromosomes by flow sorting provided an opportunity to dissect the rye genome into individual chromosomes. This is an attractive way of reducing DNA complexity in a lossless manner to facilitate various downstream molecular analyses and genome sequencing (Zwyrtková et al. 2021). However, the ability to sort only one (1R) out of the seven rye chromosomes would preclude this application. The solution was provided by Kubaláková et al. (2003) who showed that the remaining rye chromosomes 2R–7R could be sorted from wheat–rye disomic addition lines with purities ranging from 80 to 90%. Historically, the first use of flow-sorted rye chromosomes to support rye genome analysis and sequencing involved the construction of a bacterial artificial chromosome (BAC) library from the short arm of chromosome 1R (1RS). This arm is present in many cultivars of wheat, mainly in a form of the 1BL.1RS translocation, and confers resistance to wheat against diseases and may also improve the adaptation to unfavorable environments and increase its yield.

Šimková et al. (2008a) purified the 1RS arm by flow sorting from a wheat-rye ditelosomic addition line (Fig. 3.10). In total 10.3×10^6 1RS chromosome arms were sorted with the average purity of 86% as estimated by FISH with 45S rDNA and pSc200. The 1RS-specific library was the first BAC library for rye and consisted of 103,680 clones. Almost one third of this library had inserts larger than 100 kb with an average insert size of 73 kb. The library represents 14 equivalents of 1RS and the probability of finding any 1RS sequence in the library is 99.9%. This genomic resource was used by Bartoš et al. (2008) to provide until then the largest amount of genomic sequence data for rye and the team initiated systematic analysis of the DNA sequence composition of its genome. Sanger sequencing of BAC ends of 1,536 clones generated about 2 Mbp of BAC end sequences (BES) accounting for 0.5% DNA of 1RS. Repetitive sequences represented over 84% of 1RS with the Gypsy LTR retrotransposons forming almost 50% of 1RS DNA. Based on the identification of 93 genic sequences in 1RS BES, the authors estimated 36,000 genes for the entire rye genome. To support marker-assisted crop improvement, Bartoš et al. (2008) used 1RS BES

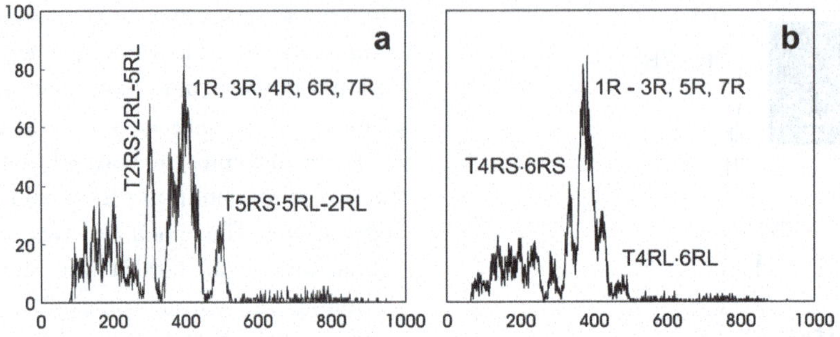

Fig. 3.8 Flow karyotyping of two chromosome translocation lines of rye. **a** A line carrying translocation chromosomes T2RS·2RL–5RL and T5RS·5RL–2RL. Both chromosomes could be easily discriminated. Analysis of their relative peak areas confirmed stability of the chromosomes in the population. **b** A line carrying translocation chromosomes T4RS·6RS and T4RL·6RL. Only some seeds contained the translocation chromosomes, which was reflected by smaller relative peak areas. *x* axis, relative DAPI fluorescence intensity; *y* axis, number of events. From Kubaláková et al. (2003), with permission

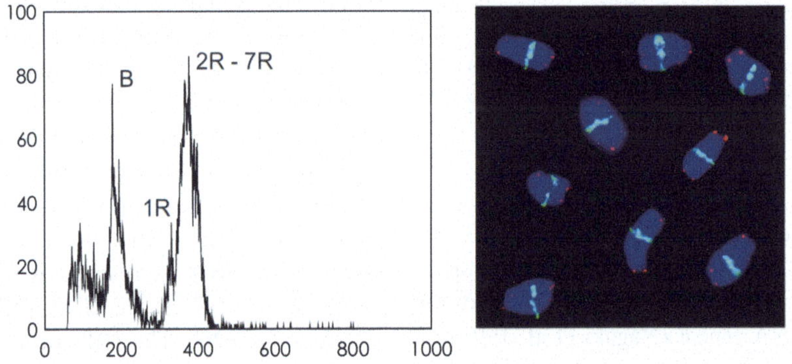

Fig. 3.9 Left panel: flow karyotype of rye cv. Adams with a peak representing B chromosomes. *x* axis, relative DAPI fluorescence intensity; *y* axis, number of events. Modified from Kubaláková et al. (2003), with permission. Right panel: Examples of sorted Bs after fluorescent labeling of Afa DNA repeats (yellow-green) and Arabidopsis-type telomere repeat (red) using FISH. Chromosomes were counterstained by DAPI (blue). Modified from Martis et al. (2013), with permission

to develop 1RS-specific molecular markers. These included insertion site-based polymorphism (ISBP) markers and also microsatellite (SSR) markers. A comparison with published data demonstrated the efficiency of the targeted approach for marker development from particular genome regions. Kofler et al. (2008) used a total of 2778 1RS BES to identify additional 216 microsatellites. After testing 138 primer pairs, they developed fourteen 1RS-specific and polymorphic markers. As an alternative approach for targeted marker development, Kofler et al. (2008) flow-sorted 30,000 1RS arms from a wheat-rye ditelosomic addition line, amplified their DNA using Phi29 multiple displacement amplification according to Šimková et al. (2008b). The amplified DNA was used to construct SSR clone libraries enriched for four different nucleotide motifs. In total, 603 microsatellites were identified after sequencing 1290 clones and testing 569 primer pairs identified 57 1RS-specific polymorphic markers. The attrition rates for the BES and SSR-enriched libraries were similar. However, the latter approach did not rely on the availability of BAC libraries, which are laborious to prepare.

The protocol for multiple displacement amplification of chromosomal DNA (Šimková et al. 2008b) enabled next generation chromosome shotgun sequencing (Mayer et al. 2011), avoiding the construction of DNA clone libraries. Thus, Fluch et al. (2012) sequenced flow-sorted 1RS by Roche 454 technology and obtained about 200 Mbp sequence reads representing over 0.4- fold coverage of 1RS. The results provided the first large-scale insight into the sequence structure and composition of 1RS and the rye genome in general. The results agreed with the findings of Bartoš et al. (2008) and the annotation of 1RS sequence reads revealed over 3000 gene loci and at least 1882 different gene functions on this arm. Examination of the conservation of homologous genes and gene order between 1RS, the genomes of rice and *Brachypodium* and the short arm of barley chromosome 1H revealed that 50% of 1RS genes corresponded to the distal end of the short arm of rice chromosome 5, the proximal region of the *Brachypodium* chromosome 2 long arm and high conservation of 1RS genes on 1HS of barley. Repetitive DNA was estimated to represent 72% of the 1RS sequence with the Gypsy/Sabrina LTR retrotransposons being most abundant. Moreover, the authors identified the presence of chloroplast insertions in 1RS DNA, as well as more than 4000 SSR loci for possible marker development.

The ability to flow-sort rye B chromosomes (Kubaláková et al. 2003) and suitability of amplified chromosomal DNA for next generation sequencing (Mayer et al. 2011) enabled Martis et al. (2013) to obtain the first shotgun sequence of a B chromosome. The study improved dramatically the knowledge on molecular organization and evolution of B chromosomes (see Chap. 4). In a follow-up study focusing on transcribed genes on rye B chromosome, Ma et al. (2017) used DNA amplified from flow-sorted rye B chromosomes to prepare a paired-end library and three mate-pair libraries, which were sequenced using Illumina HiSeq2000 and Illumina MiSeq platforms to 24-fold and 2—sixfold coverage, respectively. The identification of B-derived transcripts and comparative transcriptome analysis showed that 1954 and 1218

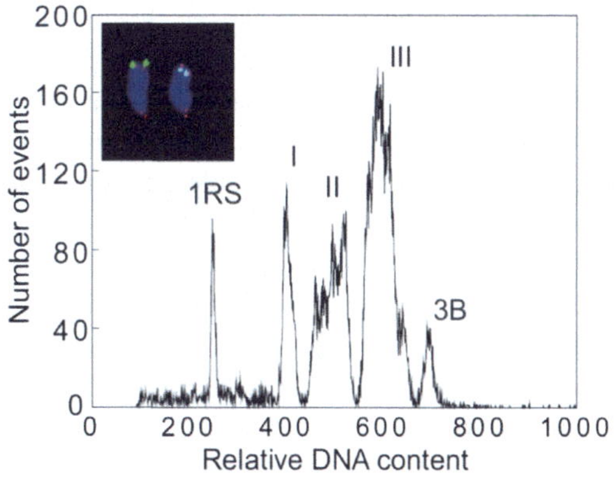

Fig. 3.10 Flow karyotype of the wheat-rye 1RS telosome addition line. The karyotype contains four peaks representing the chromosomes of wheat (labeled I, II, III, and 3B) and a peak of the telocentric chromosome 1RS. Inset: Images of the flow-sorted chromosome 1RS after FISH with probes for telomeric sequences (red color) and pSc200 repeat (green color). The chromosomes were counterstained with DAPI (blue color). From Šimková et al. (2008a), with permission

B-derived transcripts with an open reading frame were expressed in generative and vegetative tissues of a rye plant, respectively (see Chap. 4).

The possibility to dissect the rye genome into individual chromosomes and to sequence DNA from flow-sorted chromosomes by next generation sequencing (NGS) technologies set the stage for producing a draft genome of this crop. In this endeavor, Martis et al. (2013) purified 20,000 copies of rye chromosomes in two or three batches each by flow sorting from cv. Imperial (1RS) and from cv. Chinese Spring-cv. Imperial wheat-rye disomic chromosome addition lines (2R–7R) (Fig. 3.11).

The purity of sorted fractions ranged from 91.19% (5R) to 96.82% (3R) as determined by FISH with pSc119.2 and 5S rDNA. Chromosomal DNA was amplified, DNA samples from each chromosome were pooled to reduce DNA amplification bias and the pooled samples were sequenced by Roche 454 technology. In total, 8.25 Gb sequence was obtained representing 0.93-fold to 1.17-fold coverage for each rye chromosome (average 1.04-fold). These sequences and the sequences of clones selected from 1RS-specific BAC library (Šimková et al. 2003) were later used by Evtushenko et al. (2016) to characterize the arrays of the pSc200 and pSc250 tandem repeat families in subtelomeric heterochromatin of 1RS and to identify their adjacent DNA sequences across the whole genome. The authors demonstrated that the large blocks of subtelomeric heterochromatin originated from amplification of transposable elements and from expansion of tandem repeats.

Martis et al. (2013) compared repeat-masked shotgun sequences from the seven rye chromosomes with the set of barley genes and full gene sets of rice, *B. distachyon*, and sorghum and identified a set of 31,008 non-redundant rye genes. To establish a hypothetical order of the identified genes on the seven rye chromosomes, the authors used the concept of the 'genome zipper' by integrating chromosome sequence data with dense gene-based marker maps and conserved synteny information from sequenced grass genomes of *B. distachyon*, rice, and sorghum (Mayer et al. 2009, 2011). The tentative positioning of a total of 22,426 genes along the rye genome (72% of all detected genes) enabled Martis et al. (2013) to identify 17 segments representing the rye genome, which exhibit conserved synteny to the barley genome. While the rye chromosome 1R was collinear to barley chromosome 1H, the remaining chromosomes 2R–7R were composed of two to four segments corresponding to individual regions on the barley genome. These findings allowed Martis et al. (2013) to propose a revised model of the rye genome evolution (Fig. 3.12).

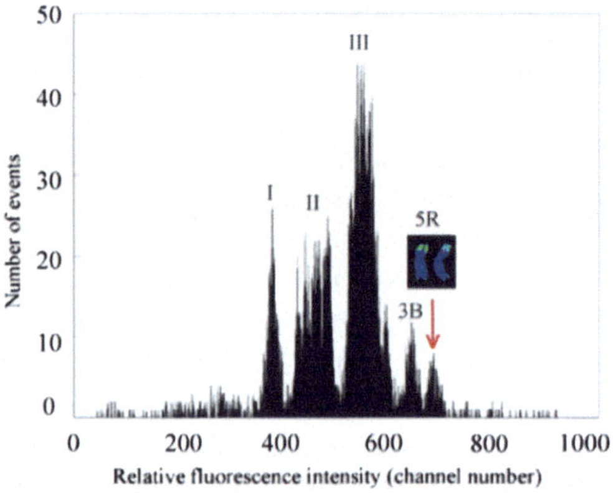

Fig. 3.11 Example of the use of wheat-rye chromosome addition lines to purify chromosomes 2R–7R using flow sorting. As the chromosomes could not be sorted directly from rye cv. Imperial, they were sorted from wheat-rye disomic addition lines. Rye chromosomes 2R–7R are larger than the largest wheat chromosome 3B and can be discriminated on histograms of relative fluorescence intensity (flow karyotype) obtained after the analysis of DAPI-stained mitotic metaphase chromosomes. As an example, a flow karyotype of a wheat-5R disomic addition line is shown. Wheat chromosomes form three composite peaks I–III, representing groups of chromosomes, and one peak representing chromosome 3B. Chromosome 5R forms a well discriminated peak (indicated by arrow) and can be easily sorted. Inset: sorted chromosomes identified by FISH with pSC119.2 (red) and 5S rDNA (green). From Martis et al. (2013), with permission

Following the construction of a draft genome sequence of rye (Martis et al. 2013), chromosome sorting also contributed to the development of the chromosome-scale genome assembly of the species (Rabanus-Wallace et al. 2021). In this project, flow-sorted rye chromosomes were used to prepare 500 bp insert Illumina sequencing libraries, which were pair-end sequenced (2 × 150 bp) on the Illumina GAIIx platform. The sequences obtained thus were used to validate the assignment of DNA sequence scaffolds from the whole genome shotgun assembly to a chromosome. As demonstrated by Šimková et al. (2003) intact and high-molecular weight (HMW) DNA can be prepared from flow-sorted chromosomes. Due to the superior quality, DNA prepared this way permits the development of high-quality optical maps with long contigs (Staňková et al. 2016). Rabanus-Wallace et al. (2021) flow-sorted 10.5 million rye chromosomes ($\sim$22 μg DNA) and used HMW DNA prepared from them to produce a rye genome optical map with a total length of 6660 Gbp and a map N50 of 1.671 Mbp. The optical map was used to order, orient, and curate DNA sequence scaffolds during the production of the rye reference genome (see Chap. 7).

Acknowledgments The authors are grateful to Ekaterina D. Badaeva for providing figures with C-banded and FISH probed mitotic chromosomes of different rye species. JD was supported by ERDF project "Plants as a tool for sustainable global development" (No. CZ.02.1.01/0.0/0.0/16_019/0000827).

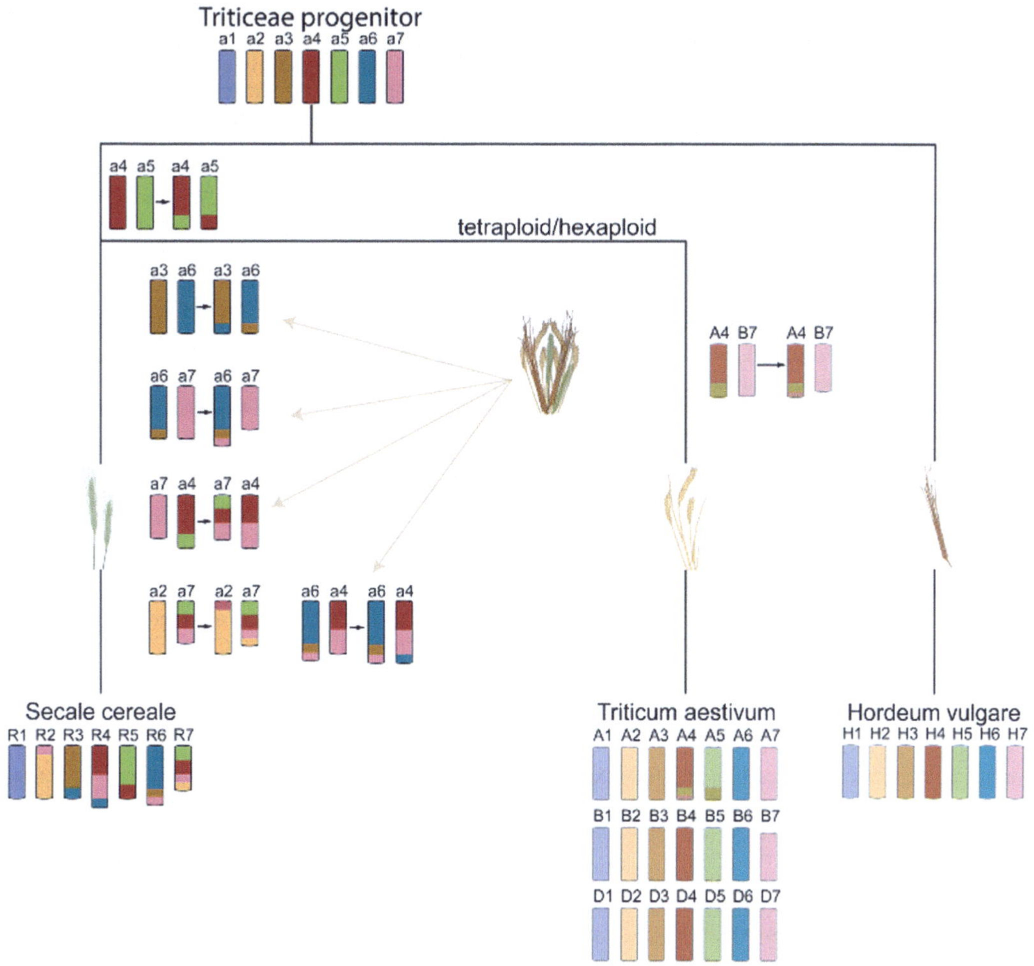

Fig. 3.12 Genome reorganization and translocation events that accompanied speciation within the tribus Triticeae and rye genome evolution. The chromosomes are shown as colored rectangles. All species within the tribe have a basic chromosome number $x = 7$ and the chromosomes of a common ancestor are labeled a1–a7. Rye genome reorganizations occurring in the common ancestor of rye and wheat (translocation between chromosomes a4 and a5) and divergence of the two lineages are postulated. Three of the five translocations that occurred after the split of wheat (a3–a6, a6–a7, a7–a4) can be ordered, while for two (a2–a7, a6–a4) the order cannot be deduced. They may have occurred in parallel or consecutively. From Martis et al. (2013), with permission

References

Aase HC (1935) Cytology of cereals. Bot Rev 1:467–496. https://doi.org/10.1007/BF02869986

Abbo MW, Dunford RP, Foote T, Reader SM, Flavell RB, Moore G (1995) Organisation of retroelement and stem-loop repeat families in the genomes and nuclei of cereals. Chromosome Res 3(1):5–15. https://doi.org/10.1007/BF00711156

Ahloowalia BS (1963) Study of translocation in tetraploid rye. Genetica 33:207–221. https://doi.org/10.1007/BF01725762

Alkhimova AG, Heslop-Harrison JS, Shchapova AI, Vershinin AV (1999) Rye chromosome variability in wheat-rye addition and substitution lines. Chromosome Res 7(3):205–212. https://doi.org/10.1023/a:1009299300018

Aragon-Alcaide L, Miller T, Schwarzacher T, Reader S, Moore G (1996) A cereal centromeric sequence. Chromosoma 105(5):261–268. https://doi.org/10.1007/BF02524643

Atayeva DM, Iordansky AB, Ajzatulina KS (1982) Karyotype polymorphism of cultivated rye. Dokl Acad Nauk SSSR 264(1):234–237 (in Russian, Abstr. Engl.)

Avdulow N (1931) Karyosystematische Untersuchung der Familie Gramineen." Bull Appl Bot Genet Pl Breed Monograph. Supplement 44-th to the Bulletin of Applied Botany, of Genetics and Plant-Breeding. Leningrad (in Russian, German Summary)

Badaeva ED, Ruban A, Aliyeva-Schnorr L, Municio C, Hesse S, Houben A (2017) In situ hybridization to plant chromosomes. In: Fluorescence in situ hybridization (FISH), pp 477–494. https://doi.org/10.1007/978-3-662-52959-1_49

Bartoš J, Paux E, Kofler R, Havránková M, Kopecký D, Suchánková P, Šafář J, Šimková H, Town CD, Lelley T, Feuillet C, Doležel J (2008) A first survey of the rye (*Secale cereale*) genome composition through BAC end sequencing of the short arm of chromosome 1R. BMC Plant Biol 8:95

Bennett MD, Smith JB (1976) Nuclear DNA amounts in angiosperms. Philos Trans R Soc Lond B Biol Sci 274 (933):227–274. https://doi.org/10.1098/rstb.1976.0044 PMID: 6977

Bhattacharyya NA, Jenkins BC (1960) Karyotype analysis and chromosome designations for *Secale cereale* L. "Dakold". Can J Genet Cytol 2(3):268–277. https://doi.org/10.1139/g60-027

Bowman JG, Rajhathy T (1977) Fusion of chromocenters in premeiotic interphase of *Secale cereale* and its possible relationship to chromosome pairing. Can J Genet Cytol 19(2):313–321. https://doi.org/10.1139/g77-034

Candela M, Fiqueiras AM, Lacadena JR (1979) Maintenance of interchange heterozygosity in cultivated rye Secale Cereale l. Heredity 42:283–289. https://doi.org/10.1038/hdy.1979.31

Clark M, Karp A, Archer S (1989) Physical mapping of the B-hordein loci on barley chromosome 5 by in situ hybridization. Genome 32(5):925–929. https://doi.org/10.1139/g89-531

Cuadrado A, Jouve N (2002) Evolutionary trends of different repetitive DNA sequences during speciation in the genus Secale. J Hered 93(5):339–345. https://doi.org/10.1093/jhered/93.5.339

Cuadrado A, Ceoloni C, Jouve N (1995) Variation in highly repetitive DNA composition of heterochromatin in rye studied by fluorescence in situ hybridization. Genome 38(6):1061–1069. https://doi.org/10.1139/g95-142

Danilova TV, Friebe B, Gill BS (2014) Development of a wheat single gene FISH map for analyzing homoeologous relationship and chromosomal rearrangements within the Triticeae. Theor Appl Genet 127(3):715–730. https://doi.org/10.1007/s00122-013-2253-z

Darvey NL, Gustafson JP (1975) Identification of rye chromosomes in wheat-rye addition lines and Triticale by heterochromatin bands. Crop Sci 15(2):239–243. https://doi.org/10.2135/cropsci1975.0011183X001500020029x

Darvey NL (1973) Genetics of seed shriveling in wheat and Trilicales. In: Sears ER, Sears LMS (eds) Proceeding of 4th wheat genetics symposium, Columbia, pp 155–160

Lima-de-Faria A (1953) Chromomere analysis of the chromosome complement of rye. Chromosoma 5 (1):1–68. https://doi.org/10.1007/BF01271482

Doležel J, Binarová P, Lucretti S (1989) Analysis of nuclear DNA content in plant cells by flow cytometry. Biol Plant 31:113–120

Doležel J, Greilhuber J, Lucretti S, Meister A, Lysák MA, Nardi L, Obermayer R (1998) Plant genome size estimation by flow cytometry: inter-laboratory comparison. Ann Bot 82 (Suppl. A):17–26

Doležel J, Čížková J, Šimková H, Bartoš J (2018) One major challenge of sequencing large plant genomes is to know how big they really are. Int J Mol Sci 19 (11):3554. https://doi.org/10.3390/ijms19113554

Eilam T, Anikster Y, Millet E, Manisterski J, Sagi-Assif O, Feldman M (2007) Genome size and genome evolution in diploid Triticeae species. Genome 50 (11):1029–1037. https://doi.org/10.1139/g07-083 PMID: 18059548

Evans LE, Jenkins BC (1960) Individual *Secale cereale* chromosome additions to *Triticum aestivum* I. The addition of individual "Dakold" fall rye, chromosomes to "Kharkov" winter wheat and their subsequent identification. Can J Genet Cytol 2(3):205–215. https://doi.org/10.1139/g60-021

Evans GM, Rees H, Snell CL, Sun S (1972) The relation between nuclear DNA amount and the duration of the mitotic cycle. Chromosomes Today 3:24–31. Longmans, London

Evtushenko EV, Levitsky VG, Elisafenko EA, Gunbin KV, Belousov AI, Šafář J, Doležel J, Vershinin AV (2016) The expansion of heterochromatin blocks in rye reflects the co-amplification of tandem repeats and adjacent transposable elements. BMC Genomics 17:337. https://doi.org/10.1186/s12864-016-2667-5

Fluch S, Kopecký D, Burg K, Šimková H, Taudien S, Petzold A, Kubaláková M, Platzer M, Berenyi M, Krainer S, Doležel J, Lelley T (2012) Sequence composition and gene content of the short arm of rye (*Secale cereale*) chromosome 1. PLoS One 7(2): e30784. https://doi.org/10.1371/journal.pone.0030784

Friebe B, Mukai Y, Dhaliwal HC, Martin TJ, Gill BC (1991) Identification of alien chromatin specifying resistence to wheat streak mosaic and greenbug in wheat germ plasm by C-banding and in situ hybridization. Theor Appl Genet 81(3):381–389. https://doi.org/10.1007/BF00228680

Fu S, Chen L, Wang Y, Li M, Yang Z, Qiu1 L, Yan B, Ren Z, Tang Z (2015) Oligonucleotide probes for ND-FISH analysis to identify rye and wheat chromosomes. Sci Rep 5:10552. https://doi.org/10.1038/srep10552

Fujigaki J, Tsuchiya T (1990) Chromosome identification of seven primary trisomics in inbred rye (*Secale cereale* L.) by acetocarmine Giemsa C-banding. Jpn J Gene 65:209–219. https://doi.org/10.1266/jjg.65.209

Gerlach WL, Dyer TA (1980) Sequence organization of the repeating units in the nucleus of wheat which contain 5S rRNA genes. Nucleic Acids Res 8 (21):4851–4865. https://doi.org/10.1093/nar/8.21.4851

Gill BS, Kimber G (1974) The Giemsa C-banded karyotype of rye. Proc Natl Acad Sci USA 71:1247–1249. https://doi.org/10.1073/pnas.71.4.1247

Giraldes R, Orellana I (1979) Metaphase I bonds, crossing over frequency and genetic length of specific chromosome arms of rye. Chromosoma 72(3):377–385. https://doi.org/10.1007/BF00331096

Giraldes R, Cermeño MC, Orellana J (1979) Comparison of C-banding pattern in the chromosomes of inbred lines and open pollinated varieties of rye. Z Pflanzenzucht 83(1):40–48

Gupta PK (1971) Homoeologous relationship between wheat and rye chromosomes. Present status. Genetica 42(2):199–213. https://doi.org/10.1007/BF00121902

Gustafson J, Lukaszewski AJ, Bennett MD (1983) Somatic deletion and redistribution of telomeric heterochromatin in the genus *Secale* and in *Triticale*. Chromosoma 88:293–298. https://doi.org/10.1007/BF00292906

Gustafson JP, Evans LE, Josifek K (1976) Identification of chromosomes in *Secale montanum* and individual *S.montanum* chromosome additions to 'Kharkov' wheat by heterochromatin bands and chromosome morphology. Can J Genet Cytol 18(2):339–343. https://doi.org/10.1139/g76-041

Hajdera I, Siwinska D, Hasterok R, Maluszynska I (2003) Molecular cytogenetic analysis of genome structure in *Lupinus angustifolius* and *Lupinus cosentinii*. Theor Appl Genet 107(6):988–996. https://doi.org/10.1007/s00122-003-1303-3

Hasegawa N (1934) A cytological study on 8-chromosome rye. Cytologia 6(1):68–77. https://doi.org/10.1508/cytologia.6.68

Heneen WK (1962) Chromosome morphology in inbred rye. Hereditas 48(1–2):182–200. https://doi.org/10.1111/j.1601-5223.1962.tb01806.x

Heneen WK, Caspersson T (1973) Identification of the chromosomes of rye by distribution patterns of DNA. Hereditas 74(2):259–279. https://doi.org/10.1111/j.1601-5223.1973.tb01127.x

Heslop-Harrison JS (1992) Molecular cytogenetics, cytology and genomic comparisons in the Triticeae. Hereditas 116(s1):93–99. https://doi.org/10.1111/j.1601-5223.1992.tb00805.x

Hrishi NJ, Müntzing A (1960) Structural heterozygosity in *Secale kuprijanovii*. Hereditas 46(3–4):745–752. https://doi.org/10.1111/j.1601-5223.1960.tb03114.x

Iordansky AB, Zurabishvili TG, Badaev NS (1978) Linear differentiation of cereal chromosomes. 3. Rye, triticale and 'Aurora' variety. Theor Appl Genet 51 (6):281–288. https://doi.org/10.1007/BF00274816

Jenkins GM, Mikhailova EI, Langdon T, Tikholiz OA, Sosnikhina SP, Jones RN (2005) Strategies for the study of meiosis in rye. In: Puertas MJ, Naranjo T (eds) "Plant Cytogenetics"—a special volume of Cytogenetic and Genome research, vol 109, no 1–3, pp 21–227. https://doi.org/10.1159/000082404

Jones GH (1978) Giemsa C-banding of rye meiotic chromosomes and the nature of "terminal" chiasmata. Chromosoma 66(1):45–57. https://doi.org/10.1007/BF00285815

Jouve N, Mc Intyre CL, Gustafson JP (1991) Chromosome preparations from protoplasts: in situ hybridization banding pattern of a dispersed DNA sequence in rye (*Secale cereale* L.). Genome 34(4):524–527. https://doi.org/10.1139/g91-080

Kamanoi M, Jenkins BC (1962) Trisomics in common rye *Secale cereale* L. Seiken ziho Rep Kihara Inst Biol Res 13:118–123

Kofler R, Bartoš J, Gong L, Stift G, Suchánková P, Šimková H, Berenyi M, Burg K, Doležel J, Lelley T (2008) Development of microsatellite markers specific for the short arm of rye (*Secale cereale* L.) chromosome 1. Theor Appl Genet 117(6):915–926. https://doi.org/10.1007/s00122-008-0831-2

Koláčková V, Perničková K, Vrána J, Duchoslav M, Jenkins G, Phillips D, Turkosi E, Šamajová O, Sedlářová M, Šamaj J,Doležel J, Kopecký D (2019) Nuclear disposition of alien chromosome introgressions into wheat and rye using 3D-FISH. Int J Mol Sci 20:4143. https://doi.org/10.3390/ijms20174143

Koller OL, Zeller FJ (1976) The homologous relationships of rye chromosomes 4R and 7R with wheat chromosomes. Genet Res 28(2):177–188. https://doi.org/10.1017/S0016672300016852

Kranz AR (1976) Karyotype analysis in meiosis. Giemsa banding in the genus *Secale* L. Theor Appl Genet 47 (3):101–107

Kubaláková M, Valárik M, Bartoš J, Vrána J, Číhalíková J, Molnár-Láng M, Doležel J (2003) Analysis and sorting of rye (*Secale cereale* L.) chromosomes using flow cytometry. Genome 46(5):893–905. https://doi.org/10.1139/g03-054

Langdon T, Seago C, Mende M, Leggett M, Thomas H, Forster JW, Jones RN, Jenkins G et al (2000) Retrotransposon evolution in diverse plant genomes. Genetics 156(1):313–325. PMID: 10978295; PMCID: PMC1461242

Leitch AR, Schwarzacher T, Jackson D, Leitch IJ (1994) In situ hybridization: a practical guide. BIOS Scientific Publishers Limited, Oxford, UK

Lelley T, Josifek K, Kaltsikes PJ (1978) Polymorphism in the Giemsa C-banding pattern of rye chromosomes. Can J Genet Cytol 20(3):307–312. https://doi.org/10.1139/g78-034

Lewitsky GA (1931) The morphology of chromosomes (Russian and Engl.). Bull Appl Bot Genet Plant Breed 24:19–174

Lukaszewski AJ (1992) A comparison of physical distribution of recombination in chromosome 1R in diploid rye and in hexaploid triticale. Theor Appl Genet 83(8):1048–1053

Lukaszewski AJ, Gustafson JP (1987) Cytogenetics of Triticale. In: Janick J (ed) Plant breeding reviews, vol 5. pp 41–93. https://doi.org/10.1002/9781118061022.ch3

Ma W, Gabriel TS, Martis MM, Gursinsky T, Schubert V, Vrána J, Doležel J, Grundlach H, Altschmied L,

Scholz U, Himmelbach A, Behrens S-E, Banaei-Moghaddam AM, Houben A (2017) Rye B chromosomes encode a functional Argonaute-like protein with *in vitro* slicer activities similar to its A chromosome paralog. New Phytol 213(2):916–928. https://doi.org/10.1111/nph.14110

Martis MM, Klemme S, Banaei-Moghaddam AM, Blattner FR, Macas J, Schmutzer T, Scholz U, Gundlach H, Wicker T, Šimková H, Novák P, Neumann P, Kubaláková M, Bauer E, Haseneyer G, Fuchs J, Doležel J, Stein N, Mayer KFX, Houben A (2012) Selfish supernumerary chromosome reveals its origin as a mosaic of host genome and organellar sequences. Proc Natl Acad Sci USA 109:13343–13346. https://doi.org/10.1073/pnas.1204237109

Martis MM, Zhou R, Haseneyer G, Schmutzer T, Vrána J, Kubaláková M, König S, Kugler KG, Scholz U, Hackauf B, Korzun V, Schön CC, Doležel J, Bauer E, Mayer KFX, Stein N (2013) Reticulate evolution of the rye genome. Plant Cell 25(10):3685–3698. https://doi.org/10.1105/tpc.113.114553

Mayer KFX, Taudien S, Martis M, Šimková H, Suchánková P, Gundlach H, Wicker T, Petzold A, Felder M, Steuernagel B, Scholz U, Graner A, Platzer M, Doležel J, Stein N (2009) Gene content and virtual gene order of barley chromosome 1H. Plant Physiol 151(2):496–505. https://doi.org/10.1104/pp.109.142612

Mayer KFX, Martis M, Hedley PE, Šimková H, Liu H, Morris JA, Steuernagel B, Taudien S, Roessner S, Gundlach H, Kubaláková M, Suchánková P, Murat F, Felder M, Nussbaumer T, Graner A, Salse J, Endo TR, Sakai H, Tanaka T, Itoh T, Sato K, Platzer M, Matsumoto T, Scholz U, Doležel J, Waugh R, Stein N (2011) Unlocking the barley genome by chromosomal and comparative genomics. Plant Cell 23(4):1249–1263. https://doi.org/10.1105/tpc.110.082537

Melz G, Schlegel R, Sybenga J (1988) Identification of the chromosomes in the 'Esto' set of rye trisomics. Plant Breed 100(3):169–172. https://doi.org/10.1111/j.1439-0523.1988.tb00236.x

Merker A (1973) Giemsa technique for rapid identification of chromosomes in *Triticale*. Hereditas 75 (2):280–282. https://doi.org/10.1111/j.1601-5223.1973.tb01171.x

Merker A (1976) The cytogenetic effect of heterochromatin in hexaploid *Triticale*. Hereditas 83(2):215–222. https://doi.org/10.1111/j.1601-5223.1976.tb01586.x

Mikhailova EI, Sosnikhina SP, Michurina TG (1993) Identification of rye chromosomes by means of conjugation test in meiosis. Genetika 29(6):978–989 (in Russian, Abstr. Engl.)

Mikhailova EI, Sosnikhina SP, Nerusheva GV, Fyong FT (1994) The use of heterochromatin markers in genetic analysis of rye (*Secale cereale* L.). Russ J Genet 30 (1):76–82

Mikhailova EI, Naranjo T, Shepherd K, van Wennekes-Eden J, Heyting C, de Jong JH (1998) The effect of the wheat Ph1 locus on chromatin organisation and meiotic chromosome pairing analysed by genome painting. Chromosoma 107(5):339–350. https://doi.org/10.1007/s004120050316

Mikhailova EI, Phillips D, Sosnikhina SP, Lovtsyus AV, Jones RN, Jenkins G (2006) Molecular assembly of meiotic proteins Asy1 and Zyp1 and pairing promiscuity in rye (*Secale cereale* L.) and its synaptic mutant sy10. Genetics 174(3):1247–1258. https://doi.org/10.1534/genetics.106.064105

Miller TE (1984) The homoeologous relationship between the chromosomes of rye and wheat current status. Can J Genet Cytol 26(5):578–589. https://doi.org/10.1139/g84-091

Moshkovich AM, Chebotar AA (1976) Rye (karyology, embriology, cytogenetics). Kishinev: Shtiintza (in Russian)

Müntzing A, Prakken R (1941) Chromosomal aberrations in rye populations. Hereditas 27(3–4):273–308. https://doi.org/10.1111/j.1601-5223.1941.tb03261.x

Nagaki K, Tsujimoto H, Isono K, Sasakuma T (1995) Molecular characterization of a tandem repeat, Afa family, and its distribution among Triticeae. Genome 38(3):479–486. https://doi.org/10.1139/g95-063

Oinuma T (1953) Karyomorphological studies on the origin of 8–9 chromosome rye plants. Jpn J Genet 28 (1):28–34 (Japanese, English Summary). https://doi.org/10.1266/jjg.28.28

Pilch I (1978) Cytological and morphological characteristics of primary trisomics in rye (*Secale cereale* L.). Genet Polon 19:137–152

Rabanus-Wallace et al (2021) (in press)

Rao PM (1975) Homoeologous relationship of Imperial rye chromosomes C with wheat chromosome 4A. Cereal Res Commun 3:103–109

Rees H (1961) The consequences of interchange. Evolution 15(2):145–152. https://doi.org/10.1111/j.1558-5646.1961.tb03138.x

Riley R, Chapman V (1958) The production and phenotypes of wheat-rye chromosome addition lines. Heredity 12(3):301–315. https://doi.org/10.1038/hdy.1958.31

Riley R (1965) Cytogenetics and plant breeding. Genetics today. Proc XI Intern Congr Genet 3:681–688

Said M, Hřibová E, Danilova TV, Karafiátová M, Čížková J, Friebe B et al (2018) The *Agropyron cristatum* karyotype, chromosome structure and cross-genome homoeology as revealed by fluorescence in situ hybridization with tandem repeats and wheat single-gene probes. Theor Appl Genet 131:2213–2227. https://doi.org/10.1007/s00122-018-3148-9

Sarma NP, Natarajan AT (1973) Identification of heterochromatic regions in the chromosomes of rye. Hereditas 74(2):233–238. https://doi.org/10.1111/j.1601-5223.1973.tb01124.x

Schlegel R, Fridrich J (1975) Erste Untersuchungen zum meiotischen Paarungs- verhalten Giemsa-markierter Chromosomen des diploiden Roggens (*Secale cereale*). Biol Rdsch 13:300–305

Schlegel R, Gill BS (1984) N-banding analysis of rye chromosomes and the relationship between N-banded

and C-banded heterochromatin. Can J Genet Cytol 26 (6):765–769. https://doi.org/10.1139/g84-121
Schlegel R, Mettin D (1982) The present status of chromosome recognition and gene localization in rye. Tagunsber Acad Landwirtschaftwiss DDR 198:131–152
Schlegel R, Melz G, Mettin D (1986) Rye cytology, cytogenetics and genetics current status. Theor Appl Genet 72:721–734. https://doi.org/10.1007/BF00266535
Schlegel R, Melz G, Nestrowicz R (1987) A universal reference karyotype in rye Secale cereale l. Theor Appl Genet 74:820–826. https://doi.org/10.1007/BF00247563
Schwarzacher T, Anamthawat-Jonsson K, Harrison GE, Islam AKMR, Jia JZ, King JP, Leitch AR, Miller TE, Reader SM, Rogers WJ, Shi M, Heslop-Harrison JS (1992) Genomic in situ hybridization to identify alien chromosomes and chromosome segments in wheat. Theor Appl Genet 84:778–786. https://doi.org/10.1007/BF00227384
Sears ER (1952) Homoeologous chromosomes in *Triticum aestivum*. Genetics 37:624
Sears ER (1966) Nullisomic-tetrasomic combinations in hexaploid wheat. In: Riley R, Lewis KR (eds) Chromosome manipulations and plant genetics. Springer, Boston, MA. https://doi.org/10.1007/978-1-4899-6561-5_4
Semenov VI, Semenova EV (1982) Polymorphism of rye chromosomes by heterochromatin blocks in some rye varieties and triticale forms. Genetika 18(11):1856–1867 (in Russian, Abstr. Engl.)
Shchapova AI, Kobylyansky VD (1976) Differential staining of plant chromosomes. II. Distant rye hybrids. Tsitologia 18(2):156–160 (in Russian, Abstr. Engl.)
Shchapova AI (1974) Differential staining of plant chromosomes. I. *Secale cereale* L. Tsitologia 16 (3):370–372 (in Russian, Abstr. Engl.)
Silkova OG, Dobrovol'skaia OB, Dubovets NI, Adonina IG, Kravtsova LA, Roder MS, Salina EA, Shchapova AI, Shumnyĭ VK (2006) Production of wheat-rye substitution lines and identification of chromosome composition of karyotypes using C-banding,GISH, and SSR markers. Russ J Genet 42:645–653. https://doi.org/10.1134/S1022795406060093
Šimková H, Číhalíková J, Vrána J, Lysák MA, Doležel J (2003) Preparation of HMW DNA from plant nuclei and chromosomes isolated from root tips. Biol Plant 46:369–373
Šimková H, Šafář J, Suchánková P, Kovářová P, Bartoš J, Kubaláková M, Janda J, Číhalíková J, Mago R, Lelley T, Doležel J (2008a) A novel resource for genomics of Triticeae: BAC library specific for the short arm of rye (*Secale cereale* L.) chromosome 1R (1RS). - BMC Genomics 9:237. https://doi.org/10.1186/1471-2164-9-237
Šimková H, Svensson JT, Condamine P, Hřibová E, Suchánková P, Bhat PR, Bartoš J, Šafář J, Close TJ, Doležel J (2008b) Coupling amplified DNA from flow-sorted chromosomes to high-density SNP mapping in barley. BMC Genomics 9:294
Singh RJ, Röbbelen G (1975) Comparison of somatic Giemsa banding pattern in several species of rye. Z Pflanzenz Bd 75, 270–285.
Singh RJ, Röbbelen G (1977) Identification by Giemsa technique of the translocations separating cultivated rye from three wild species of *Secale*. Chromosoma 59:217–225. https://doi.org/10.1007/BF00292779
Smirnov VG, Sosnikhina SP (1984) Genetics of Rye. Leningrad University Press, Leningrad (in Russian)
Sosnikhina SP, Mikhailova EI, Tikholiz OA, Priyatkina SN, Smirnov VG, Dadashev SY, Kolomiets OL, Bogdanov YF (2005) Meiotic mutations in rye *Secale cereale* L. In: Puertas MJ, Naranjo T (eds) "Plant Cytogenetics"—a special volume of Cytogenetic and Genome research, vol 109, pp 215–220
Staňková H, Hastie AR, Chan S, Vrána J, Tulpová Z, Kubaláková M, Visendi P, Hayashi S, Luo M, Batley J, Edwards D, Doležel J, Šimková H (2016) BioNano genome mapping of individual chromosomes supports physical mapping and sequence assembly in complex plant genomes. Plant Biotechnol J 14:1523–1531. https://doi.org/10.1111/pbi.12513
Sybenga J (1959) Some sources of errors in the determination of chromosome length. Chromosoma 10(1–6):355–364. https://doi.org/10.1007/BF00396578
Sybenga J, van Eden J, van der Meijs QG, Roeterdink BW (1985) Identification of the chromosomes of the rye translocation tester set. Theor Appl Genet 69:313–316. https://doi.org/10.1007/BF00662451
Sybenga J, Wolters AHG (1972) The classification of chromosomes of rye (*Secale cereale* L.). A translocation tester set. Genetica 43:453–464. https://doi.org/10.1007/BF00156140
Sybenga J (1965) The quantitative analysis of chromosome pairing and chiasma formation based on the relative frequencies of MI configurations. II. Primary trisomics. Genetica; 36(1):339–350. https://doi.org/10.1007/BF01557167
Sybenga J (1983) Rye chromosome nomenclature and homoeology relationships. Workshop report. Z Pflanzenziichtg 90:297–304
Tarkowski C, Stefanowska G (1972) Chromosome morphology in the genome of rye *Secale cereale* L. and in Triticale 6x and 8x. Genet Polon 13:83–89
Thomas JB, Kaltsikes PJ (1976) A bouquet-like attachment plate for telomeres in leptotene of rye revealed by heterochromatin staining. Heredity 36(2):155–162. https://doi.org/10.1038/hdy.1976.19
Thompson JB, Rees H (1956) Selection for heterozygotes during inbreeding. Nature 177:385—386
Tikhonovich IA, Fadeyeva TS (1976) The study of heterochromatin in karyotypes of the forms of genetic collections of rye. Genetika 12:5–14 (in Russian, Abstr. Engl.)
Tikhonovich IA (1975) Chromosomes polymorphism in rye inbred lines and populations. Dokl Acad Nauk SSSR 224(4):944–946 (in Russian, Abstr. Engl.)

Tikhonovich IA, Dubovskaya AG, Sosnikhina SP (1987) The use of heterochromatin markers of chromosomes in cytogenetic studies in rye (*Secale cereale* L.). Part 1. The behaviour of chromosomes 1 (7R) and 5 (4R) in meiosis. Genetika 23:838–844 (in Russian, Abstr. Engl.)

Tsujimoto H, Gill BS (1991) Repetitive DNA sequences from polyploid *Elymus trachycaulus* and the diploid progenitor species: detection and genomic affinity of *Elymus* chromatin added to wheat. Genome 34 (5):782–789. https://doi.org/10.1139/g91-122

Unfried I, Gruendler P (1990) Nucleotide sequence of the 5.8S and 25S rRNA genes and the internal transcribed spacers from *Arabidopsis thaliana*. Nucleic Acids Res 18(13):4011. https://doi.org/10.1093/nar/18.13.4011

Verma SC, Rees H (1974) Giemsa staining and distribution of heterochromatin in rye chromosomes. Heredity 32:118–122. https://doi.org/10.1038/hdy.1974.12

Veronika KK, Jan P, Martin V, Glyn D, Dylan J, Edina P, Olga T,Michaela Š, Jozef S, Jaroslav Š, Doležel David K (2019) Nuclear Disposition of Alien Chromosome Introgressions into Wheat and Rye Using 3D-FISH. Int J Mol Sci 20(17) 4143. https://doi.org/10.3390/ijms20174143

Vershinin AV, Schwarzacher T, Heslop-Harrison JS (1995) The large-scale genomic organization of repetitive DNA families at telomeres of rye chromosomes. Plant Cell 7(11):1823–1833. https://doi.org/10.1105/tpc.7.11.1823

Vosa CG (1977) Heterochromatic patterns and species relationship. Nucleus 20:33–41

Vosa CG (1974) The basic karyotype of rye (*Secale cereale* L.) analysed with Giemsa and fluorescence methods. Heredity 33(3):403–408. https://doi.org/10.1038/hdy.1974.106

de Vries JM, de Sybenga J (1976) Identification of rye chromosomes: the Giemsa banding pattern and the translocations tester set. Theor Appl Genet 48(1):35–43. https://doi.org/10.1007/BF00282408

Weimarck A (1975) Heterochromatin polymorphism in the rye karyotype as detected by the Giemsa C-banding technique. Hereditas 79(2):293–300. https://doi.org/10.1111/j.1601-5223.1975.tb01486.x

Weimarck A (1974) Elimination of wheat and rye chromosomes in a strain of octoploid *Triticale* as revealed by Giemsa banding technique. Hereditas 77 (2):281–286. https://doi.org/10.1111/j.1601-5223.1974.tb00940.x

Xi W, Tang S, Du H, Luo J, Tang Z, Fu S (2020) ND-FISH-positive oligonucleotide probes for detecting specific segments of rye (*Secale cereale* L.) chromosomes and new tandem repeats in rye. Crop J 8:171–181. https://doi.org/10.1016/j.cj.2019.10.003

Xiaofeng Y, Zili Z (1988) In situ induction of G-bands and macrocoils in plant chromosomes. Hereditas 109 (1):45–51. https://doi.org/10.1111/j.1601-5223.1988.tb00180.x

Xie J, Zhao Y, Yu L, Liu R, Dou Q (2020) Molecular karyotyping of Siberian wild rye (*Elymus sibiricus* L.) with oligonucleotide fluorescence in situ hybridization (FISH) probes. PLoS One 15(1):e0227208. https://doi.org/10.1371/journal.pone.0227208

Zeller FJ, Kimber G, Gill BS (1977) The identification of rye trisomics by translocations and Giemsa staining. Chromosoma 62(3):279–289. https://doi.org/10.1007/BF00286049

Zonneveld BJ, Leitch IJ, Bennett MD (2005) First nuclear DNA amounts in more than 300 angiosperms. Ann Bot 96(2):229–244

Zwyrtková J, Šimková H, Doležel J (2021) Chromosome genomics uncovers plant genome organization and function. Biotech Adv 41:107659

The B Chromosome of Rye

4

A. Houben, W. Ma, and A. M. Banaei-Moghaddam

Abstract

Dispensable B chromosomes are the most enigmatic components of many eukaryotic genomes, including thousands of plant species. Since their discovery more than a century ago, the origin, genetic content and the mechanisms by which they remain in populations have been long-standing questions. In this regard, the B chromosome of rye (*Secale cereale*) was the first for which these questions have been addressed using a combination of modern laboratory and bioinformatic tools. In this review, our current understanding of how the rye B chromosome originated and the cellular mechanisms of its post-meiotic drive will be summarized. Besides, the DNA and chromatin composition, the transcriptional activity of the rye B and its effects on the host genome are discussed. Potential applications of the rye B are evaluated.

4.1 Introduction

The supernumerary B chromosome (B) is one of the most exciting components of the rye genome. B chromosomes are dispensable parts of the genome. They occur in a wide range of taxa, from fungi to plants and animals. To date, Bs were found in 2087 plants [53% monocots and 47% eudicots, see B chromosome database: http://www.bchrom.csic.es) (D'Ambrosio et al. 2017)]. The number of Bs varies between species and individuals of a population. Although dependent on the species, they may vary in size, structure and chromatin properties, they share certain features that make them distinguishable from other types of chromosome polymorphisms like, for example, aneuploidy. B chromosomes do not recombine with the standard set of A chromosomes (As) at meiosis and therefore have their own evolutionary pathway (reviewed in, e.g. Jones 1995; Camacho et al. 2000; Houben et al. 2013). The presence of a low number of Bs is associated with mild or no apparent phenotypes. However, increased numbers of Bs cause phenotypic differences and reduced fertility (reviewed in Bougourd and Jones 1997). Because most Bs do not confer any advantages to the organisms that harbour them, they are considered

A. Houben (✉)
Leibniz Institute of Plant Genetics and Crop Plant Research (IPK) Gatersleben, 06466 Seeland, Germany
e-mail: houben@ipk-gatersleben.de

W. Ma
Science Center for Future Foods, Jiangnan University, 1800 Lihu Road, Wuxi 214122, Jiangsu, China

A. M. Banaei-Moghaddam
Laboratory of Genomics and Epigenomics (LGE), Department of Biochemistry, Institute of Biochemistry and Biophysics (IBB), University of Tehran, Tehran, Iran

© The Author(s), under exclusive license to Springer Nature Switzerland AG 2021

M. T. Rabanus-Wallace and N. Stein (eds.), *The Rye Genome*, Compendium of Plant Genomes,
https://doi.org/10.1007/978-3-030-83383-1_4

as parasitic, selfish elements that persist in populations by making use of the cellular machinery required for the inheritance and function of A chromosomes. The de novo formation of a B is probably a rare event because the occurrence of similar B chromosome variants within related species suggests that they arose from a single origin either from the same or from a related species (Muñoz-Pajares et al. 2011; Marques et al. 2013). This review focuses on the origin, composition, behaviour and effects of rye B chromosomes. An earlier review written by Jones and Puertas (1993) contains additional information about the B of rye.

4.2 Rye B Chromosome Occurrence and Structural Variability

The B chromosome of rye was discovered unknowingly by M. Nakao more than 100 years ago. Looking at the meiotic chromosomes of rye (*Secale cereale*), wheat (*Triticum aestivum*) and barley (*Hordeum vulgare*), he stated that 'the number of chromosomes is 8 in wheat and rye, and 7 in barley' (Nakao 1911). Without being aware, Nakao analysed plants of rye carrying B chromosomes. Thirteen years later, another Japanese scientist, K. Gotoh, published a study in which he investigated rye with two small extra 'k-chromosomes' (Gotoh 1924). Later, Randolph (1928) classified them as B chromosomes.

Cultivated rye (*Secale cereale* subsp. *cereale*) tolerates up to eight additional Bs (Jones and Rees 1982) (Fig. 4.1a). The number of Bs in all somatic plant tissues is stable. Each standard B of rye adds ~580 Mbp to the normal complement of seven pairs of A chromosomes (1C–7917 Mbp) (Martis et al. 2012). Bs can be found in landraces and old cultivars, but are unlikely present in modern rye cultivars. In addition, Bs were found in perennial weedy rye (*Secale cereale* subsp. *segetale*) (Akita and Sakamoto, 1982; Jones and Puertas 1993; Niwa and Sakamoto 1995; Niwa and Sakamoto 1996). Based on similar morphology and meiotic pairing of Bs derived from weedy and cultivated rye lines in F_1 hybrids, a monophyletic origin for the rye B is likely (Niwa and Sakamoto 1995). Structural rearrangements of the Bs occurred at about 2% per generation in experimental crosses (Jimenez et al, 1995) giving rise mainly to B isochromosomes and truncated Bs (Jones and Puertas 1993).

The overall structure of the B of cultivated rye and weedy rye (subsp. *afghanicum* and subsp. *segetale*, respectively) from different origins was analysed by FISH using probes specific for the

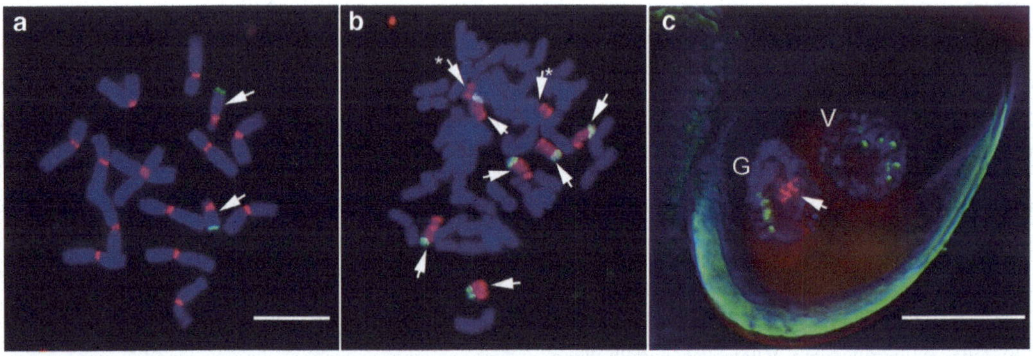

Fig. 4.1 Rye B chromosomes (arrowed) characterized by fluorescence in situ hybridization. **a** Mitotic metaphase of rye + Bs labelled with the rye centromere-specific repeat Bilby (in red) and the B-specific repeat E3900 (in green). **b** Mitotic metaphase of hexaploid wheat line with additional rye Bs labelled with the rye genome-specific repeat Revolver (in red) and the B-specific repeat E3900 (in green). * indicates truncated Bs. **c** Binucleated pollen of rye + Bs labelled with the rye centromere-specific repeat Bilby (in green) and the B-specific probe ScCl11 (in red). As a result of nondisjunction, Bs accumulate in the generative nucleus (G). Vegetative nucleus (V) contains only A chromosomes. Chromatin is counterstained with DAPI (in blue). Bars = 10 μm

pericentromeric region (ScCl11 and mitochondrial DNA), the nondisjunction control region of the long arm (D1100, E3900, Sc26c38 and Sc9c130 repeats) and the interstitial region (Sc36c82, Sc55c1, Sc63c34, chloroplast and mitochondrial DNA) of Bs (Marques et al. 2013). Regardless of the origin, Bs of both subspecies showed the same distribution of hybridization signals of ScCl11, Sc36c82 and D1100 in the pericentromeric, interstitial and terminal regions, respectively. In contrast, the A chromosomes of different origins and subspecies revealed distribution polymorphisms for ScCl11 and Sc36c82. The retroelements *Revolver* and *Sabrina* revealed genotype-independent distribution patterns along A and B chromosomes. The distribution of high-copy repeats in the B of rye is summarized in Fig. 4.2.

FISH with labelled organellar DNA and a $(CAA)_{10}$-type microsatellite allowed the identification of two structural variants of Bs. Likely an inversion of the pericentromeric region was responsible for the origin of the observed B chromosome polymorphism (Marques et al. 2013). Besides, early analysis of pachytene chromosomes suggested some minor variations in the B chromosome structure (Lima-De-Faria 1963). Thus, apart from these minor differences, the distribution of most repetitive sequences on the rye Bs is still conserved at the chromosomal level, regardless of the geographic origin of the populations.

The high degree of chromosome conservation was unexpected since the nonessential nature of Bs should allow structural polymorphisms. Indeed, a higher rate of mutations of single nucleotide polymorphisms has been found in genic sequences of the cultivated rye B compared to the homologous sequences of As, reflecting reduced selective pressure for these sequences on the Bs (Martis et al. 2012). Also, in many other species, B chromosome structural variants derived from a single type have been reported (for review see Jones 1995).

Rare spontaneous A/B chromosome translocations were reported for rye (Schlegel and Pohler 1994). To generate a high number of A/B translocations, an experiment involving radiation was performed. Although high frequencies of structural chromosome rearrangements were detected in M1 plants, none were present at meiosis or in somatic cells of their progeny. This observation may explain why this form of chromosome mutation is mostly absent from natural populations (Hasterok et al. 2002).

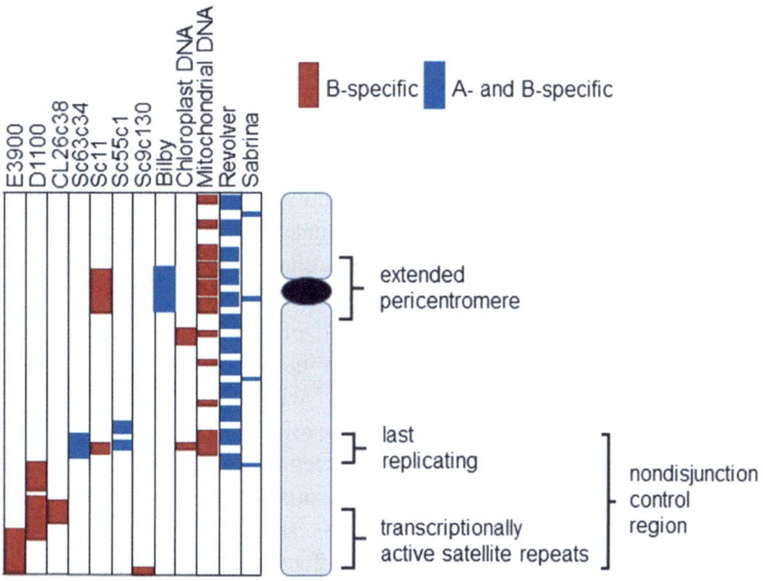

Fig. 4.2 Rye B chromosomes features. Distribution of B- and A/B-specific repeats along the mitotic metaphase chromosome

4.3 Rye B Chromosome Origin

The question of the origin of B chromosomes in rye arose with the first discovery of this supernumerary genome component and remains unanswered for the majority of Bs in other species. Generally, it is assumed that Bs are derived from standard A chromosomes, either from the same or from a related species. The path of their formation likely differs between different species and B chromosome types (reviewed in Marques et al. 2018; Martins and Ahmad 2019). Recent advances in genome sequencing have revolutionized our understanding of the B chromosome origin and have shown that Bs are more complex than was previously imagined (reviewed in Ruban et al. 2017).

The rye B originated approximately 1.1–1.3 million years ago, 0.4–0.6 million years after the formation of the genus *Secale* (Martis et al. 2012). Using a combination of flow sorting, next-generation sequencing and bioinformatic tools, the B of rye was shown to have originated from multiple A chromosomes. To trace the origin of the B, the positional information of genic sequence reads of the B on the respective A chromosomes were depicted by mapping them against the virtual gene map of rye (Martis et al. 2012). The rye B shows extended sequence overlaps with the rye A chromosomes 3R and 7R, along with thousands of short genic sequence stretches derived from all over other A chromosomes. Likely, a proto-B chromosome was initially formed by segmental genome duplication and interspecies hybridization events, followed by reductive chromosome translocations, unbalanced segregation of a small translocation chromosome and subsequent sequence insertions. Alternatively, the centromeric 'by-product' of a Robertsonian translocation between two nonhomologous acrocentric chromosomes with breakpoints close to centromeres could evolve into a proto-B. After proto-B formation, recombination with donor As became restricted, probably due to multiple rearrangements and illegitimate recombination involving different As, which precluded extended pairing with the formerly homologous A chromosome regions. This restriction of recombination may be considered as the starting point for the independent evolution of the B. At the same time, drive or a fitness benefit of the nascent B is a prerequisite for an evolving B. It is tempting to speculate that Bs originated as a by-product of A chromosome rearrangement events.

Because an increased gene dosage may affect gene expression, the expression of B-located paralogues genes might have been reprogrammed potentially through epigenetic mechanisms early during the evolution of the Bs. Thus, proto-B genes might first have been transcriptionally suppressed and then degenerated due to mutations, including the insertion of sequences derived from other A chromosomal regions and organellar genomes. Selection pressure is only acting on regions responsible for the drive and thus for the maintenance of B chromosomes.

In addition, compared to As, the rye B was found to accumulate large amounts of specific repeats and organellar DNA. A similar sequence composition was found for the B of *Aegilops speltoides* (Ruban et al. 2020). It seems that the B chromosome acts like a 'sponge' which collects nucleus, chloroplast- and mitochondria-derived sequences (Fig. 4.3). However, an entirely annotated B chromosome sequence is not yet

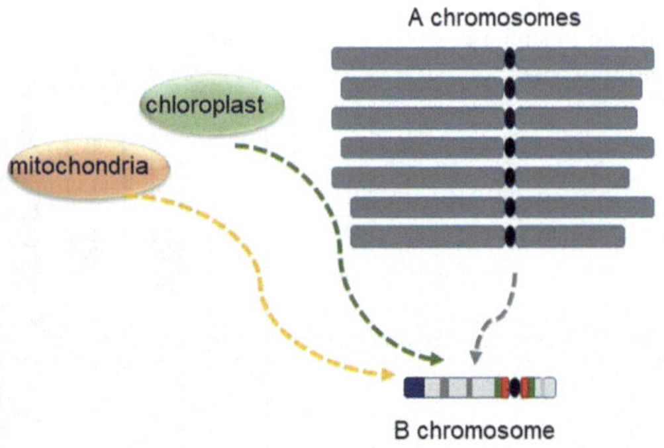

Fig. 4.3 Model of the evolution of the rye B chromosome. Accumulation of mitochondria-, chloroplast- and standard A chromosome-derived DNA fragments, amplification of B-specific repeats, erosion and inactivation of A-derived genes and gain of chromosome drive resulted in the B chromosome

available for any organism. The mosaic-like structure of Bs suggests analogies between the process of chromothripsis [chromothripsis results in simultaneous fragmentation of distinct chromosomal regions and then subsequent imperfect reassembly by DNA repair pathways, reviewed by Leibowitz et al. (2015)] and the mechanism behind the formation of B chromosomes.

4.4 DNA and Chromatin Composition of Rye Bs

Despite the different transcription activities of As and Bs, both types of chromosomes exhibit a similar interphase organization in meristematic and differentiated interphase nuclei (Schubert et al. 2016). Applying fluorescence in situ hybridization (FISH), both rye A and B chromosomes added to hexaploid wheat showed in meristematic interphase nuclei, a string-like shape and clear centromeric clustering (Rabl orientation). In 4C-differentiated leaf nuclei, a more relaxed chromatin structure, round-shaped chromosome territories and a less pronounced Rabl configuration were found. The sister chromatid exchange frequency of Bs did not differ from that of A chromosomes (Schubert et al. 2016).

Although the chromatin structure is increasingly seen as playing an essential role in different aspects of chromosome function, little information is available on the chromatin composition of Bs, and whether it differs from that of the standard A chromosomes. Based on classical cytological observations (e.g. Giemsa-banding), an early survey suggested that the Bs in about half of the B-positive plant species are heterochromatic (Jones 1975).

The distribution patterns along A and B chromosomes observed for the heterochromatin marks histone H3K9me1, 2 were mainly uniform (Houben et al. 2003). The terminal heterochromatic regions of As and Bs showed little H3K27me1 but were enriched in di- and trimethylated H3K27 (Carchilan et al. 2007). The late-replicating, heterochromatic nondisjunction control region of the B is characterized by a unique combination of histone methylation marks (Carchilan et al. 2007). In contrary to the heterochromatic regions of the A chromosomes, this domain is simultaneously marked by trimethylated histone H3K4 and trimethylated H3K27. In addition, this domain shows a dark Giemsa band at mitosis, but undergoes decondensation during interphase and reveals transcription of B-specific high-copy repeat families.

4.5 Meiotic Behaviour of Rye Bs

Meiotic pairing and synapsis depend on the numbers of Bs. At pachytene, the organization of Bs and As is comparable (Müntzing and Lima-de-Faria 1949). Self-synapsis occurs in the case of one B only. Plants with 2 Bs perform a regular synaptonemal complex assembly and bivalent formation was observed. Plants carrying 3 Bs showed different modes: 84% of their meiocytes had one bivalent and one univalent, in 13% a trivalent of all 3Bs was formed, and only 3% of the cells contained three univalents. In 4B plants, 64.5% of meiocytes contained only bivalents, 30% multivalents including all Bs and 6% had one bivalent plus two univalents (Hesse et al. 2019). The synaptonemal complex formation of Bs in plants with odd numbers of Bs may be impaired. They form intrachromosomal synaptonemal complexes ranging from small clusters to long synaptonemal complex stretches which were observed for univalent Bs. But in general, rye Bs form a similar synaptonemal complex as based on the immunolocalization of conserved synaptonemal complex proteins like NSE4A, HEI10, ASY1 and ZYP1 (Hesse et al. 2019).

At metaphase I, rye Bs can form univalents, bivalents or multivalents depending on the number of Bs present. When the plant has only one B, it is frequently included in one pole at anaphase I, and therefore, the loss of the B univalent is prevented. Univalents are often eliminated at the second division. In the case of multivalents, they give irregular chromosome segregations so that overall there is a certain amount of meiotic elimination (Jones 1993). Rye genotypes for high and low B transmission rates have been selected. In low transmission plants, 2

Bs form bivalents only in 20% of the metaphase I cells. When B univalents divide equatorially at anaphase I, they are subsequently eliminated as micronuclei. Conversely, Bs in high transmission plants form bivalents in nearly 90% of the pollen mother cells and they are present in 85% of the pollen grains (Jiménez et al. 1997). Therefore, it can be concluded that rye B transmission and population polymorphism mainly depend on the Bs and that a regular meiotic behaviour is essential for a B chromosome to be maintained in the long term.

4.6 Drive Mechanism of the Rye B Chromosome

When transmission rates of chromosomes are higher than 0.5, not obeying the Mendelian law of equal segregation, the resulting transmission advantage is collectively referred to as 'drive'. Drive is the key to the evolutionary success of Bs. Depending on the species, B chromosome drive is due to pre-meiotic, meiotic or post-meiotic events (reviewed in Jones 1991, 2018; Houben 2017).

The post-meiotic drive of rye B is one of the best-analysed mechanisms among Bs. The behaviour of Bs during first pollen mitosis in rye was first studied by Hasegawa (1934), who described '…the two split halves (sister chromatids) of the extra chromosomes are in most cases included in the generative nucleus in late anaphase'. He observed that the two chromatids of the B do not separate at anaphase of the first pollen grain mitosis, and in most cases, both chromatids were included in the generative nucleus (Fig. 4.1c, 4.4). At second pollen mitosis, B-sister chromatids divide normally like standard chromosomes.

The frequency of nondisjunction at first pollen mitosis depends on the genotype (Puertas et al. 1998, 2000). The drive of Bs also occurs in female gametophytes during the first post-meiotic division (Müntzing 1945; Håkanson 1948). A similar drive of Bs during the first pollen mitosis was found in *Ae. speltoides* and *Ae. mutica* (Mendelson and Zohary 1972; Ohta 1996; Wu et al. 2019).

Chromosome drive works equally well when the rye B was introduced into *S. vavilovii* (Puertas et al. 1985), hexaploid wheat (Müntzing 1970; Endo et al. 2008) or triticale (X *Triticosecale* Wittmark) (Kishikawa and Suzuki 1982). Hence, the B chromosome controls the nondisjunction process by itself (Matthews and Jones 1983; Romera et al. 1991).

The end of the long B chromosome arm is controlling the process of chromosome drive. Rye Bs lacking the so-called nondisjunction control region, undergo normal disjunction at first pollen anaphase (Müntzing 1945, 1948; Håkanson 1959; Endo et al. 2008). Also, in somatic interphase nuclei the nondisjunction element is active to keep sister (peri)centromeres of Bs together (Schubert et al. 2011). The control region acts *in trans* because post-meiotic nondisjunction occurs for both the standard and the deficient B if the terminal region of the long arm (Endo et al. 2008) or a standard B (Lima-de-Faria 1962) coexist in the same cell containing a deficient B without the nondisjunction control region.

The nondisjunction control region is heterochromatic, late-replicating and enriched in B-specific satellite DNAs (Sandery et al. 1990; Blunden et al. 1993; Houben et al. 1996; Carchilan et al. 2007; Pereira et al. 2009; Klemme et al. 2013). Despite the heterochromatic nature of the nondisjunction control region, this region is enriched in the euchromatin-specific histone modification mark H3K4me3 and produces long-noncoding RNA predominantly in anthers (Carchilan et al. 2007). A possible involvement of nondisjunction control region-derived noncoding RNAs in maintaining the cohesion of sister chromatids, causing nondisjunction, is a likely option.

The (peri)centromere of the B is the second component involved in the chromosome drive process. The repeat composition of rye A and B (peri)centromeres differs and (peri)centromere of Bs is more extended (Banaei-Moghaddam et al. 2012). In addition to repeats shared by both types of chromosomes, the pericentromere of Bs contains B-specific ScCl11 repeats and mitochondrion-derived DNA. Hence, the

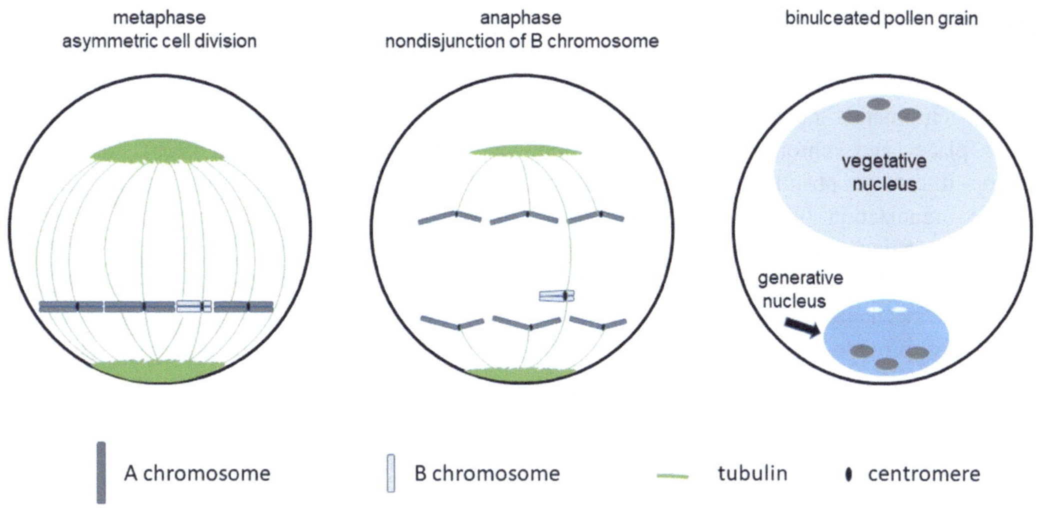

Fig. 4.4 Post-meiotic drive of rye Bs. Rye Bs accumulate by directed nondisjunction in both male and female gametophytes. In pollen grains, this nondisjunction happens in the first pollen mitosis, where the cell division is asymmetric. During metaphase, both A- and B-centromeres are active and attach to the microtubules. At anaphase stages, however, while A-sister chromatids are separated, the B-sister chromatids show extended cohesion and remain unseparated. After anaphase and by the formation of the nuclear envelope, as the placement of Bs is to the vicinity of the generative pole, lagging Bs are included in the generative nucleus. By this mechanism, the number of Bs is doubled

centromere of the B originated from an A chromosome and new sequences accumulated in the B centromere afterwards. However, ScCl11 and mitochondrial sequences are not part of the centromere active CENH3-containing chromatin (Banaei-Moghaddam et al. 2012). CENH3 is a centromere-specific histone H3 variant which marks the active centromeres of most plant species (Talbert and Henikoff 2020). A comparable distinct repeat composition was also reported for the B-centromeres of *Zea mays* and *Ae. speltoides* (Jin et al. 2005; Wu et al. 2019). The B-specific accumulation of repeats in the pericentromere is likely involved in the formation of pericentric heterochromatin. Heterochromatin has been suggested to play a role in chromosome segregation (Yamagishi et al. 2008).

Why does the B migrate preferentially to the generative nucleus at the first pollen grain mitosis? During the first pollen mitosis, the microtubule spindle is asymmetrical and the asymmetry of this division plays a critical role in the determination and subsequent fate of the two unequal daughter cells, the vegetative and the generative one. Considering the asymmetric geometry of the spindle at first pollen mitosis, it is likely that, as suggested by Jones (1991), the inclusion of Bs in the generative nucleus is caused by the fact that the equatorial plate is closer to the generative pole and lagging Bs are passively included in the generative nuclei. It seems that the asymmetrical spindle is another key component of the post-meiotic drive of rye and *Ae. speltoides* Bs (Banaei-Moghaddam et al. 2012; Wu et al. 2019).

Strikingly, comparing the cellular process of B chromosome drive with the process of B chromosome elimination, a process taking place in *Ae. speltoides,* revealed high similarity (Wu et al. 2019; Ruban et al. 2020). In both, nondisjunction of Bs occurs despite centromere activity and centromere–tubulin interaction. However, the spindle symmetry differs between both types of processes. While during the first pollen mitosis an asymmetric cell division occurs (Banaei-Moghaddam et al. 2012), the spindle in root cells is symmetric. As a consequence, in roots of *Ae. speltoides*, lagging Bs form micronuclei and undergo elimination. In contrast, due to the asymmetric geometry of the spindle at first pollen mitosis, the inclusion of the lagging joint B chromatids in the generative nucleus takes place and chromosome accumulation occurs. It can be postulated that the type of spindle organization (symmetric versus asymmetric) determines whether drive or elimination of B chromosomes follows. Understanding of the molecular mechanism behind B chromosome drive may provide clues about chromosome nondisjunction, which is a major cause of genetic diseases across species.

4.7 Effects of Rye B Chromosomes

Since each rye B contains the equivalent of 7% of the DNA of a rye genome comprising seven pairs of A chromosomes (Martis et al. 2012), adding an extra $\sim$580 Mbp with each additional B changes the genome size, and it finally produces a spectrum of physiological phenotypes. The genetical and physiological effects caused by the rye Bs are quantitative and proportional. The presence of Bs has only mild or slight phenotypic effects if their number is low. Conversely, excessive amounts of Bs can reduce the fertility and fitness of the host (reviewed in Jones and Rees 1982; Jones 1995; Bougourd and Jones 1997). Generally, Bs are not beneficial from the perspective of physiological effects under normal growth conditions.

4.8 Morphological Changes Associated with Bs

The effects of Bs on vegetative growth have been thoroughly investigated. Approximately proportional to the number of Bs present, rye plants were reduced in weight, height, seed weight and tiller numbers (Müntzing 1943, 1963; Moss 1966). Besides the effects on vegetative growth, plants with Bs also show generative effects. Puertas and Carmona (1976) found that rye plants with 2 Bs have the highest number of germinating pollen and a higher speed of pollen tube growth than 0B plants. But these effects were reduced in plants with 4 Bs (Håkanson 1957; Puertas and Carmona 1976). Also, plants with Bs showed a delay in seed germination (Moss 1966) and flowering (Jones and Rees 1967), decreased seed weight (Moss 1966), reduced seed fertility (Müntzing 1943) and seed set (Müntzing 1943, 1963). Hence, under normal growth conditions, the presence of Bs is not beneficial for the host. To date, no direct evidence exists that rye Bs can confer a selective advantage to survival or fitness under specific environmental conditions. However, rye plants with Bs showed heat-induced positive effects indicating that Bs have implications on heat tolerance and may protect meiocytes against heat stress-induced damage (Pereira et al. 2017). The effects of Bs on the nuclear physiology and the so-called odd-and-even effect of Bs have been reviewed by Jones (1995).

4.9 Effects of Bs on Chiasma Formation and Homologous Chromosome Association

Rye Bs affect the meiotic behaviour of As. In a synthetic population, obtained by crossing *S. cereale* with *S. vavilovii* carrying two Bs, Jones and Rees (1967) demonstrated that the presence of Bs influenced the distribution of chiasma formation of As. Zečević and Paunović (1969) working with two wild rye populations from Yugoslavia showed that the mean chiasma frequency increased with an increasing number of Bs. Otlowska-Miazga (1974) using Transbaikal rye showed that the Bs affect the mean chiasma frequency. The chromosome association at metaphase I was compared between rye 0B and 2B plants using the Giemsa C-banding technique. It was found that the presence of iso-Bs increased the homologous association which was measured by the frequencies of chiasmata at metaphase I (Alvarez et al. 1991).

4.10 Transcriptional Effects of Rye B Chromosomes

To assay the transcriptional activity of rye Bs at the cytological level, the localization of active RNA polymerase II (phosphorylated at serine 2) was visualized in interphase nuclei of rye and wheat with Bs by immunostaining. In both species, the RNAPIISer2P signals were closely associated with chromatin of As and Bs (Ma et al. 2017). The transcriptional activity of Bs was confirmed first by comparative cDNA-AFLP analysis using plants with and without Bs from three isogenic rye lines (Carchilan et al. 2009). Using 63 primer combinations, 112 extra bands (4.9% of the average number of bands per rye line) were found in plants with Bs of either 1, 2 or all 3 +B rye lines. However, only 16 B-extra bands (0.7% of the average number of bands per rye line) of different intensities were consistently found in all three rye lines. Assuming that the transcriptome of a 0B rye plant is equally derived from all seven A pairs, each A chromosome-type could encode around 14% of the transcriptome. Not unexpectedly, this number is much higher than the 0.7% of B-associated transcripts and confirms a low transcriptional activity of B chromosomes. However, due to the limited number of primer combinations tested for AFLP and the likely high sequence similarity of A- and B-derived transcripts, analysis was an underestimate of the B-encoded transcripts.

Indirect proof that the B is influencing the transcription of A-located genes was provided by silver staining of metaphase chromosomes to evaluate the activity of the A-located 45S rDNA. A single B had no measurable effect, but, as the number of Bs increased, there were significant changes in the physical dimensions of the A chromosome metaphase NORs, reflecting reduced levels of their activity earlier in the cell cycle and also in the condensation patterns of the interphase rDNA loci (Morais-Cecilio et al. 1996).

Banaei-Moghaddam et al. (2013) analysed 15 pseudogene-like fragments to address whether rye B-located genic fragments exhibit transcriptional activity. The results showed that besides transcriptional activity, they could modulate their paralogs on As in a tissue- and genotype-dependent manner. Some of these B-located genic fragments showed alternative splicing patterns, suggesting that at least for some B-located genic sequences, the regulatory elements remained functional (Banaei-Moghaddam et al. 2013).

More recently, to determine the transcriptional activity of the rye B as well as its possible influences on the gene expression pattern of the host genome, a comparative transcriptome analysis was conducted at the genome-wide level. Therefore, polyadenylated mRNA from vegetative (roots and leaves) and generative tissue (anthers) of rye with and without Bs were sequenced using next-generation sequencing (Ma et al. 2017). In total, 1954 and 1218 B-located and expressed genic sequences with an open reading frame of ≥ 180 bp in generative and vegetative tissues were considered as B-located genes. Sixteen per cent (308) of the B-originated transcripts in anther and 20% (247) of them in vegetative tissue showed similarities to canonical protein-coding genes. These B-located genes belonged to various biological processes including photosynthetic and respiratory processes, cell division, and negative regulation of gene expression. The enrichment of transcripts having a role in photosynthesis and respiratory processes was attributed to the ongoing insertion of chloroplast DNA into the rye B. The overrepresentation of negative regulatory function was thought, to some extent, to explain the detrimental effects associated with the high-copy number of Bs and the observed downregulation of A-located genes in the presence of Bs. So far, there is no direct evidence for the function of any rye B-located genes in the drive mechanism of the rye B. However, enrichment analysis in this study highlighted the overrepresentation of 'microtubule-based movement', 'mitosis' and 'cell division'.

Interestingly the GO enrichment for anther and vegetative B-transcripts showed a different pattern that indirectly implies that B-located genes have tissue-specific gene expression patterns. In addition to determining the B-originated transcripts, statistically significant differentially expressed genes in the presence of rye Bs were determined. A higher number of genes (1569) in generative tissues than vegetative tissues (916) showed dysregulation in the presence of Bs. As 92% of these dysregulated genes were downregulated in 4B plants compared to the less than 10% of upregulated ones, the authors concluded that Bs had affected *in trans* the regulation of A-located genes (Ma et al. 2017).

To test whether B-derived transcripts finally translate to proteins in vivo, a comparative mass spectrometry study was performed using the protein samples isolated from shoots of rye plants with and without Bs. 319 out of 16,776 quantified features were found in at least three out of five +B plants but not in 0B plants. 31 out of 319 features were identified as B-associated peptide features. Thus, the existence of B chromosomes can alter the composition of the rye proteome (Ma et al. 2021).

4.11 B Chromosome Located Genes and Their Evolution

Due to the dispensable nature of Bs, which put them under less selection pressure, it was expected that the B-located genic sequences would accumulate mutations with a higher rate as compared to the As, except for those necessary for the drive mechanism of rye Bs. Indeed, comparison of rye B-genic sequences to their A-located counterparts showed an elevated level of sequence polymorphisms. Nevertheless, the extent of these dissimilarities varied among different B-genic sequences. Highly degraded B-located genic sequences may represent those that were located on B very early at the onset of B evolution. Alternatively, they may represent dosage-sensitive genes which are rapidly getting inactivated upon their duplication by the formation of the B. Silencing or activity of B-located genes likely depends on how their activity interfered with the expression and function of their A-located paralogs. In this regard, it would be expected that some of the dosage-sensitive B-located genes are silenced epigenetically at the onset of B formation (Banaei-Moghaddam et al. 2013). The mechanisms by which these genic sequences were inserted in Bs remains unclear. However, the observation of intron-containing genes from different As on rye Bs suggested that they may have inserted via hitchhiking on transposable elements or during double-strand break repair.

Most of the identified expressed B-located genic sequences (54% in anthers and up to 41% in vegetative tissues out of total transcripts) belonged to uncharacterized, often short-length sequences without any similarity to the functionally annotated proteins (Ma et al. 2017). These are likely products of the pseudogenization process (Fig. 4.5).

Three of the rye B-located genes were characterized in more detail. One of these B-specific transcripts showed homology to Argonaute 4 of *Arabidopsis thaliana* (ScAGO4B). Sequence alignment of ScAGO4B showed that compared to the A-originated transcript, the B-transcript had deletions that were divisible by three, and none of the point mutations caused a non-sense mutation. In vitro analysis of the A and B chromosome encoded AGO4B protein variants demonstrated that both possess RNA slicer activity. Thus, B-encoded genes may provide an additional level of gene control and complexity in combination with their related A-located genes. Hence, physiological effects, associated with the presence of B chromosomes, can partly be explained by the activity of B-located (pseudo)genes (Ma et al. 2017).

4.12 Potential Applications of B Chromosomes

Plant B chromosomes have been employed in mapping the standard genome, modulating meiotic recombination, and exploring the structure of the centromere and the process of nondisjunction, as discussed by Jones et al. (2008a, b), Birchler et al. (2009), Harper et al. (2018) and Jones and Ruban (2018). In future, Bs could become even crucial for the generation of minichromosome-based vectors for gene transfer.

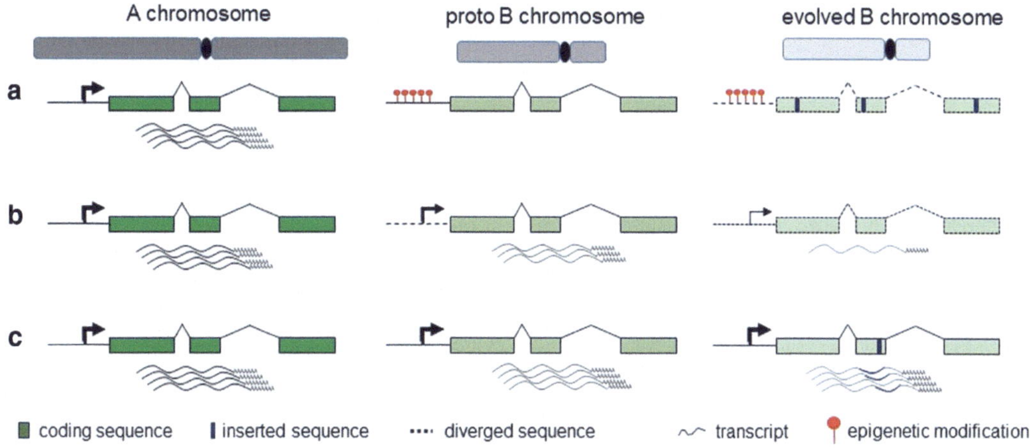

Fig. 4.5 Possible evolutionary fate of rye B-located genes. The B of rye descended from multiple As, and therefore, proto-B still shows sequence similarity to the parental As. **a** Dosage-sensitive B-located genes are getting suppressed at proto-B stage by, for example, epigenetic modifications. After inactivation, they became pseudogenized by accumulating mutations, mobile and satellite DNAs. Genes on Bs show a higher rate of sequence polymorphisms compared to their A-located paralogs. **b** Non-dosage-sensitive *B*-located genes that have no benefit for their host or maintenance of Bs, remain active upon the formation of Bs. However, as the Bs follow their evolutionary path, they gradually collect mutations. Depending on the position of mutations, these genes have expressions level ranging from similar to parental genes to the silenced ones. **c** Genes on Bs that are necessary for the maintenance of Bs remain active. Their sequences are under selection pressure; nevertheless, due to possible neofunctionalization, some sequence polymorphism on them is expected

Telomere-mediated chromosome truncation has been used for the chromosomes of maize, wheat, barley and *A. thaliana* (reviewed in Mette and Houben 2015; Birchler and Swyers 2020). Concerning the possible use of Bs as a vector for transgenes, it is important to recall that Bs have little or no effect on an individual's phenotype, and this issue is only of concern where a high number of Bs can reduce vigour (Puertas 2002). Constitutive transgene expression from maize B chromosome-derived minichromosomes suggests that inactivation of transgenes on Bs (Yu et al. 2007), if it occurs, is at least not a rapid process. Because of the intrinsic post-meiotic drive of intact rye Bs, a B chromosome-derived vector might potentially reveal an increase of transmission frequency above Mendelian expectation.

Acknowledgements We thank our colleagues and reviewers for their helpful suggestions, insightful comments and discussions on this review. This work was supported by the Deutsche Forschungsgemeinschaft DFG (HO1779/30-1) and China CSC Scholarship.

References

Akita T, Sakamoto S (1982) Cytogenetical studies of B-chromosomes in *Secale afghanicum*, a weed species closely related tp cultivated *S. cereale*. Report of Plant Germ-plasm Institute, Faculty of Agriculture, Kyoto University, Kyoto, Japan, 5:53–64

Alvarez M, Fominaya A, De La Vega MP (1991) A possible effect of B-chromosomes on metaphase I homologous chromosome association in rye. Heredity 67:123–128

Banaei-Moghaddam AM, Schubert V, Kumke K et al (2012) Nondisjunction in favor of a chromosome: the mechanism of rye B chromosome drive during pollen mitosis. Plant Cell 24:4124–4134

Banaei-Moghaddam AM, Meier K, Karimi-Ashtiyani R, Houben A (2013) Formation and expression of pseudogenes on the B chromosome of rye. Plant Cell 25:2536–2544

Birchler JA, Gao Z, Han F (2009) A tale of two centromeres–diversity of structure but conservation of function in plants and animals. Funct Integr Genomics 9:7–13

Birchler JA, Swyers NC (2020) Engineered minichromosomes in plants. Exp Cell Res 388:111852

Blunden R, Wilkes TJ, Forster JW et al (1993) Identification of the E3900 family, a second family of rye B

chromosome specific repeated sequences. Genome 36:706–711
Bougourd SM, Jones RN (1997) B chromosomes: a physiological enigma. New Phytol 137:43–54
Camacho JPM, Sharbel TF, Beukeboom LW (2000) B-chromosome evolution. Philos Trans R Soc Lond Ser B-Biol Sci 355:163–178
Carchilan M, Delgado M, Ribeiro T et al (2007) Transcriptionally active heterochromatin in rye B chromosomes. Plant Cell 19:1738–1749
D'Ambrosio U, Alonso-Lifante MP, Barros K et al (2017) B-chrom: a database on B-chromosomes of plants, animals and fungi. New Phytol
Endo TR, Nasuda S, Jones N et al (2008) Dissection of rye B chromosomes, and nondisjunction properties of the dissected segments in a common wheat background. Genes Genet Syst 83:23–30
Gotoh K (1924) Über die Chromosomenzahl von *Secale cereale* L. Jap J Genet 7:135–152
Håkanson A (1948) Behaviour of accessory rye chromosomes in the embryo-sac. Hereditas 34:35–59
Håkanson A (1957) Meiosis and pollen mitosis in rye plants with many accessory chromosomes. Hereditas 43:603–620
Håkanson A (1959) Behaviour of different small accessry rye chromosomes at pollen mitosis. Hereditas 45:623–631
Harper J, Phillips D, Thomas A et al (2018) B chromosomes are associated with redistribution of genetic recombination towards lower recombination chromosomal regions in perennial ryegrass. J Exp Bot 69:1861–1871
Hasegawa N (1934) A cytological study on 8-chromosome rye. Cytologia 6:68–77
Hasterok R, Jenkins G, Langdon T, Jones RN (2002) The nature and destiny of translocated B-chromosome-specific satellite DNA of rye. Chromosome Res 10:83–86
Hesse S, Zelkowsk M, Mikhailova EI et al (2019) Ultrastructure and dynamics of synaptonemal complex components during meiotic pairing and synapsis of standard (A) and accessory (B) rye chromosomes. Front Plant Sci 10:773
Houben A (2017) B Chromosomes—a matter of chromosome drive. Front Plant Sci 8:210
Houben A, Kynast RG, Heim U et al (1996) Molecular cytogenetic characterization of the terminal heterochromatic segment of the B-chromosome of rye (*Secale cereale*). Chromosoma 105:97–103
Houben A, Demidov D, Gernand D, Meister A et al (2003) Methylation of histone H3 in euchromatin of plant chromosomes depends on basic nuclear DNA content. Plant J 33:967–973
Houben A, Banaei-Moghaddam AM, Klemme S, Timmis JN (2013) Evolution and biology of supernumerary B chromosomes. Cell Mol Life Sci 71:467–478
Jimenez MM, Romera F, Gallego A, Puertas MJ (1995) Genetic control of the rate of transmission of rye B chromosomes. II. 0Bx2B crosses. Heredity 74:518–523
Jiménez MM, Romera F, González-Sánchez M, Puertas MJ (1997) Genetic control of the rate of transmission of rye B chromosomes III. Male Meiosis and Gametogenesis. Heredity 78:636–644
Jin WW, Lamb JC, Vega JM et al (2005) Molecular and functional dissection of the maize B chromosome centromere. Plant Cell 17:1412–1423
Jones RN (1975) B chromosome systems in flowering plants and animal species. Int Rev Cytol 40:1–100
Jones RN (1991) B-chromosome drive. Am Nat 137:430–442
Jones RN (1995) Tansley review n° 85—B chromosomes in plants. New Phytol 131:411–434
Jones RN (2018) Transmission and drive involving parasitic B chromosomes. Genes 9:388
Jones RN, Rees H (1967) Genotypic control of chromosome behaviour in rye XI. The influence of B chromosomes on meiosis. Heredity 22:333–347
Jones RN, Rees H (1982) B chromosomes, 1st edn. Academic Press, London New York
Jones N, Ruban A (2018) Are B chromosomes useful for crop improvement? Plants People Planet 1:84–92
Jones RN, Gonzalez-Sanchez M, Gonzalez-Garcia M et al (2008a) Chromosomes with a life of their own. Cytogenet Genome Res 120:265–280
Jones RN, Viegas W, Houben A (2008b) A century of b chromosomes in plants: so what? Ann Bot 101:767–775
Jones RN, Puertas MJ (1993) The B-chromosomes of rye (*Secale cereale* L.). In: Dhir KK, Sareen TS (eds) Frontiers in plant science research. Dehli (India): Bhagwati Enterprises, pp 81–112
Jones R (1993) The B-chromosomes of rye (*Secale cereale* L.). Fron Plant Sci Res
Kishikawa H, Suzuki A (1982) Cytological study on hypo-pentaploid Triticale with four B chromosomes of rye. Jpn J Genet 57:17–24
Klemme S, Banaei-Moghaddam AM, Macas J et al (2013) High-copy sequences reveal distinct evolution of the rye B chromosome. New Phytol 199:550–558
Leibowitz ML, Zhang CZ, Pellman D (2015) Chromothripsis: a new mechanism for rapid karyotype evolution. Annu Rev Genet 49:183–211
Lima-de-Faria A (1962) Genetic interaction in rye expressed at chromosome phenotype. Genetics 47:1455–1462
Lima-De-Faria A (1963) The evolution of the structural pattern in a rye B-chromosome. Evolution 17:289–295
Ma W, Gabriel TS, Martis MM et al (2017) Rye B chromosomes encode a functional argonaute-like protein with in vitro slicer activities similar to its A chromosome paralog. New Phytol 213:916–928
Ma W, Liu ZJ, Beier S, Houben A, Carpentier S (2021) Identification of rye B chromosome-associated peptides by mass spectrometry. New Phytologist (in press)
Marques A, Banaei-Moghaddam AM, Klemme S et al (2013) B chromosomes of rye are highly conserved and accompanied the development of early agriculture. Ann Bot 112:527–534

Marques A, Klemme S, Houben A (2018) Evolution of plant B chromosome enriched sequences. Genes 9:515

Martins C, Ahmad SF (2019) The modern view of B chromosomes under the impact of high scale omics analyses. Cells 8:156

Martis MM, Klemme S, Banaei-Moghaddam AM et al (2012) Selfish supernumerary chromosome reveals its origin as a mosaic of host genome and organellar sequences. Proc Nat Acad Sci USA 109:13343–13346

Matthews RB, Jones RN (1983) Dynamics of the B chromosome polymorphism in rye II. Estimates of parameters. Heredity 50:119–137

Mendelson D, Zohary D (1972) Behavior and transmission of supernumerary chromosomes in *Aegilops speltoides*. Heredity 29:329–339

Mette MF, Houben A (2015) Engineering of plant chromosomes. Chromosome Res 23:69–76

Morais-Cecilio L, Delgado M, Jones RN, Viegas W (1996) Painting rye B chromosomes in wheat: interphase chromatin organization, nuclear disposition and association in plants with two, three or four Bs. Chromosome Res 4:195–200

Moss J (1966) The adaptive significance of B-chromosomes in rye

Muñoz-Pajares A, Martinez-Rodriguez L, Teruel M et al (2011) A single, recent origin of the accessory B chromosome of the grasshopper *Eyprepocnemis plorans*. Genetics 187:853–863

Müntzing A (1943) Genetical effects of duplicated fragment chromosomes in rye. Hereditas 29:91–112

Müntzing A (1945) Cytological studies of extra fragment chromosomes in rye. II. Transmission and multiplication of standard fragments and iso-fragments. Hereditas 31:457–477

Müntzing A (1948) Cytological studies of extra fragment chromosomes in rye. V. A new fragment type arisen by deletion. Hereditas 34:435–442

Müntzing A (1963) Effects of accessory chromosomes in diploid and tetraploid rye. Hereditas 49:371–426

Müntzing A (1970) Chromosomal variation in the Lindstroem strain of wheat carrying accessory chromosomes of rye. Hereditas 66:279–285

Müntzing A, Lima-de-Faria A (1949) Pachytene analysis of standard fragments and large iso-fragments in rye. Hereditas 35:263–268

Nakao M (1911) Cytological studies on the nuclear division of the pollen mother cells of some cereals and their hybrids. J Coll Agr Tohoku Imp Univ 4:173–190

Niwa K, Sakamoto S (1995) Origin of B-chromosomes in cultivated rye. Genome 38:307–312

Niwa K, Sakamoto S (1996) Detection of B chromosomes in rye collected from Pakistan and China. Hereditas 124:211–215

Ohta S (1996) Mechanisms of B-chromosome accumulation in *Aegilops mutica* Boiss. Genes Genet Syst 71:23–29

Otlowska-Miazga D (1974) Influence of B chromosomes on the chiasma frequency in the bivalents of the diploid rye, Transbaikal. Hodowla Rosl Aklim Inasiennic (in Polish)

Pereira HS, Barao A, Caperta A et al (2009) Rye Bs disclose ancestral sequences in cereal genomes with a potential role in gametophyte chromatid segregation. Mol Biol Evol 26:1683–1697

Pereira HS, Delgado M, Viegas W et al (2017) Rye (*Secale cereale*) supernumerary (B) chromosomes associated with heat tolerance during early stages of male sporogenesis. Ann Bot 119:325–337

Puertas MJ (2002) Nature and evolution of B chromosomes in plants: a non-coding but information-rich part of plant genomes. Cytogenet Genome Res 96:198–205

Puertas M, Carmona R (1976) Greater ability of pollen tube growth in rye plants with 2B chromosomes. Theor Appl Genet 47:41–43

Puertas MJ, Romera F, Delapena A (1985) Comparison of B-chromosome effects on *Secale cereale* and *Secale vavilovii*. Heredity 55:229–234

Puertas MJ, González-Sánchez M, Manzanero S et al (1998) Genetic control of the rate of transmission of rye B chromosomes. IV. Localization of the genes controlling B transmission rate. Heredity 80:209–213

Puertas MJ, Jiménez G, Manzanero S et al (2000) Genetic control of B chromosome transmission in maize and rye. In: Chromosomes today. Birkhäuser, Basel, pp 79–92

Randolph L (1928) Types of supernumerary chromosomes in maize. Anat Rec 41:102

Romera F, Jimenez MM, Puertas MJ (1991) Factors controlling the dynamics of the B chromosome polymorphism in Korean rye. Heredity 67:189–195

Ruban A, Schmutzer T, Scholz U, Houben A (2017) How next-generation sequencing has aided our understanding of the sequence composition and origin of B chromosomes. Genes 8:294

Ruban A, Schmutzer T, Wu DD et al (2020) Supernumerary B chromosomes of *Aegilops speltoides* undergo precise elimination in roots early in embryo development. Nat Commun 11:2764

Sandery MJ, Forster JW, Blunden R, Jones RN (1990) Identification of a family of repeated sequences on the rye B-chromosome. Genome 33:908–913

Schlegel R, Pohler W (1994) Identification of an a B chromosome translocation in diploid rye (*Secale cereale* L). Breed Sci 44:279–283

Schubert V, Meister A, Tsujimoto H et al (2011) Similar rye A and B chromosome organization in meristematic and differentiated interphase nuclei. Chromosome Res 19:645–655

Schubert V, Zelkowski M, Klemme S, Houben A (2016) Similar Sister Chromatid Arrangement in Mono- and Holocentric Plant Chromosomes. Cytogenet Genome Res 149:218–225

Talbert P, Henikoff S (2020) What makes a centromere? Exp Cell Res 389:111895
Wu DD, Ruban A, Fuchs J et al (2019) Nondisjunction and unequal spindle organization accompany the drive of *Aegilops speltoides* B chromosomes. New Phytol 223:1340–1352
Yamagishi Y, Sakuno T, Shimura M, Watanabe Y (2008) Heterochromatin links to centromeric protection by recruiting Shugoshin. Nature 455:251–255
Yu W, Han F, Gao Z et al (2007) Construction and behavior of engineered minichromosomes in maize. Proc Natl Acad Sci USA 104:8924–8929
Zečević L, Paunović D (1969) The effect of B-chromosomes on chiasma frequency in wild populations of rye. Chromosoma 27:198–200

Dissection of the Rye Genome by the Gametocidal System

5

Takashi R. Endo

Abstract

Dissection of the rye genome is desirable for the genomics research of rye. Some alien chromosomes called "gametocidal chromosomes" induce chromosomal rearrangements in gametes lacking the alien chromosome when they are introduced into common wheat. This genetic system was named "gametocidal system." The gametocidal system also induces chromosomal rearrangements in rye chromosomes added to common wheat and has been used for generating rye chromosomal segments in the background of common wheat. Common wheat lines with single rye chromosomal segments, which are called "dissection lines," have proved useful for the cytological mapping of the rye chromosomes.

Rye *(Secale cereale* L.) has a large genome (1Cx = 7,917 Mbp; Bartoš et al. 2008) with seven pairs of chromosomes ($2n = 2x = 14$). The rye chromosomes (1R–7R) have separately been introduced to common wheat (*Triticum aestivum* L., $2n = 6x = 42$), and the seven wheat-rye chromosome addition lines have become available. Also, each of the rye chromosome arms has been added to common wheat and many of the possible wheat-rye telosomic addition lines have been developed (for more information, visit the website of the Wheat Genetic Resource Center at Kansas State University, USA: https://www.k-state.edu/wgrc/). These wheat-rye addition lines are useful for allocating rye chromosome-specific genes and DNA markers. Moreover, flow cytometry allows individual rye chromosomes to be flow-sorted from the wheat-rye addition lines for the systematic sequencing of the rye genome (Martis et al. 2013). Although the whole-genome shotgun method is a routine procedure in genome projects nowadays, sequences of flow-sorted rye chromosomes were used to gain higher specificity of the constructed rye whole-genome sequencing assembly (Bauer et al. 2017). Thus, the rye chromosome addition lines are still useful in the era of whole-genome sequencing.

Dissection of the rye genome into smaller segments is obviously desirable. There are some genetic systems to manipulate the rye chromosomes at the sub-arm level in the background of common wheat. One is the homoeologous pairing (*Ph*) system, which controls homoeologous chromosome pairing between wheat and rye chromosomes. For example, using the *ph* system, Lukaszewski (2000) induced homoeologous recombination between the short arms of rye chromosome 1R and wheat chromosome arm 1B and successfully removed the *Sec 1* locus from the short arm of chromosome 1R. The *Sec 1* gene produces the rye seed storage protein secalin,

T. R. Endo (✉)
3-16-7 Sakuragaoka, Seika-cho, Soraku-gun, Kyoto 619-0232, Japan
e-mail: endo.takashi.2e@gmail.com

© The Author(s), under exclusive license to Springer Nature Switzerland AG 2021

M. T. Rabanus-Wallace and N. Stein (eds.), *The Rye Genome*, Compendium of Plant Genomes,
https://doi.org/10.1007/978-3-030-83383-1_5

which reduces the bread-making quality. Another is the gametocidal system, which utilizes chromosomal rearrangements induced by particular alien chromosomes in common wheat. Details of the gametocidal system for the rye chromosomes are described below.

5.1 The Gametocidal System

Particular chromosomes of the genus *Aegilops* were found to cause partial sterility in common wheat if added to wheat in an unusual manner (Endo and Tsunewaki 1975; Maan 1975). Such alien chromosomes, which are called gametocidal chromosomes, induce chromosomal rearrangements in gametes lacking the gametocidal chromosome (Finch et al. 1984) (Fig. 5.1). The gametocidal system is a genetic system in which the gametocidal chromosome induces chromosomal rearrangements in common wheat. Such chromosomal rearrangements are either lethal or semilethal to the gametes depending on the type of the added gametocidal chromosome and also the cultivar of common wheat (Endo 2007). Thanks to the hexaploid nature of common wheat, some gametes carrying chromosomal rearrangements develop into viable zygotes after fertilization, and the rearranged chromosomes such as deletions continue to exist for generations. The gametocidal system was successfully applied to produce the deletion stocks in the common wheat cultivar Chinese Spring (Endo 1988; Endo and Gill 1996). The gametocidal system also induced rearranged barley chromosomes added to Chinese Spring (Endo 2009).

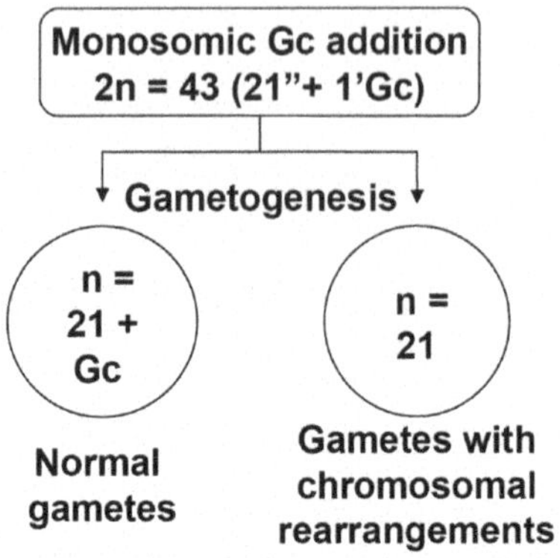

Fig. 5.1 Schematic diagram of the gametocidal system in common wheat. Twenty-one bivalents (symbolized by 21" in the figure) represent the entire wheat genome. The gametocidal chromosome in monosomic condition (1'Gc) induces chromosomal rearrangements only in gametes without itself

5.2 Dissection of Rye Chromosomes in Common Wheat by the Gametocidal System

Figure 5.2 illustrates a general scheme to induce chromosomal rearrangements in rye chromosomes in the genetic background of common wheat. The gametocidal system induces randomly semilethal chromosomal rearrangements in wheat and rye chromosomes in the rye chromosome addition lines of common wheat; therefore, the chromosomal rearrangements result in wheat-rye translocations, as well as terminal deletions of rye and wheat chromosomes. So far, two types of gametocidal chromosomes have been used to induce chromosomal rearrangements in rye chromosomes. One is chromosome 2C from *Ae. cylindrica*. This chromosome has been used in the gametocidal system for rye chromosomes added to Chinese Spring, namely chromosome 1R of *S. cereale* cv. Imperial (Tsuchida et al. 2008), chromosome 1R of *S. montanum* (Li et al. 2015), Imperial rye chromosomes 2R–7R (Li et al. 2013), and rye B chromosomes (Endo et al. 2008). The other is chromosome 3C derived from *Ae. triuncialis*. This chromosome has been used in the gametocidal system for chromosome 1R in a common wheat cv. Burgas 2 with a 1R(1B) substitution (Endo et al. 1994; Masoudi-Nejad et al. 2002). Chromosomal rearrangements induced by chromosome 3C are nearly lethal in Chinese Spring but semilethal in some Japanese common wheat cultivars carrying a dominant gene (*Igc1*) that mitigates the gametocidal effect (Tsujimoto and Tsunewaki 1985). Therefore, Endo et al. (1994) crossed wheat Burgas 2 with a disomic 3C addition line of wheat Norin 26 carrying *Igc1*.

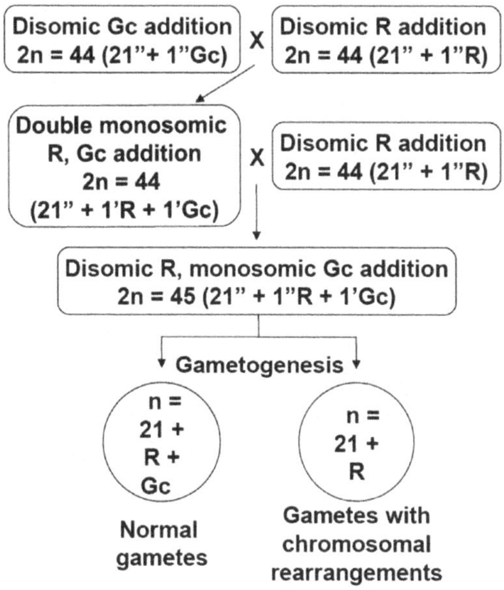

Fig. 5.2 General scheme to induce chromosomal rearrangements in rye chromosomes in the genetic background of common wheat. First, double monosomic alien chromosome addition ($2n$ = 44) with a rye chromosome and a gametocidal chromosome (represented by "R" and "Gc" in the figure, respectively) is produced by a cross between a disomic R addition line and a disomic Gc addition line. Then, the $2n$ = 44 plant is backcrossed to the disomic R addition line, and then $2n$ = 45 plants (disomic for R and monosomic for Gc) are selected from the backcrossed progeny. Chromosomal rearrangements are to occur in the progeny of the $2n$ = 45 plant

Another gametocidal chromosome $3C^{SAT}$, a satellite chromosome derived from *Ae. triuncialis*, will be suitable for the gametocidal system of rye chromosomes in Chinese Spring (Endo 2007); indeed, it induced chromosomal rearrangements in barley chromosome 2H added to Chinese Spring (Joshi et al. 2011).

5.3 Production of Dissection Lines of Rye Chromosomes

The most reliable way to detect rye chromosomal rearrangements in common wheat is a cytological analysis using genomic in situ hybridization (GISH), which clearly distinguishes rye from wheat chromosomes (Endo 2011). Tsuchida et al. (2008) selected 106 plants carrying one or more rearranged 1R chromosomes from 572 plants of the selfed progeny of the 45-chromosome plants with disomic Imperial 1R addition and monosomic 2C addition (21″ + 1″1R + 1′2C) lines. Further examining the progeny of the 106 plants, they finally established 55 common wheat lines carrying single rearranged 1R chromosomes in either hemizygous or homozygous condition with a success rate of 51.9%. Lines with rearranged rye chromosomes were called dissection lines. The dissection of chromosome 1R from Burgas 2 started with 340 pre-screened plants carrying one or more rearranged 1R chromosomes. These plants were selected by C-banding or GISH from 1934 plants of the progeny of a common wheat line with disomic substitution 1R (1B) and monosomic 3C addition ($2n$ = 43, 20″ + 1″1R(1B) + 1′3C) (Endo 2003). Gyawali et al. (2009, 2010) examined the chromosome constitutions of the progeny of 201 plants chosen from the pre-screened plants and established a total of 120 dissection lines carrying single rearranged 1R chromosomes in either hemizygous or homozygous condition with a success rate of 59.7%. Thus, in the first round of screening, cytological analysis detected plants with rearranged 1R chromosomes induced by the gametocidal system with high rates, 18.5% (106 out 572) for Imperial chromosome 1R and 17.6% (340 out of 1934) for Burgas 2 chromosome 1R. Many of the rearranged 1R chromosomes in the pre-screened plants were not established as dissection lines. Consequently, the percentage of the number of established dissection lines to the number of plants initially examined is 9.6% (=18.5% × 51.9%) for Imperial chromosome 1R and 10.4% (= 17.6% × 59.7%) for Burgas 2 chromosome 1R. The failure in the retrieval of rearranged 1R chromosomes in the next generation may be attributed to some technical problems in the cytological analysis or to the actual absence of rearranged 1R chromosomes in the germline of the pre-screened plants, which is possible as suggested below.

Joshi et al. (2013) selected 81 plants cytologically by their normal-appearing chromosome

2H and no 2C chromosomes from the progeny of a cross between the wheat–barley 2H addition line with monosomic 2C addition ($2n = 45$, 1″ 2H + 1′2C) and euploid common wheat. They found by PCR analysis that six of the 81 plants lacked some of the 2H-specific EST markers and then confirmed the presence of rearranged 2H chromosomes in the progeny of the six plants by GISH analysis. This fact suggests that the gametocidal system sometimes induces chromosomal rearrangements only in aerial parts of the plant including the germline, however, not in root tissues. The converse may also be true. That is, the gametocidal system sometimes induces chromosomal rearrangements only in root tissues. If this is the case, it is explicable that part of the rearranged 1R chromosomes detected by the cytological analysis was not retrieved in the next generation.

PCR is an alternative method for screening dissection lines of rye chromosomes. As illustrated in Fig. 5.3, the lack of one or more of single-locus DNA markers on a rye chromosome suggests the loss of certain segments of the rye chromosome in the progeny of the 45-chromosome plants (disomic for the rye chromosome and monosomic for a gametocidal chromosome) backcrossed to euploid common wheat. Cytological analysis can identify the types of chromosomal rearrangement in the progeny of the plants with chromosomal rearrangements predefined by PCR. However, PCR analysis cannot detect the occurrence of chromosomal rearrangements, for instance, when single plants carry reciprocal translocations between rye and wheat chromosomes because no part of the rye chromosome is lost in them. Nevertheless, the PCR-based screening is a more comfortable and reliable way to identify dissection lines, although not yet practiced with rye chromosomes.

The PCR-based screening involving the nulli-/tetrasomic lines of common wheat has proven to be effective in detecting deletions induced by the gametocidal system in common wheat. The nulli-/tetrasomic lines lack one pair of chromosomes and possess an extra pair of homoeologous chromosomes. The progeny (2041 plants) from a cross between the nullisomic 6B-tetrasomic 6A line as a female parent and the monosomic 2C addition line was examined by PCR using 6B-specific DNA markers, and 102 plants were found to have a 6B deletion (Endo 2015). Likewise, Svačina et al. (2020) established 113 deletion lines from the progeny of a cross between the nullisomic 3D-tetrasomic 3A (or 3B) line as a female parent and the monosomic 2C addition line.

5.4 Types of Chromosomal Rearrangement Induced by the Gametocidal System

The gametocidal system induces various types of rearranged 1R chromosome with breakpoints in either the chromosome arms or the centromere (Fig. 5.4). There are three forms of translocations involving chromosome 1R: those with the rye centromere, those with the wheat centromere, and those of Robertsonian type. Chromosome 2C induced more deletions than translocations and chromosome 3C induced more translocations than deletions, whereas both gametocidal chromosomes induced chromosomal rearrangements in chromosome 1R with almost the same frequency (Table 5.1). The occurrence of chromosome breaks is relatively high in the centromere despite its narrow region and seems to be directly proportional to the length of the 1R chromosome arms (Table 5.2). It is noteworthy that the gametocidal system induces chromosome breaks as frequently in the proximal regions, where chromosomal crossovers are restricted, as in the distal region (Tsuchida et al. 2008; Gyawali et al. 2009).

5.5 Use of the Dissection Lines

The rye dissection lines are useful in mapping the rye chromosomes because they have various segments of each of the rye chromosomes. So far, cytological mapping using the dissection lines has been conducted only for chromosome 1R (Tsuchida et al. 2008; Gyawali et al. 2009, 2010; Li et al. 2015). A common method of

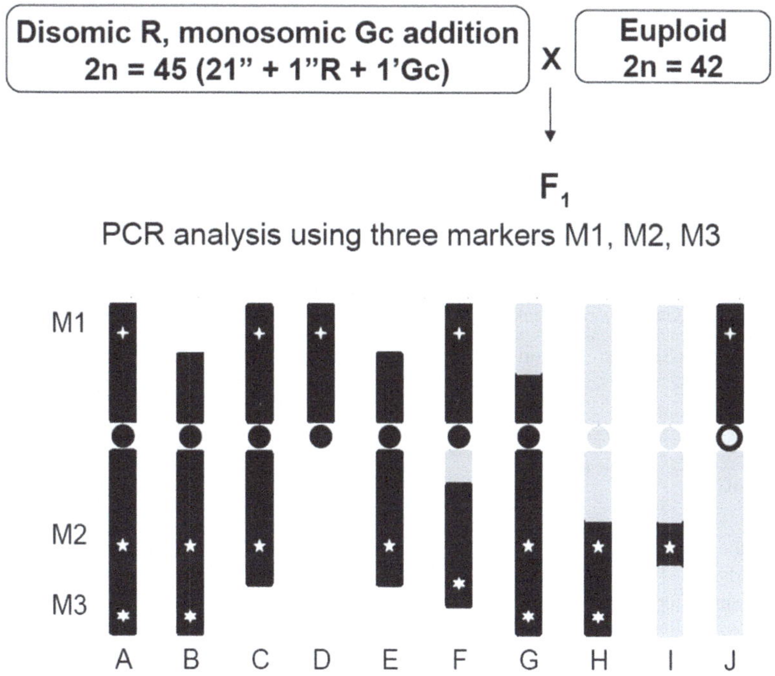

Fig. 5.3 PCR-based screening for chromosomal rearrangements induced by the gametocidal system in a rye chromosome. The absence of some of the markers (M1–M3) suggests the occurrence of chromosomal rearrangements in the rye chromosome. **a** No rearrangement. **b** Deletion in the short arm. **c** Deletion in the long arm. **d** Deletion of the entire long arm. **e** Deletion in both arms. **f** Interstitial deletion in the long arm. **g** Translocation of a wheat chromosomal segment to the rye chromosome. **h** Translocation of a segment of the rye chromosome to a wheat chromosome. **i** Interstitial translocation of a segment of the rye chromosome to a wheat chromosome. **j** Robertsonian translocation between the rye chromosome and a wheat chromosome (the centromere could be derived from the rye or wheat chromosome, or from both chromosomes). Black and gray bars represent the rye and wheat chromosomes, respectively, and circles stand for the centromere

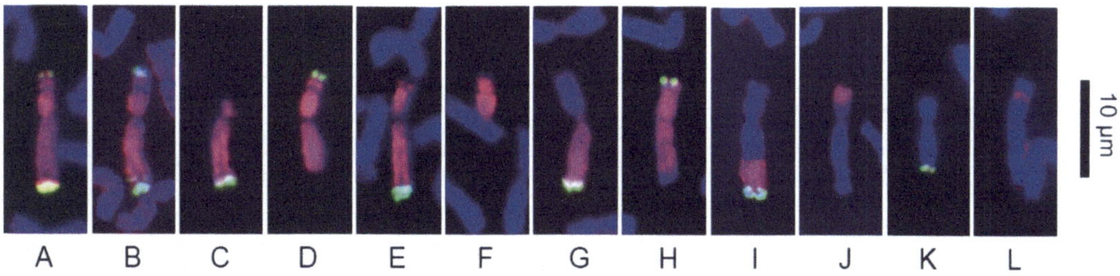

Fig. 5.4 Representative chromosomal rearrangements induced by the gametocidal system in chromosome 1R derived from rye cultivar Imperial ($1R^{i}$) and common wheat cultivar Burgas 2 ($1R^{B}$). **a** Normal $1R^{i}$. **b** Normal $1R^{B}$. **c** $1R^{i}$-5, deletion in the short arm. **d** $1R^{i}$-63, deletion in the long arm. **e** $1R^{i}$-47, deletion in the satellite. **f** $1R^{i}$-20, double deletion in the centromere and satellite. **G** $1R^{i}$-7, Robertsonian translocation between the $1R^{i}$ long arm and a wheat chromosome arm. **h** $1R^{i}$-48, translocation of a wheat chromosomal segment to the $1R^{i}$ long arm. **i** $1R^{B}$-135, translocation of a $1R^{B}$ long-arm segment to a wheat chromosome. **j** $1R^{i}$-60, translocation of a $1R^{i}$ interstitial segment to a wheat chromosome. **k** $1R^{B}$-20, translocation of the $1R^{B}$ satellite segment to a wheat chromosome. **l** $1R^{B}$-119, interstitial translocation of a $1R^{B}$ segment to a wheat chromosome. Dissection lines carrying these 1R rearrangements were used in the previous studies (Masoudi-Nejad et al. 2002; Tsuchida et al. 2008; Gyawali et al. 2009; Li et al. 2015)

Table 5.1 Frequency of deletions and translocations involving chromosome 1R in the dissection lines

Source of 1R	Gc[a]	Deletion	Translocation	Total	References
Imperial rye	2C	41	14[b]	55	Tsuchida et al. (2008)
Burgas 2 wheat	3C	31	64[c]	95	Gywali et al. (2009)

[a]Type of the gametocidal chromosome
[b]Two translocations with the 1R centromere, five with the wheat centromere, and seven Robertsonian translocations
[c]Twenty-four translocations with the 1R centromere, 31 with the wheat centromere, and 9 Robertsonian translocations

Table 5.2 Frequency of chromosome breaks in different sites of chromosome 1R in the dissection lines

Source of 1R	Gc[a]	Satellite	Short arm[b]	Long arm	Centromere	?	Total	References
Imperial rye	2C	4	10	29	15	0	58[c]	Tsuchida et al. (2008)
Burgas 2 wheat	3C	9	26	51	13	2	99[d]	Gywali et al. (2009)

[a]Type of the gametocidal chromosome
[b]The satellite is excluded
[b]Two breaks are counted for each of the three double deletions
[d]Two breaks are counted for each of the two interstitial translocations and the four translocations with deletions

cytological mapping is first to conduct PCR analysis with the 1R dissection lines and then to arrange the DNA markers in the proper order based on the presence or absence of the DNA markers in the dissection lines (Fig. 5.5). The cytological mapping using dissection lines sometimes corrected the order of the markers determined by genetic linkage mapping (e.g., Gyawali et al. 2010).

The gametocidal system induces chromosome breaks at random in any chromosome, and therefore, controlled deletion or introgression of a target region of a rye chromosome is impossible. Nevertheless, the gametocidal system is

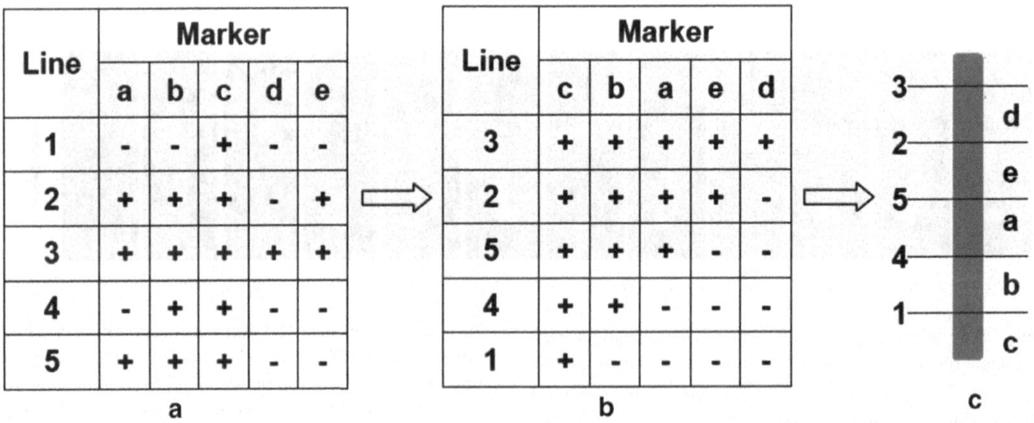

Fig. 5.5 Illustrative example of the construction of a deletion map using five dissection lines and five markers. **a** Raw data matrix of the result of PCR analysis. **b** Processed matrix in which the dissection lines are arranged in the order of the numbers of markers retained in them. **c** Marker order along a chromosome. The horizontal lines represent the breakpoints of the chromosomal rearrangements in the dissection lines and the markers are placed in bins flanked by the breakpoints

efficacious enough to produce rye chromosomes with specific deletions or wheat chromosomes with specific rye chromosomal segment introgressions. For example, Masoudi-Nejad et al. (2002) attempted to separate the *Sec-1* gene from a cluster of rust resistance genes that are located distal to the *Sec-1* gene on the satellite of chromosome 1R. Analysis of the rearranged 1R chromosomes induced by the gametocidal system resulted in two wheat chromosomes carrying 1R satellite segments containing the rust resistance genes but not the *Sec-1* gene (one of the two chromosomes concerned is $1R^{B}$-20 shown in Fig. 5.4k).

Whole-genome sequencing has become a reality for the genome of rye, thanks to the next-generation sequencing technologies and bioinformatics. Chromosomal landmarks are indispensable to assemble sequence contigs into supercontigs and further into chromosomes. The gametocidal system will provide almost limitless chromosomal landmarks to the rye genome in the form of chromosomal rearrangements in each of the rye chromosomes added to common wheat.

References

Bartoš J, Paux E, Kofler R et al (2008) A first survey of the rye (Secale cereale) genome composition through BAC end sequencing of the short arm of chromosome 1R. BMC Plant Biol 8:95. https://doi.org/10.1186/1471-2229-8-95

Bauer E, Schmutzer T, Barilar I et al (2017) Towards a whole-genome sequence for rye (*Secale cereale* L.). Plant J 89:853–869. https://doi.org/10.1111/tpj.13436

Endo TR (1988) Induction of chromosomal structural changes by a chromosome *Aegilops cylindrica* L. in common wheat. J Hered 79:366–370. https://doi.org/10.1093/oxfordjournals.jhered.a110529

Endo TR (2003) Wheat stocks carrying alien chromosomal segments induced by the gametocidal system. In: Pogna NE, Romano M, Pogna EA et al (eds) Proceeding 10th international wheat symposium, Paestum, Italy, pp 69–72

Endo TR (2007) The gametocidal chromosome as a tool for chromosome manipulation in wheat. Chromosome Res 15:67–75. https://doi.org/10.1007/s10577-006-1100-3

Endo TR (2009) Cytological dissection of barley genome by the gametocidal system. Breeding Sci 59:481–486. https://doi.org/10.1270/jsbbs.59.481

Endo TR (2011) Cytological Dissection of the Triticeae Chromosomes by the gametocidal system. In: Birchler JA (ed) Plant chromosome engineering: methods and protocols, Methods in molecular biology, vol 701, pp 247–257. https://doi.org/10.1007/978-1-61737-957-4_14

Endo TR (2015) New aneuploids of common wheat. In: Ogihara Y, Takumi S, Handa H (eds) Advances in wheat genetics: from genome to field, pp 73–80. https://doi.org/10.1007/978-4-431-55675-6_8

Endo TR, Gill BS (1996) The deletion stocks of common wheat. J Hered 87:295–307. https://doi.org/10.1093/oxfordjournals.jhered.a023003

Endo TR, Tsunewaki K (1975) Sterility of common wheat with *Aegilops triuncialis* cytoplasm. J Hered 66:13–18. https://doi.org/10.1093/oxfordjournals.jhered.a108562

Endo TR, YamamotoM MY (1994) Structural changes of rye chromosome 1R induced by a gametocidal chromosome. Jpn J Genet 69:13–19. https://doi.org/10.1266/jjg.69.13

Endo TR, Nasuda S, Jones N et al (2008) Dissection of rye B chromosomes, and nondisjunction properties of the dissected segments in a common wheat background. Genes Genet Syst 83:23–30. https://doi.org/10.1266/ggs.83.23

Finch RA, Miller TE, Bennett MD (1984) Cuckoo" *Aegilops* addition chromosome in wheat ensures its transmission by causing chromosome breaks in meiospores lacking it. Chromosoma 90:84–88. https://doi.org/10.1007/BF00352282

Gyawali YP, Nasuda S, Endo TR (2010) A cytological map of the short arm of rye chromosome 1R constructed with 1R dissection stocks of common wheat and PCR-based markers. Cytogenet Genome Res 129:224–233. https://doi.org/10.1159/000314556

Gyawali YP, Nasuda S, Endo TR (2009) Cytological dissection and molecular characterization of chromosome 1R derived from 'Burgas 2' common wheat. Genes Genet Syst 84:407–416. https://doi.org/10.1266/ggs.84.407

Joshi GP, Nasuda S, Endo TR (2011) Dissection and cytological mapping of barley chromosome 2H in the genetic background of common wheat. Genes Genet Syst 86:231–248. https://doi.org/10.1266/ggs.86.231

Joshi GP, Endo TR, Nasuda S (2013) PCR and sequence analysis of barley chromosome 2H subjected to the gametocidal action of chromosome 2C. Theor Appl Genet 126:2381–2390. https://doi.org/10.1007/s00122-013-2142-5

Li J, Nasuda S, Endo TR (2013) Dissection of rye chromosomes by the gametocidal system. Genes Genet Syst 88:321–327. https://doi.org/10.1266/ggs.88.321

Li J, Gyawali YP, Stein N et al (2015) Comparative study of the structure of chromosome 1R derived from *Secale montanum* and *Secale cereale*. Plant Breeding. https://doi.org/10.1111/pbr.12314

Lukaszewski AJ (2000) Manipulation of the 1RS.1BL Translocation in wheat by induced homoeologous

recombination. Crop Sci 40:216–225. https://doi.org/10.2135/cropsci2000.401216x
Maan SS (1975) Exclusive preferential transmission of an alien chromosome in common wheat. Crop Sci 15:287–292. https://doi.org/10.2135/cropsci1975.0011183X001500030002x
Martis MM, Zhou R, Haseneyer G et al (2013) Reticulate evolution of the rye genome. Plant Cell 25:3685–3698. https://doi.org/10.1105/tpc.113.114553
Masoudi-Nejad A, Nasuda S, McIntosh RA et al (2002) Transfer of rye chromosome segments to wheat by a gametocidal system. Chromosom Res 10:349–357. https://doi.org/10.1023/A:1016845200960
Svačina R, Karafiátová M, Malurová M et al (2020) Development of deletion lines for chromosome 3D of bread wheat. Front Plant Sci 10:1756. https://doi.org/10.3389/fpls.2019.01756
Tsuchida M, Fukushima T, Nasuda S et al (2008) Dissection of rye chromosome 1R in common wheat. Genes Genet Syst 83:43–53. https://doi.org/10.1266/ggs.83.43
Tsujimoto H, Tsunewaki K (1985) Gametocidal genes in wheat and its relatives. II. Suppressor of the chromosome 3C gametocidal gene of *Aegilops triuncialis*. Can J Genet Cytol 27:178–185. https://doi.org/10.1139/g85-027.doi:10.1139/g85-027

Evolution and Domestication of Rye

6

Mona Schreiber, Hakan Özkan, Takao Komatsuda, and Martin Mascher

Abstract

More than 12,000 years ago, with the beginning of the Neolithic Age, one of the most fundamental innovations in human history took place: the beginning of agriculture. The plants and animals that became an indispensable aspect of human life began to differ more and more from their wild relatives, and in the course of time, adaptive traits advantageous for living together with humans were fixed. The history of cultivated rye (*Secale cereale* subsp. *cereale*) has long been mysterious, and many theories have been put forward as to its origin. The putative wild progenitor *Secale cereale* subsp. *vavilovii* still occurs in Southwest Asia, as do the close relatives of rye, wheat and barley. But in contrast to the latter two species, rye was not used as a cereal crop until the European Bronze Age, several millennia after start of the Neolithic revolution. Hence, it is not among the founder crops of agriculture. Archaeobotanical studies, genetic kinship analyses as well as the cultural history and etymology support the hypothesis of a secondary domestication origin from a weed in wheat and barley fields. In northern and central Europe, its exceptional winter hardiness and its tolerance to grow on poor soils may have enabled it to be grown as a grain crop in its own right. Yet still, in some regions of the world, e.g. Southwest Asia, rye is considered a weed growing among other cereals. In this chapter, we review the current knowledge of the taxonomy of the small genus *Secale*, the basics of the Neolithic transition and the domestication of cereals, as well as the peculiarities of the cultural history of rye as a secondary domesticate collating evidence from multiple disciplines.

M. Schreiber (✉) · M. Mascher (✉)
Leibniz Institute of Plant Genetics and Crop Plant Research (IPK) Gatersleben, Seeland, Germany
e-mail: mona@besenkunst.de

M. Mascher
e-mail: mascher@ipk-gatersleben.de

M. Schreiber
Department of Biology, University of Marburg, Marburg, Germany

H. Özkan
Department of Field Crops, Faculty of Agriculture, Çukurova University, Adana, Turkey

T. Komatsuda
Institute of Crop Science, National Agriculture and Food Research Organization (NARO), Tsukuba, Japan

T. Komatsuda
Crop Research Institute, Shandong Academy of Agricultural Sciences (SAAS), Jinan, Shandong, China

M. Mascher
German Centre for Integrative Biodiversity Research (iDiv) Halle-Jena-Leipzig, Leipzig, Germany

© The Author(s), under exclusive license to Springer Nature Switzerland AG 2021

M. T. Rabanus-Wallace and N. Stein (eds.), *The Rye Genome*, Compendium of Plant Genomes,
https://doi.org/10.1007/978-3-030-83383-1_6

6.1 Rye—A Remarkable Cereal

Although from a global perspective rye, *Secale cereale* ssp. *cereale* is a crop of minor economic interest, in Central and Northern Europe it is, after wheat, the second most important cereal for the production of bread (FAO 2018). Furthermore, rye is also utilised as a fodder plant as well as a cover crop and is used in the production of alcoholic beverages such as rye whisky.

Rye is a member of the genus *Secale*, belonging to the monocots and among these to the Poaceae, the true grasses. Together with wheat (*Triticum* spp.) and barley (*Hordeum vulgare*), rye belongs to the tribe of Triticeae and is generally adapted to a wider range of environmental and climatic conditions than its close relatives. Rye has a remarkable tolerance to various biotic and abiotic stress conditions, grows on sandy and rocky soils and requires a lower nutrient intake. The members of the small genus *Secale* can be found in the temperate regions of six continents, including dry and cold habitats as well as higher elevations in tropical and subtropical latitudes, where other cereals cannot thrive (Tang et al. 2011).

6.2 The Small Genus *Secale*

The genus *Secale* includes perennial and annual, self-incompatible and outcrossing species (Vences et al. 1987). Following the taxonomic revision of Frederiksen and Petersen (1998) confirmed by various genetic and genomic analyses (Shang et al. 2006; Ren et al. 2011; Hagenblad et al. 2016; Maraci et al. 2018; Schreiber et al. 2019) the genus is composed of three species (Table 6.1), all of which are diploid ($2n$ = 14). The haploid genome size of *S. cereale* is ~8 Gb (Rabanus-Wallace et al. 2021), with slightly smaller size for *S. strictum* (7.8 Gb) and *S. sylvestre* (7.6 Gb).

In general, all three species can be crossed with each other. A notable example is *S.* x *derzhavinii*, as an artificial fertile hybrid between *S. cereale* and *S. strictum*, that is, currently explored as potential source of a perennial growth habit (as in *S. strictum*) in cultivated rye (Tsvelev 1973; Clayton et al. 2006; Barkworth 2007). Efforts were undertaken to establish *S. strictum* (synonym *S. montanum*) as a forage crop in Australia (Oram 1996).

Systematic cross-pollination studies reported a reduced fertility in hybrids, presumably because of chromosomal translocations (Stutz 1957; Singh 1977). Among *Secale* taxa, crossability with domesticated rye was lowest in *S. sylvestre* and highest with *S. strictum* (Khush and Stebbins 1961; Khush 1962). It is interesting to note that pollen fertility of *S. cereale* × *S. vavilovii* (19%) was lower than that of *S. cereale* × *S. strictum* (31%).

The origin of the outstanding winter hardiness of rye is evident in the ability of *S. strictum* and *S. cereale* subsp. *vavilovii* to grow on field margins and forest clearings up to altitudes of 2500 m in Southwest Asia (Schlegel 2013). The divergence times of *Secale* species are most likely much more recent than in *Hordeum* (barley) and *Triticum* (wheat) taxa (Bernhardt et al. 2017). The chloroplast phylogeny of Skuza et al. 2019 did not reflect species assignments of samples, indicating that low evolutionary rates of plastid genomes (Clegg et al. 1994) and recent divergence have prevented the fixation of species-specific polymorphisms. A recent split time between *Secale* taxa is in concordance with the extent of incomplete lineage sorting found in the marker data from the nuclear genome. Using genome-wide marker data, Schreiber et al. (2019) arrived at phylogenetic relationships that supported the conspecific of cultivated rye and *S. cereale* subsp. *vavilovii* and indicated the *S. sylvestre* split earlier from *S. cereale* than *S. strictum* (Fig. 6.1). This topology is supported by principal component analysis and model-based ancestry estimation from genome-wide marker data. Nevertheless, Maraci et al. (2018) reported incomplete lineage sorting (i.e. ancestral variants still segregating in multiple *Secale* taxa) for one 667 bp nuclear locus. This observation was confirmed by the analysis of Schreiber et al. (2019), who reported a substantial proportion of

Table 6.1 Key characteristics of the three species within the genus *Secale*

	S. cereale	*S. strictum*	*S. sylvestre*
Subspecies	*Vavilovii, cereale, rigidum, dighoricum, afghanicum, segetale, ancestrale*	*Africanum, anatolicum, strictum, kuprijanovii*	
Lifecycle	Annual	(mostly) Perennial	Annual
Pollination	Outcrossing	Self-compatible/outcrossing	Self-compatible
Geographic distribution	Natural distribution of subsp. *vavilovii*: Southwest Asia	From the Mediterranean Sea to central Asia, distinct subspecies in South Africa (subsp. *africanum*)	East-Europe, Asia
Domestication status	Wild, weedy, feral and domesticated	Wild	Wild
Synonyms	*S. vavilovii* (wild), *S. cereale* (domesticated/weedy)	*S. montanum*	*S. silvestre, S. fragile*

shared segregating sites between all Secale taxa. The degree of incomplete lineage sorting was consistent with the proposed phylogenetic distance of the three species *S. cereale*, *S. strictum* and *S. sylvestre*.

To conclude, the recent split times and partial interfertility of *Secale* taxa make species delimitation blurry. However, recent genome-wide marker data made it possible to establish a species-level phylogeny (Schreiber et al. 2019), which however, can be violated at individual loci, due to incomplete lineage sorting (Maraci et al. 2018; Skuza et al. 2019).

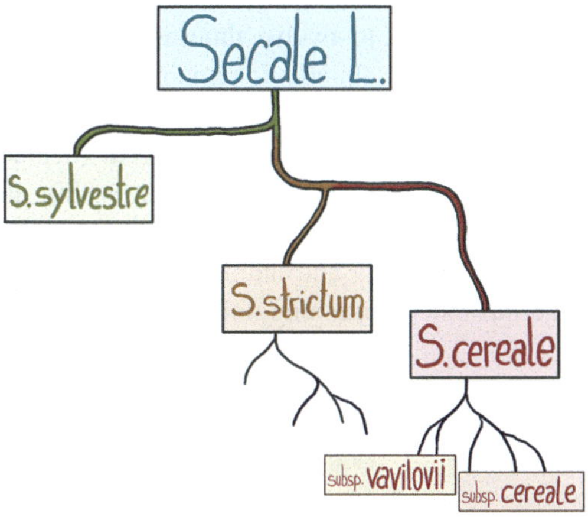

Fig. 6.1 Schematic phylogeny of the small genus *Secale*. Wild *S. sylvestre* is the taxon most distinct from domesticated rye. *S. strictum* is divided into several subspecies. *S. cereale* comprises domesticated and weedy rye (subsp. *cereale*) and its wild progenitor subsp. *vavilovii* (see Table 6.1 for further information). The topology of the tree is based on the results of Schreiber et al. (2019). Branch lengths are not to scale

6.3 The Dynamic Multi-stage Model of Plant Domestication

It is tempting, when looking at the characteristics within the diversity of our cultivated plants today, to view domestication as a straightforward process. We should always keep in mind that the development of settlements and the cultivation of plants, and later also animals, occurred over a time span of thousands of years, i.e. dozens of human generations. Even as hunter-gatherers, humans used wild plants, and the path to domestication was long. At the same time, the characteristics and properties that distinguish our cultivated plants today did not necessarily arise from human intention, but were certainly in many cases rather by-products of the joint life of plants and humans in the course of cultivation. Archaeological finds indicate, for example, that the increase in seed size may have been a by-product of the necessary soil cover to protect the seeds from being eaten by birds and other animals (Shennan 2018).

The dynamic and multi-stage model of plant domestication as a protracted process based on consecutive steps was first described by Harris and Hillman (1989) and is now widely accepted on the basis of extensive archaeobotanical evidence (Fuller 2007). This model describes four stages on the path from wild plants to fully domesticated cultivated plants and was further modified by Fuller in 2007 (illustrated in Fig. 6.2).

Stage 1 describes the process of collecting wild plants as hunters and gatherers did for thousands of years. In many cases, the wild plants collected in the stage were to become the wild ancestors of our current crop plants, as the example of Einkorn shows (Zohary et al. 2012). In the archaeobotanical context, there is much greater diversity among the specimen collected from the wild, than there is among the species later grouped together in the Neolithic package. Hence, the question arises at which point in the domestication process "lost crops" were abandoned and how their influence on the overall process should be evaluated (Abbo et al. 2013).

In **stage 2**, the first signs of cultivation become apparent in the form of the targeted growth of wild plants as well as their harvest, storage and initial dispersal movements.

Stage 3 follows the systematic cultivation of wild plants that were morphologically still wild. This includes the necessary soil preparation and the appearance of weeds.

Stage 4 is the subsequent replacement of the first crop plants by their fully domesticated descendants and thus the beginning of agriculture. This process is accompanied by a steady increase in the amount of work invested, larger and increasingly dense settlements in which more and more people had to be supplied (Harris and Hillman 1989; Fuller 2007).

In the following, we discuss cereal, and in particular rye, domestication, with a focus on the origin and dispersal of crops and the characteristics of domesticates. We will show that rye evolution fits the paradigm of protracted evolution in some aspects, and followed a separate trajectory in others, mainly through an intervening phase of weediness. The history of rye will also illustrate that domestication is not a one-way street. Plants can change between wild, domesticated and wild status in the course of their evolution, and rye is a prime example for this.

6.4 Cereal Domestication in the Fertile Crescent

Domestication is generally described as the transformation of a wild species by artificial selection as it adapts to a new ecological niche: the human-made agricultural environment. This is the foundation of an interdependent relationship between humans and their domesticated species. Domestication should be understood as a mutually inclusive long-term process, of which the first clear evidence is found in the Neolithic period. The oldest archaeological remains of domesticated plants and animals date back as far as to around ten to twelve thousand years ago with the most abundant remains found in Southwest Asia in an area (highlighted in Fig. 6.3) known as the Fertile Crescent (Childe 1928, 1936; Harris et al. 1998), but within an astonishingly short period of time, independently domesticated plants and animals were found in numerous other places on (almost) all continents.

Fig. 6.2 Protracted process of cereal domestication and subsequent spread. Human use of plants started with gathering of wild plants. Early cultivation and field preparation led over time to the establishment of fully domesticated species, which spread to other geographic regions and adapted to diverse climatic and environmental conditions

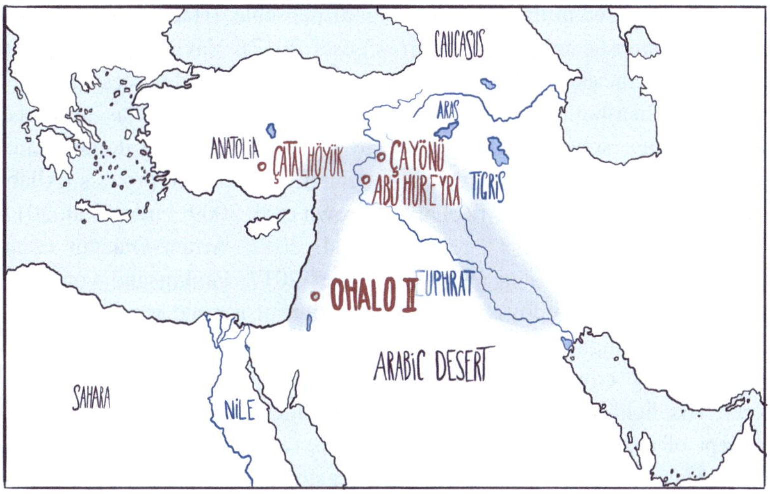

Fig. 6.3 Selection of some archaeological sites in Southwest Asia relevant for the establishment of agriculture. Depicted in grey is the Fertile Crescent, the region where the oldest evidence for the transition to a sedentary lifestyle was found. Shown are Ohalo II, dated to the upper Palaeolithic, Abu Hureyra with Epipalaeolithic and Neolithic remains, the Neolithic site Çatalhöhük as well as Çayönü spanning to the outgoing Neolithic. A recent analysis of plant remains from early agricultural sites was done by Wallace et al. (2019)

Among other parallels, the centres of diversity of the respective wild progenitors of rye, barley and wheat are found in the same geographical area, which extends across the Levant and Anatolia (Zohary et al. 2012). The taxon of the wild ancestor of cultivated rye, *Secale cereale* subsp. *vavilovii*, was unclear for a long time. The hypothesis had been put forward that cultivated rye is a hybrid of *S. cereale* subsp. *vavilovii* and *S. strictum* (Schlegel 2013). However, taxonomic (Frederiksen and Petersen 1998) and phylogenetic analyses (Schreiber et al. 2019) have shown that accessions of wild *S. cereale* subsp. *vavilovii* are closely related to domesticated rye, ruling out substantial *S. strictum* ancestry. Schreiber et al. (2019) did not rule out gene flow due to natural crosses between sympatric *Secale* taxa. In fact, they found a specific example for possible gene flow between *S. cereale* and *S. strictum* in Armenia. However, global ancestry analysis did not support the notion that domesticated rye possess a major ancestry component tracing back to *S. strictum.*

An interesting observation puts the domestication history of rye into the spotlight, as rye is not being considered to be one of the so-called founder crops, like wheat and barley, although the respective wild species are all native to the same geographic area. Fascinatingly, in the pre-pottery Neolithic, archaeobotanical remains from the northern Levant show that rye grains are present alongside with barley and wheat, but always only together with at least one of the other two cereals and with uncertain domestication status as ascertained from the abscission zone discernible from charred grains (Hillman 1978; Nesbitt 2002). As humans settled into a sedentary way of life, and by the time the pottery Neolithic began, rye had disappeared from the archaeobotanical record, suggesting it had ceased to be among the crops of early Near Eastern civilisations.

6.5 Centres of Diversity and Wild Ancestors

We are familiar with most of the wild ancestors of our modern cultivated plants, as well as their current distribution ranges and basic ecological preferences. One underlying assumption is that the wild progenitors of the founder crop plants still have their diversity centres in the same area where they were first domesticated. Apart from barley and flax, the wild ancestors of our crop plants have very limited distribution ranges. Wild emmer (*Triticum turgidum* subsp. *dicoccoides*) and wild chickpeas (*Cicer reticulatum*) are limited to the area of the Fertile Crescent (Zohary et al. 2012).

The Russian botanist, geneticist and research traveller Nikolai Ivanovič Vavilov (*1887, †1943) pioneered the investigation of how the genetic variability of our crops is distributed around the world. His field studies led him to develop the concept of *centres of origin* of our cultivated plants, which he placed in those areas of the world where the genetic diversity in related wild species is highest. Based on these centres of diversity, he precisely characterised many wild ancestors of our cultivated plants (Vavilov and Dorofeev 1992). The review of McElroy (2014) gives a good introduction to Vavilov's life, his tragic conflict with the Soviet authorities and his impact on weed science.

Assuming that there has been no dramatic change in these areas of distribution over the past millennia, domestication of these cereals and pulses could only have taken place in this geographical area. In particular, the endemic distribution of wild emmer in the Fertile Crescent and its immense historical importance as a cereal far beyond this small geographical area prove that Neolithic agriculture could only have begun there, in the Fertile Crescent (Zohary et al. 2012). Although the ancestors of einkorn, lentil, bitter vetch and pea also occur outside the Fertile Crescent, they are most diverse in this very region. Vavilov used such diversity centres as the decisive feature for determining the wild ancestors of our cultivated plants (Vavilov and Dorofeev 1992).

Today, in many cases the wild species proposed by Vavilov have been confirmed as direct ancestors by genetic studies (Zohary et al. 2012). However, the hypothesis of centres of diversity has been controversially discussed at least since Harlan and de Wet (1971), who considered the inclusion of human settlement and migration history in the study of plant domestication to be indispensable (Harlan and de Wet 1971; Bar-Yosef 2017). Vavilov's assumption that domestication was a rapid process limited to those small geographical regions has now been challenged by the hypothesis that domestication was a protracted, multicentric process (Allaby et al. 2008; Brown et al. 2009; Fuller et al. 2011, 2012; Abbo et al. 2012; Arranz-Otaegui et al. 2016; Bar-Yosef 2017; Pankin and von Korff 2017). The question of a rapid versus a protracted domestication process is still under debate (Heun et al. 2012).

Besides lending his name to *S. cereale* subsp. *vavilovii* (initially as *S. vavilovii*), the wild ancestor of cultivated rye, Vavilov developed the hypothesis of a secondary domestication origin of rye. Although he found the greatest diversity of *S. vavilovii* in Southwest Asia, he saw it mainly in weedy rye, which persisted within barley and wheat fields; a phenomenon which can still be observed today (McElroy 2014, Fig. 6.4). Vavilov assumed that wild rye had taken advantage of early farming practices and grew as a weed in fields of the other fully domesticated cereals. He considered the advantages of human care to be so substantial that mimicking the phenotypic characteristics of the target crop was a plausible evolutionary trajectory for a weed (Vavilov and Dorofeev 1992). These traits include the non-shattering rachis, as well as the adjustment of phenology and grain

Fig. 6.4 *Secale cereale* growing in a field of durum wheat around the province of Gaziantep in Turkey

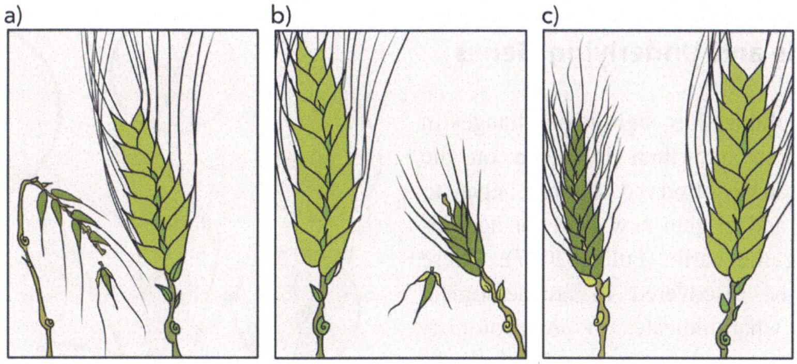

Fig. 6.5 **Vavilovian mimicry**. **a** In the wild, plants have to survive, reproduce and disperse seed. **b** In early farming, farmers deliberately cultivated plants. However, in the early farmers field, also some wild plants adapted to human-made habitats and become "weeds" as beneficiaries of cultivation. Weeds were unwelcome guests. But since they were barely distinguishable from the cultivated plants (**c**), they escaped removal by human hands

size so that the weed was sown and harvested together with the actual crop (highlighted and described in b). Following his observations, this form of imitation, which Vavilov called plant mimicry, is now known as "Vavilovian mimicry" (Vavilov and Dorofeev 1992; McElroy 2014), and rye is recognised as important example (Fig. 6.5).

6.6 Archaeobotany Dates Onset of Cereal Domestication

Archaeozoology and archaeobotany, two subdisciplines of archaeology, are concerned, among other topics, with the Neolithic transition and signs of the initial domestication of animals and plants, as well as their subsequent spread and development over time.

Archaeobotany investigates plant remains from the archaeological context, so-called macrofossils. These were preserved by mechanisms such as charring, humidity, mineralisation or desiccation and allow us to make inferences about agricultural history. In quantitative terms, most plant remains are charred. In the process of charring, a kind of carbon skeleton is created by combustion in the absence of oxygen through carbonisation, which protects the plant remains from further decay but destroys DNA and other biomolecules (Moffett 2009).

If we examine the archaeobotanical remains of Neolithic settlements, the chaff and seeds of various weeds make up the majority of the findings (Fuller 2007). The remains of processed food, remnants of the processing itself, fire materials, animal dung, structural elements and ritual items allow the archaeologist glimpses into the beginning of sedentary life (van der Veen et al. 2007).

Of course, cereals are not only used to make bread, but bread has been a basic source of nutrition for thousands of years and is often one of the first associations when thinking about cereals. The processing of wild cereals began long before settlements developed, as the findings from Ohalo II, a site in Israel dated to 23,500–22,500 years before the present have demonstrated most convincingly (Piperno et al. 2004). For a long time, it was only possible to speculate indirectly about the products into which these wild cereals were processed by means of associated finds (Carretero et al. 2017). Recently, however, Arranz-Otaegui et al. (2018) found 14,400-year-old charred flatbread made from einkorn, barley and oats in Jordan, thus providing the first direct evidence of the pre-Neolithic production of bread.

6.7 Characteristics of Domesticated Cereals and Underlying Genes

During the Neolithisation, significant changes in human behaviour and their influence on the environment can be observed. People began to cultivate soils and to gain new land for agriculture by tilling the earth (Fuller 2007). These changes can be discovered in archaeological contexts. But what indicates to archaeologists that they have found the remains of a domesticated plant, rather than of a wild relative? What do they have to look for to investigate the beginning of domestication, and how can wild and domesticated species be distinguished from each other? The changes through cultivation that lead to domestication of a species are often described as the "domestication syndrome" (illustrated for cereals in Fig. 6.6) and define the characteristic aspects of cultivated plants through the targeted selection of certain features (Harlan and de Wet 1971; Harlan 1992; Hammer 1984; Vavilov and Dorofeev 1992).

These features differ according to the crop. Farmers have selected—consciously or unconsciously—for very different traits in fruit trees, tubers, vines, legumes and cereals (Harlan and de Wet 1971; Fuller 2007; Zohary et al. 2012). In the case of cereals, the features relevant to archaeobotany are the size of grains, the adherence of husks to grains, and even more importantly, the loss of brittleness of the inflorescence stem (the "rachis"), which is a hallmark for human control of seed dispersal (Zohary et al. 2012).

In general, the characteristics selected for the domestication of plants were those that enabled humans to sow, cultivate and harvest them reliably. This applies to the growth of the plant, where the cultivated forms are usually more erect and more uniform, especially in cereals.

Furthermore, it is advantageous for the management by humans if all plants in a field germinate, grow and reach maturity at the same time, in order to facilitate cultivation and harvesting. In addition, an increase in seed production and fruit size is obvious advantages, as these characteristics directly increase yield. Awns and glumes became thinner, and in addition, a transition to naked or free-threshing grains and adaptations to different climatic conditions can be observed, which facilitated direct handling and further processing. The spread to colder regions also led to adaptations as the need for vernalisation, i.e. a cold period after germination, which, depending on the climatic zone, stimulates tillering and flowering of a plant (Zohary et al. 2012).

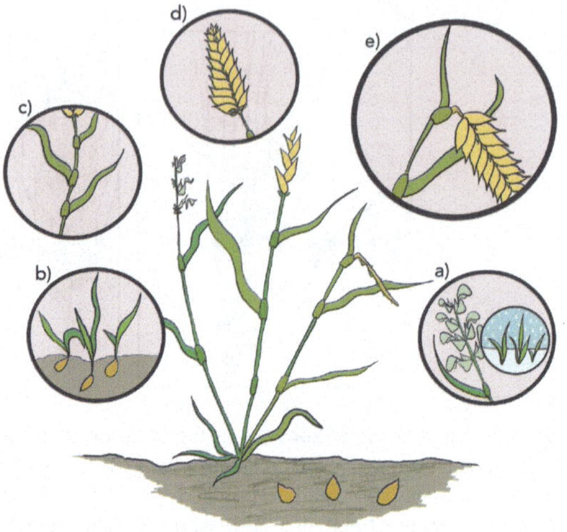

Fig. 6.6 Key traits during cereal domestication. Adapted from Lenser and Theißen (2013) showing a schematic wild grass in the centre and the characteristics of domesticated cereals in the surrounding circles. Clockwise from the bottom right: **a** the adaptation to different climatic conditions (e.g. the need for vernalisation, or loss thereof), **b** reduced seed dormancy, **c** a more regular plant morphology, **d** more and larger seeds, as well as **e** the loss of the brittleness of the spike. The latter serves as the main trait to differentiate between wild and domesticated taxa in archaeobotany, since this trait is observable in archaeobotanical remains. However, it is necessary to keep in mind that the mere separation of wild and domesticated is agnostic as to weediness

Little is known about the molecular basis of domestication characters in rye. This is in contrast to the situation in wheat and barley. Molecular genetic studies have found that the *Non-brittle rachis 1* (*btr1*) and *Non-brittle rachis 2* (*btr2*) genes are conserved among wheat and barley and that mutations in these genes underlie the abolition of grain shattering in both crops

(Pourkheirandish et al. 2015; Avni et al. 2017). Likewise, allelic variation in three key regulators *VERNALIZATION1*, *VERNALIZATION2* and *VERNALIZATION3* (*VRN1*, *VRN2*, *VRN3*; Distelfeld et al. 2009) underlies variation in flowering time and vernalisation response in both wheat and barley. Lastly, genes balancing seed dormancy and preharvest sprouting show allelic variation shaped by crop evolution in both wheat and barley (Nakamura 2018). By contrast, the evolution of naked or free-threshing grains is due to different genetic factors in barley and wheat: the *NUDUM* (Taketa et al. 2008) and *Q* genes (Simons et al. 2006), respectively.

The key domestication and crop improvement characters—seed shattering, loss of husk adherence, increased grain size, flowering time adaptations—are shared between Triticeae crops. Hence, it is a plausible hypothesis that the *Btr1/2* and *VRN* genes have played an important role also in rye domestication and crop evolution. Restriction fragment length polymorphism mapping revealed that quantitative loci controlling vernalisation response (Plaschke et al. 1993) are located on orthologous chromosomal regions in wheat, barley and rye. Assisted by genome sequences (Li et al. 2021; Rabanus-Wallace et al. 2021) and high-throughput marker assays (Haseneyer et al. 2011; Schreiber et al. 2019), genetic mapping and candidate genes studies may now proceed to provide conclusive links between sequence variation and domestication traits in rye.

6.8 Weeds and Crops

Weeds play an important role in the archaeobotanical context. The occurrence of cultivated plants is not the only hallmark of the beginning of agriculture and domestication: as crops rose in importance, also weeds became more diverse and abundant (De Wet and Harlan 1975; Baker 1991; Willcox et al. 2008). Already in the Palaeolithic, plants had to adapt to the environmental changes caused by humans. The use of fire, the establishment of campsites and the first fields for the cultivation of different plants as well as hunting practices that influenced the composition of local wild species (van Vuure 2005) created new environmental conditions for plants to adapt to and thrive in. These synanthropic plants were poised to spread together with humans, crops and agriculture (Snir et al. 2015). It is likely that some species transported by early migrant farmers to new habitats thrived in their new location thus spreading beyond farmers' fields, e.g. representing the early forms of invasive species (Alpert and Maron 2000). Here, also rye is an excellent example of a crop that has become wild and invasive. For example, in California, weedy rye is a massive problem (Burger et al. 2006) going back to the direct descendants of domesticated rye imported from Europe (Snir et al. 2015).

Remains of weeds can help date the onset of agriculture. In northern Syria, in addition to the well-known cereals and legumes, finds of rye and the seeds of various weeds as *Chenopodium album, Chrozophora tinctoria, Emex spinosa, Hordeum spontaneum* (wild barley), *Scorpiurus muricatus* and *Silybum marianum* (Hartmann-Shenkman et al. 2015). These weeds that even today still occur in farmer's fields have provided evidence for the beginning of intensive cultivation in the pre-pottery Neolithic, around 11,500 BC (Moore et al. 2000; Shennan 2018).

At the foothills of the Zagros Mountains in Iran, the transition from the pre-domestication cultivation of einkorn, emmer and barley to the growth of fully domesticated crops can be observed, spanning over a period of ~2,200 years: phenotypically wild cereals appeared 12,000 years ago, domesticated forms no later than 9800 years before the present. Counterpointing the domestication process of the first cereals, typical field weeds became more frequent as well. Both crops and weeds support the assumption that the cultivation of plants should be seen as a protracted process based on diverse interacting factors (Riehl et al. 2013; Purugganan 2019).

Accordingly, the evolution of weeds proceeded in parallel with the development of cultivated plants, but in the absence of intentional selection by humans, and mostly unwanted. This makes weeds an interesting model to study rapid evolutionary adaptation mechanisms subject to natural selection (Vigueira et al. 2012).

6.9 A Secondary Domestication Origin for Rye

The first archaeological evidence for domesticated rye is from the European Bronze Age (Hartyányi and Nováki 1975). With the beginning of the Iron Age, rye had become a widely used grain in Central and Northern Europe (Behre 1992). Thus, the theory of a secondary domestication origin for rye seems plausible.

In this scenario, synanthrope rye was spread unintentionally by farmers, as a stowaway, in the form of weeds in barley and wheat fields. As soon as weedy rye had arrived in regions with poorer soils and colder climatic conditions, in which it grew better than wheat and barley, it became a cultigen (Zohary et al. 2012). As a result, the domestication of rye took place in two successive steps: (i) first rye adapted to the agricultural environment as a travelling weed already in possession of some domesticated characters such as a non-shattering spike and an upright plant architecture; but only (ii) after arriving in Europe, rye was sown intentionally as a crop, becoming fully domesticated and evolving into a locally significant cereal crop (Preece et al. 2017).

The main evidence for the role of Vavilovian mimicry in the evolution of rye comes from archaeological finds, or the absence thereof. Studies of genetic diversity based on molecular markers (e.g. Schreiber et al. 2019; Hagenblad et al. 2016) observed low differentiation between populations and high intra-accession diversity, in contrast to the inbreeding species wheat and barley. These observations may also be a direct consequence of the outcrossing nature of rye and do not necessarily argue for its origin as a secondary domesticate. Future population genomic studies with sequence data of wild, weedy and domesticated accessions, and, probably more importantly, more sophisticated approaches such as coalescent simulations (Excoffier et al. 2013) and estimation of past effective population sizes (Li et al. 2011) may provide further evidence for episodes of weediness in rye crop evolution.

6.10 The Peculiar Situation on the Iberian Peninsula and a Little Excursion into Linguistics

Numerous recent studies using ancient DNA sequences retrieved from human remains show the great influence that various population movements from the more eastern parts of Eurasia had on the human gene pool in Europe. The first to arrive were the farmers and cattle breeders that brought the Neolithic from Anatolia and the Levant to Europe (Lazaridis et al. 2014; Skoglund et al. 2014, 2012; Omrak et al. 2016; Jones et al. 2017; Oms et al. 2018; Skoglund and Mathieson 2018). Cattle herders from the Pontic steppe followed in the Bronze Age (Allentoft et al. 2018; Haak et al. 2015). The Neolithic reached the Iberian Peninsula about 7500 years ago, followed by a regional diversification in the production of ceramics and stone tools in the north of the peninsula, as can be observed along the Mediterranean coast (Utrilla 2012; Alday et al. 2018).

However, Neolithic ceramics from southern Iberia (Andalusia) differ from those of the oldest sites in the north and show similarities with archaeological material from Morocco (Sánchez et al. 2012; Chocarro et al. 2013; Martínez-Sánchez et al. 2018). The possibility of North African influences on the southern part of the Iberian Peninsula has long been considered (Sánchez et al. 2012), but recent studies were not able to confirm this based on ancient human DNA sequences (Martínez-Sánchez et al. 2018; Valdiosera et al. 2018). Valdiosera et al. (2018) examined genome-wide sequencing data of 13 individuals from both the north and south of Iberia, covering a period of 4000 years (from 7500 to 3500 BC). They confirmed the assumption of previous studies that the first Iberian farmers were the descendants of Anatolian migrants (Olalde et al. 2015; Haak et al. 2015) who mixed with native hunter-gatherer populations in the following centuries while maintaining their new farming practices (Haak et al. 2015; Olalde et al. 2015; Martiniano et al. 2017; Valdiosera et al. 2018).

More evidence for a connection between the Levant and the Iberian Peninsula at some time after the beginning of settlement was provided by population genomic studies of rye (Schreiber et al. 2019), barley (Russell et al. 2016) and spelt wheat (Abrouk et al. 2018). In the case of rye, the Iberian material clustered together in diversity space to wild material, domesticated and weedy accessions from the Levant and to domesticated or weedy rye accessions there and is clearly distinct from northern and central European samples (Schreiber et al. 2019).

A similar picture, albeit less pronounced, was also found by Russell et al. (2016) in barley, whose diversity panels also included samples from the Maghreb regions, thus emphasising again the connection between Levant and Iberia through North Africa.

In addition to the genetic peculiarities of Iberian cereals, the etymology of the Spanish language is pointing towards past differences. Spanish words denoting cereals do not follow the pattern of the other Latin languages. For example, *centeno* (rye) and *cebada* (barley) differentiate clearly from Secale, which is not only the botanical name but also the relative for the other Latin languages (e.g. French: *sègle*, Italian: *segala*; see Fig. 6.7). Although the modern words in Castilian (and Portuguese) are attributed to a Latin origin, the exact etymology remains unclear, with a possible Andalusian Arabic origin, which would fit in with the close connection with the Middle East (Blake 1987). Whether this

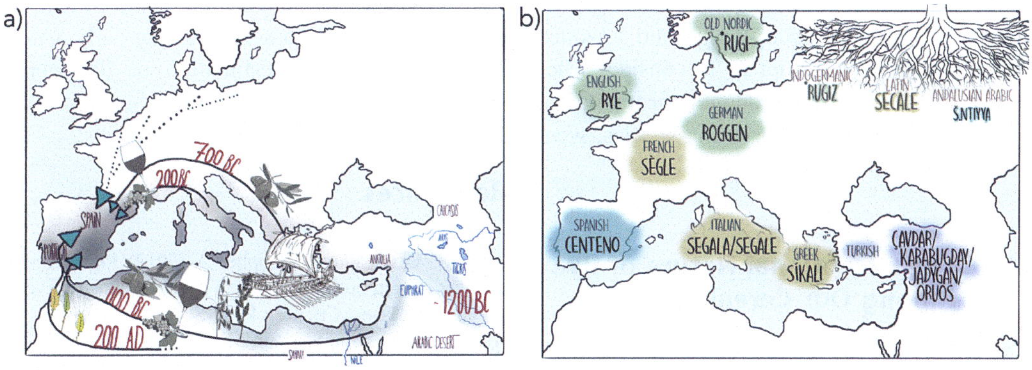

Fig. 6.7 **a** Map showing the settlement history of the Iberian Peninsula with the introduction of the respective dominant crops. Around 1200 BC, the rise of the Phoenicians as traders began in the Levant. Within a few decades, they established trade routes by sea and over land across the entire Mediterranean. By around 1100 BC, they had reached the Iberian Peninsula, whose agriculture they radically changed with the introduction of grapevine and olives. These crops were cultivated in large monocultures. Moreover, the Phoenicians introduced oats and millet as new cereals to Spain. Attracted by the local climate, colonists kept coming to the peninsula along with their crops over next few centuries. Around 700 BC, olives introduced by Greek settlers dominated the agricultural landscape of Spain. Olives, in turn, were replaced as the dominant crop by grapevine, introduced by Romans starting from 200 BC. Other groups such as Celtic and Germanic tribes (represented by the dotted lines) also arrived on the Iberian peninsula, but without bringing about significant changes in agricultural practices. Immigrants from North Africa arrived from 200 AD onwards and changed agricultural practices by bringing cereals to the fore **b** Etymology of words for rye in different European languages, coloured according to language groups and their respective etymological history: Green indicates a germanic origin as in English rye and German Roggen. In brown, the Latin group around the botanical name Secale is shown with Italian Segala/Segale, French Sègle and Greek Síkali. Spanish is presented in blue as an outlier with Centeno, which goes back to a Latin origin as well, but it is speculated that the Latin word may trace back to an Andalusian-Arabic origin. In purple are the diverse terms for rye in Turkish shown: they are all rather unspecific with meanings between "black wheat", "weed in wheat fields" or just "weed" (Kamil, 2008), indicating the common presence of rye as weed in wheat and barley fields and only minor importance as a crop plant

Andalusian Arabic origin (Buxo 2006, Chocarro et al. 2013) of the terms is due to the influence of the Phoenicians or later of the Moors remains controversial.

The lack of genetic affinity between human individuals of the Moroccan Neolithic and Iberian Neolithic (Martínez-Sánchez et al. 2018), as well as the clear relationship of the first hunter-gatherers of the peninsula with those of the rest of the European Mediterranean (Valdiosera et al. 2018), points to a later arrival of Levantine cereals. Historical sources provide us with two possible sources of a post-neolithic Levantine influence on the peninsula: the Iron Age colonisation by the Phoenicians and the Arabic rule of Al-Andalus (Freller and Vázquez 2012).

The Phoenician colonial period had a lasting influence on the local agriculture of the Iberian Peninsula, with the import of wine and olives, as well as millet and oats, the change to large monocultures and the intensification of agricultural practices (Buxó 2006; Freller and Vázquez 2012). This should prompt us to consider whether the Phoenicians also brought their traditional cereals, which, unlike the existing species, were better adapted to the new system.

6.11 The Exceptional Nature of Rye Among Our Cereal Crops

The beginning of agriculture well over ten thousand years ago is without question one of the most far-reaching changes in the human way of life. The history of our cultivated plants also begins with this change. Perhaps it was bread that led man to settle, perhaps it was beer (Katz and Voigt 1986). We do not know.

But we do know that during this long process, domesticated plants and animals became both more and more distinct from their wild relatives and an indispensable part of our human existence. The history of rye has demonstrated that the transition from wild to cultivated need not to be straightforward and that unlike the other major cereals it followed a separate and intriguing path. A detour as a weed is possible. Many aspects of the domestication history of rye have yet to be resolved, and there are still many questions to ask. But the steady improvement of genomic resources, new archaeological finds and the interdisciplinary interaction of all fields of research will certainly further our understanding of this important cereal. The recent publication of chromosome-scale reference genome sequences of two rye genotypes (Rabanus-Wallace et al. 2021; Li et al. 2021) will facilitate the mapping and cloning of "domestication genes", for example those controlling shattering and vernalisation response. Population and pan-genomics across wild and domesticated taxa will unravel the impact of human selection on patterns of sequence diversity in the rye genome, or maybe the lack thereof in this recent, outcrossing and highly diverse cultigen. Analysis of sequence data of extant weedy accession, and hopefully, access to ancient DNA, may accrue molecular genetic evidence for the hypothesis of Vavilovian mimicry. A genus-wide pangenome, i.e. chromosome-scale assemblies for all *Secale* species, will enable estimation of divergence times and rates of gene flow.

References

Abbo S, Lev-Yadun S, Gopher A (2012) Plant domestication and crop evolution in the near east: on events and processes. Crit Rev Plant Sci 31(3):241–257

Abbo S, Lev-Yadun S, Heun M, Gopher A (2013) On the 'lost' crops of the neolithic near east. J Exp Bot 64 (4):815–822

Abrouk M, Stritt C, Müller T, Keller B, Roulin A, Krattinger S (2018) High-throughput genotyping of the spelt gene pool reveals patterns of agricultural history in Europe, 481424

Alday A, Perales U, Soto A (2018) Beyond the archaeological contexts: the debate on the records of the Mesolithic and early Neolithic in the upper Ebro Basin. Quatern Int 472:195–204

Allaby RG, Fuller DQ, Brown TA (2008) The genetic expectations of a protracted model for the origins of domesticated crops. Proc Natl Acad Sci 105 (37):13982–13986

Allentoft ME et al (2018) Population genomics of Bronze Age Eurasia. Nature 522:167–172

Alpert P, Maron JL (2000) Carbon addition as a countermeasure against biological invasion by plants. Biol Invasions 2(1):33–40

Arranz-Otaegui A, Colledge S, Zapata L, Teira-Mayolini LC, Ibanez JJ (2016) Regional diversity on the timing for the initial appearance of cereal cultivation and domestication in southwest Asia. Proc Natl Acad Sci 113(49):14001–14006

Arranz-Otaegui A, Gonzalez Carretero L, Ramsey MN, Fuller DQ, Richter T (2018) Archaeobotanical evidence reveals the origins of bread 14,400 years ago in north-eastern Jordan. Proc Natl Acad Sci 115 (31):7925–7930

Avni R, Nave M, Barad O, Baruch K, Twardziok SO, Gundlach H, Hale I, Mascher M, Spannagl M, Wiebe K, Jordan KW, Golan G, Deek J, Ben-Zvi B, Ben-Zvi G, Himmelbach A, MacLachlan RP, Sharpe AG, Fritz A, Ben-David R, Budak H, Fahima T, Korol A, Faris JD, Hernandez A, Mikel MA, Levy AA, Steffenson B, Maccaferri M, Tuberosa R, Cattivelli L, Faccioli P, Ceriotti A, Kashkush K, Pourkheirandish M, Komatsuda T, Eilam T, Sela H, Sharon A, Ohad N, Chamovitz DA, Mayer KFX, Stein N, Ronen G, Peleg Z, Pozniak CJ, Akhunov ED, Distelfeld A (2017) Wild emmer genome architecture and diversity elucidate wheat evolution and domestication. Science 357(6346): 93–97. https://doi.org/10.1126/science.aan0032 PMID: 28684525

Baker HG (1991) The continuing evolution of weeds. Econ Bot 45(4):445–449

Barkworth ME (2007) Secale. In: ME Barkworth et al. (eds) Flora of North America north of Mexico, vol 24, pp 259–260.Oxford University Press, New York-Oxford

Barrett SH (1983) Crop mimicry in weeds. Econ Bot 37 (3):255–282

Bar-Yosef O (2017) Chapter 19—multiple origins of agriculture in Eurasia and Africa. In: Tibayrenc M, Ayala FJ (eds) On human nature. Academic Press, San Diego, pp 297–331

Behre K-E (1992) The history of rye cultivation in Europe. Veg Hist Archaeobotany 1(3):141–156

Bernhardt N, Brassac J, Kilian B, Blattner F (2017) Dated tribe-wide whole chloroplast genome phylogeny indicates recurrent hybridizations within Triticeae. BMC Evol Biol 17(1):141

Blake RJ (1987) New linguistic sources for old Spanish. Hisp Rev 55(1):1–12

Brown TA, Jones MK, Powell W, Allaby RG (2009) The complex origins of domesticated crops in the fertile crescent. Trends Ecol Evol 24(2):103–109

Burger J, Lee S, Ellstrand N (2006) Origin and genetic structure of feral rye in the western United States. Mol Ecol 15(9):2527–2539

Buxo R (2006) Paisajes culturales y reconstruccion historica de la vegetacion. Revista Ecosistemas 15(1)

Carretero LG, Wollstonecroft M, Fuller DQ (2017) A methodological approach to the study of archaeological cereal meals: a case study at Catalhoyuk East (Turkey). Veg Hist Archaeobotany 26(4):415–432

Childe GV (1928) The most ancient east: the oriental prelude to European prehistory. Kegan Paul, London

Childe GV (1936) Man makes himself. Watts, London

Chocarro LP, Jorda GP, Mateos JM, Rodriguez JCV (2013) Y llegaron los agricultores, agricultura y recoleccion en el occidente del Mediterraneo. Menga: Revista de prehistoria de Andalucia, pp 15–34

Clayton WD, Harman KT, Williamson H (2006) World grass species—synonymy database the board of trustees of the Royal Botanic Gardens, Kew

Clegg MT, Gaut BS, Learn GH Jr, Morton BR (1994) Rates and patterns of chloroplast DNA evolution 91:6795–6801

De Wet JMJ, Harlan JR (1975) Weeds and domesticates: evolution in the man-made habitat. Econ Bot 29 (2):99–108

Distelfeld A, Li C, Dubcovsky J (2009) Regulation of flowering in temperate cereals. Curr Opin Plant Biol 12(2):178–184. https://doi.org/10.1016/j.pbi.2008.12.010 Epub 2009 Feb 3 PMID: 19195924

Excoffier L, Dupanloup I, Huerta-Sánchez E, Sousa VC, Foll M (2013) Robust demographic inference from genomic and SNP data. PLOS Genet 9(10):e1003905. https://doi.org/10.1371/journal.pgen.1003905

Food and Agriculture Organization of the United Nations (FAO), FAOSTAT statistical database

Frederiksen S, Petersen G (1998) A taxonomic revision of *Secale* (Triticeae, Poaceae). Nord J Bot 18(4):399–420

Freller T, Vazquez MA (2012) Die Geschichte der Iberischen Halbinsel: mit landeskundlichen und kulturhistorischen Beiträgen von Miguel Vazquez. Thorbecke

Fuller DQ (2007) Contrasting patterns in crop domestication and domestication rates: recent archaeobotanical insights from the old world. Ann Bot 100(5):903–924

Fuller DQ, Willcox G, Allaby RG (2011) Cultivation and domestication had multiple origins: arguments against the core area hypothesis for the origins of agriculture in the near east. World Archaeol 43(4):628–652

Fuller DQ, Willcox G, Allaby RG (2012) Early agricultural pathways: moving outside the 'core area' hypothesis in Southwest Asia. J Exp Bot 63(2):617–633

Haak W et al (2015) Massive migration from the steppe was a source for Indo-European languages in Europe. Nature 522:207–211

Hagenblad J, Oliveira HR, Forsberg NEG, Leino MW (2016) Geographical distribution of genetic diversity in Secale landrace and wild accessions. BMC Plant Biol 16(1):23

Hammer K (1984) Das Domestikationssyndrom. Die Kulturpflanze 32(1):11–34

Harlan JR (1992) Crops and man. American society of agronomy, 2nd edn

Harlan JR, de Wet JMJ (1971) Toward a rational classification of cultivated plants. Taxon 20(4):509–517

Harris DR, Hillman G (1989) An evolutionary continuum of people-plant interaction. In: White P, Denham T (eds) Foraging and farming: the evolution of plant exploitation. Routledge, London, pp 11–26

Harris D, Damania A, Valkoun J, Willcox G, Qualset C et al (1998) The spread of Neolithic agriculture from the Levant to western Central Asia. In: The origins of agriculture and crop domestication: proceedings.

International Center for Agricultural Research in the Dry Areas, pp 65–82
Hartmann-Shenkman A, Kislev ME, Galili E, Melamed Y, Weiss E (2015) Invading a new niche: obligatory weeds at Neolithic Atlit-Yam, Israel. Veg Hist Archaeobotany 24(1):9–18
Hartyanyi BP, Novaki G (1975) Samen und Fruchtfunde in Ungarn von der Neusteinzeit bis zum 18. Jahrhundert. Agrartorteneti szemle
Haseneyer G, Schmutzer T, Seidel M et al (2011) From RNA-seq to large-scale genotyping—genomics resources for rye (Secale cereale L.). BMC Plant Biol 11:131. https://doi.org/10.1186/1471-2229-11-131
Heun M, Abbo S, Lev-Yadun S, Gopher A (2012) A critical review of the protracted domestication model for near-eastern founder crops: linear regression, long-distance gene flow, archaeological, and archaeobotanical evidence. J Exp Bot 63(12):4333–4341. https://doi.org/10.1093/jxb/ers162 Epub 2012 Jun 19 PMID: 22717409
Hillman G (1978) On the origins of domestic rye—*Secale cereale*: the finds from aceramic Can Hasan III in Turkey. Anatol Stud 28:157–174
Jones ER, Zarina G, Moiseyev V, Lightfoot E, Nigst PR, Manica A, Pinhasi R, Bradley DG (2017) The Neolithic transition in the Baltic was not driven by admixture with early European farmers. Curr Biol 27 (4):576–582
Katz S, Voigt M (1986) Bread and beer: the early use of cereals in the human diet. Expedition Magazine 28 (2):23–34
Khush GS (1962) Cytogenetic and evolutionary studies in *Secale*. II. Interrelationships of the wild species. Evolution 16
Khush GS, Stebbins GL (1961) Cytogenetic and evolutionary studies in *Secale*. I. some new data on the ancestry of *S. cereale*. Am J Botany 723–730
Lazaridis I et al (2014) Ancient human genomes suggest three ancestral populations for present-day Europeans. Nature 513:409–413
Lenser T, Theißen G (2013) Molecular mechanisms involved in convergent crop domestication. Trends Plant Sci 18(12):704–714. https://doi.org/10.1016/j.tplants.2013.08.007 Epub 2013 Sep 12 PMID: 24035234
Li H, Durbin R (2011) Inference of human population history from individual whole-genome sequences. Nature 475:493–496. https://doi.org/10.1038/nature10231
Li G, Wang L, Yang J et al (2021) A high-quality genome assembly highlights rye genomic characteristics and agronomically important genes. Nat Genet 53:574–584. https://doi.org/10.1038/s41588-021-00808-z
Maraci Ö, Özkan H, Bilgin R (2018) Phylogeny and genetic structure in the genus Secale. PLoS One 13(7): e0200825
Martinez-Sanchez RM, Vera-Rodriguez JC, Perez-Jorda G, Pena-Chocarro L, Bokbot Y (2018) The beginning of the neolithic in northwestern Morocco. Quatern Int 470:485–496
Martiniano R, Cassidy LM, O'Maolduin R, McLaughlin R, Silva NM, Manco L, Fidalgo D, Pereira T, Coelho MJ, Serra M, Burger J, Parreira R, Moran E, Valera AC, Porfirio E, Boaventura R, Silva AM, Bradley DG (2017) The population genomics of archaeological transition in west Iberia: investigation of ancient substructure using imputation and haplotype-based methods. PLoS Genet 13(7):1–24
McElroy JS (2014) Vavilovian mimicry: Nikolai Vavilov and his little-known impact on weed science. Weed Sci 62(2):207–216
Moffett L (2009) The archaeology of medieval plant foods. In: Woolgar CM, Waldron T, Serjeantson D (eds) Food in medieval England. Oxford University Press, pp 41–55
Moore AMT, Hillman GC, Legge AJ (2000) Village on the Euphrates: from foraging to farming at Abu Hureyra. Oxford University Press xvii, Oxford
Nakamura S (2018) Grain dormancy genes responsible for preventing pre-harvest sprouting in barley and wheat. Breed Sci 68(3):295–304. https://doi.org/10.1270/jsbbs.17138
Nesbitt M (2002) When and where did domesticated cereals first occur in southwest Asia, Chap. 1, pp 113–132. ex oriente Berlin, r. t. j. cappers and bottema edition
Olalde I, Schroeder H, Sandoval-Velasco M, Vinner L, Lobon I, Ramirez O, Civit S, Garcia Borja P, Salazar-Garcia DC, Talamo S, Maria Fullola J, Xavier Oms F, Pedro M, Martinez P, Sanz M, Daura J, Zilhao J, Marques-Bonet T, Gilbert MTP, Lalueza-Fox C (2015) A common genetic origin for early farmers from Mediterranean Cardial and Central European LBK cultures. Mol Biol Evol 32(12):3132–3142
Omrak A, Günther T, Valdiosera C, Svensson EM, Malmström H, Kiesewetter H, Aylward W, Stora J, Jakobsson M, Götherström A (2016) Genomic evidence establishes Anatolia as the source of the European neolithic gene pool. Curr Biol 26(2):270–275
Oms FX, Terradas X, Morell B, Gibaja JF (2018) Mesolithic-Neolithic transition in the northeast of Iberia: chronology and socioeconomic dynamics. Quatern Int 470:383–397
Oram RN (1996) Secale montanum—a wider role in Australasia? New Zealand J Agric Res 39:629–633
Pankin A, von Korff M (2017) Co-evolution of methods and thoughts in cereal domestication studies: a tale of barley (Hordeum vulgare). Curr Opin Plant Biol 36:15–21
Piperno DR, Weiss E, Holst I, Nadel D (2004) Processing of wild cereal grains in the upper palaeolithic revealed by starch grain analysis. Nature 430:670–673
Plaschke J, Börner A, Xie DX et al (1993) RFLP mapping of genes affecting plant height and growth habit in rye. Theoret Appl Genet 85:1049–1054. https://doi.org/10.1007/BF00215046
Pourkheirandish M, Hensel G, Kilian B, Senthil N, Chen G, Sameri M, Azhaguvel P, Sakuma S, Dhanagond S, Sharma R, Mascher M, Himmelbach A, Gottwald S, Nair SK, Tagiri A, Yukuhiro F, Nagamura Y, Kanamori H, Matsumoto T, Willcox G,

Middleton CP, Wicker T, Walther A, Waugh R, Fincher GB, Stein N, Kumlehn J, Sato K, Komatsuda T (2015) Evolution of the grain dispersal system in Barley. Cell 162(3):527–539. https://doi.org/10.1016/j.cell.2015.07.002 PMID: 26232223

Preece C, Livarda A, Christin P-A, Wallace M, Martin G, Charles M, Jones G, Rees M, Osborne CP (2017) How did the domestication of fertile crescent grain crops increase their yields? Funct Ecol 31(2):387–397

Purugganan M (2019) Evolutionary insights into the nature of plant domestication. Curr Biol 29(14):R705–R714

Rabanus-Wallace MT, Hackauf B, Mascher M et al (2021) Chromosome-scale genome assembly provides insights into rye biology, evolution and agronomic potential. Nat Genet 53:564–573. https://doi.org/10.1038/s41588-021-00807-0

Ren TH et al (2011) Evolutionary trends of microsatellites during the speciation process and phylogenetic relationships within the genus Secale. Genome 54(4):316–326

Riehl S, Zeidi M, Conard NJ (2013) Emergence of agriculture in the foothills of the Zagros mountains of Iran. Science 341(6141):65–67

Russell J, Mascher M, Dawson IK, Kyriakidis S, Calixto C, Freund F, Bayer M, Milne I, Marshall-Griffiths T, Heinen S, Hofstad A, Sharma R, Himmelbach A, Knauft M, van Zonneveld M, Brown JWS, Schmid K, Kilian B, Muehlbauer GJ, Stein N, Waugh R (2016) Exome sequencing of geographically diverse barley landraces and wild relatives gives insights into environmental adaptation. Nat Genet 48:1024–1030

Sánchez MC, Espejo FJJ, Vallejo MDS, Bao JFG, Carvalho AF, Martinez-Ruiz F, Gamiz MR, Flores J-A, Paytan A, Saez JAL et al (2012) The mesolithic-neolithic transition in southern Iberia. Quatern Res 77 (2):221–234

Schlegel R (2013) Rye: genetics, breeding, and cultivation. Taylor & Francis

Schreiber M, Himmelbach A, Börner A, Mascher M (2019) Genetic diversity and relationship between domesticated rye and its wild relatives as revealed through genotyping-by-sequencing. Evol Appl 12(1):66–77

Shang HY, Wei YM, Wang XR, Zheng YL (2006) Genetic diversity and phylogenetic relationships in the rye genus Secale L. (rye) based on Secale cereale microsatellite markers. Genet Mol Biol 29(4):685–691

Shennan S (2018) The first farmers of Europe. An evolutionary perspective. Cambridge University Press

Simons KJ, Fellers JP, Trick HN, Zhang Z, Tai Y-S, Gill BS, Faris JD (2006) Molecular characterization of the major wheat domestication gene Q GENETICS, vol 172, no 1 547–555. https://doi.org/10.1534/genetics.105.044727

Singh R (1977) Cross compatibility, meiotic pairing and fertility in 5 *Secale* species and their interspecific hybrids. Cereal Res Commun 5(1):67–75

Skoglund P, Mathieson I (2018) Ancient human genomics: the first decade. Annu Rev Genom Hum Genet 19(1):13–40

Skoglund P, Malmström H, Raghavan M, Storå J, Hall P, Willerslev E, Gilbert MTP, Götherström A, Jakobsson M (2012) Origins and genetic legacy of Neolithic farmers and hunter-gatherers in Europe. Science 336 (6080):466–469

Skoglund P, Malmström H, Omrak A, Raghavan M, Valdiosera C, Günther T, Hall P, Tambets K, Parik J, Sjögren K-G, Apel J, Willerslev E, Stora J, Götherström A, Jakobsson M (2014) Genomic diversity and admixture differs for stone-age Scandinavian foragers and farmers. Science 344(6185):747–750

Skuza L, Szućko I, Filip E, Strzała T (2019) Genetic diversity and relationship between cultivated, weedy and wild rye species as revealed by chloroplast and mitochondrial DNA non-coding regions analysis. PLoS One 14(2):e0213023. https://doi.org/10.1371/journal.pone.0213023. PMID: 30811487; PMCID: PMC6392296

Snir A, Nadel D, Groman-Yaroslavski I, Melamed Y, Sternberg M, Bar-Yosef O, Weiss E (2015) The origin of cultivation and proto-weeds, long before Neolithic farming. PLoS One 10(7):e0131422

Stutz HC (1957) A cytogenetic analysis of the hybrid *Secale cereale* L. x *Secale montanum* Guss. and its progeny. Genetics 42(3):199

Taketa S, Amano S, Tsujino Y, Sato Y, Saisho D, Kakeda K, Nomura M, Suzuki T, Matsumoto T, Sato K, Kanamori H, Kawasaki S, Takeda K (2008) Barley grain with adhering hulls is controlled by an ERF family transcription factor gene regulating a lipid biosynthesis pathway. Proc Nat Acad Sci 105(10): 4062–4067. https://doi.org/10.1073/pnas.0711034105

Tang ZX et al (2011) Secale. In: Kole C (ed) Wild crop relatives: genomic and breeding resources. Springer, Berlin, Heidelberg, pp 367–396

Tsvelev N (1973) Conspectus specierum tribus Triticeae Dum. familiae Poaceae in flora URSS. Nov Sist Vyssh Rast 10:19–59

Utrilla P (2012) Caminos para el neolitico aragones: la aportacion del radiocarbono y del arte rupestre. Rubricatum: revista del Museu de Gava 5:555–564

Valdiosera C, Günther T, Vera-Rodríguez JC, Ureña I, Iriarte E, Rodriguez-Varela R, Simões LG, Martinez-Sanchez RM, Svensson EM, Malmström H, Rodriguez L, Bermudez de Castro J-M, Carbonell E, Alday A, Hernandez Vera JA, Götherström A, Carretero J-M, Arsuaga JL, Smith CI, Jakobsson M (2018) Four millennia of Iberian biomolecular prehistory illustrate the impact of prehistoric migrations at the far end of Eurasia. Proc Natl Acad Sci 115 (13):3428–3433

van der Veen M, Livarda A, Hill A (2007) The archaeobotany of Roman Britain: current state and identification of research priorities. Britannia 38:181–210

Van Vuure C (2005) Retracing the Aurochs. Pensoft Publishers, History, Morphology and Ecology of an Extinct Wild Ox

Vavilov NI, Dorofeev VF (1992) Origin and geography of cultivated plants. Cambridge University Press

Vences FJ, Vaquero F, Perez de la Vega M (1987) Phylogenetic relationships in *Secale* (Poaceae): an isozymatic study. Plant Syst Evol 157:33–47

Vigueira CC, Olsen KM, Caicedo AL (2012) The red queen in the corn: agricultural weeds as models of rapid adaptive evolution. Heredity 110:303–311

Wallace M, Jones G, Charles M et al (2019) Re-analysis of archaeobotanical remains from pre- and early agricultural sites provides no evidence for a narrowing of the wild plant food spectrum during the origins of agriculture in southwest Asia. Veg Hist Archaeobot 28:449–463. https://doi.org/10.1007/s00334-018-0702-ys

Willcox G, Fornite S, Herveux L (2008) Early Holocene cultivation before domestication in northern Syria. Veg Hist Archaeobotany 17(3):313–325

Zohary D, Hopf M, Weiss E (2012) Domestication of plants in the old world: the origin and spread of domesticated plants in southwest Asia, Europe, and the Mediterranean Basin. OUP Oxford

Assembling the Rye Genome

7

M. Timothy Rabanus-Wallace, Daowen Wang, Jianping Yang, Guangwei Li, and Nils Stein

Abstract

The diploid rye genome is around 8 Gbp in size and exceptionally repetitive. Coupled with the added challenges of residual heterozygosity and the limitations of technology, the daunting task of assembling a full reference quality genome sequence was achieved only in 2021, some four years after the first reference quality Triticeae genomes appeared. The two assemblies that now exist were made possible by the integration of state-of-the-art technologies. These new resources have already begun fulfilling their promise to enhance rye's agricultural potential, and to further research into the evolution of rye and the cereals in general. We provide a general overview of the process of contemporary large crop genome assembly, describe the specific procedures used to assemble the genomes of inbred line 'Lo7' and the Chinese local variety 'Weining', briefly cover the first investigations made possible using the two assemblies, compare the assemblies, and comment on the future potential and prospects for rye genome assembly.

7.1 Introduction

Reference quality annotated genome sequences represent an indispensable resource for geneticists. To give but a handful of examples relevant to the crop sciences, the pure volume of unique sequence expedites the development of molecular markers, which can later be used by breeders to characterise the genetic diversity in a panel (e.g. Vendelbo et al. 2020; see Chap. 9), or to identify linkage between markers and loci influencing a particular phenotype (e.g. Gaikpa et al. 2020; Braun et al. 2019; see Chap. 9). Juxtaposition of linkage maps against the full genome sequence allows detailed study of recombination rates, which breeders depend upon to produce favourable combinations of genes in the production of new varieties. Evolutionary biologists might use the same markers to perform phylogenetic inference (e.g. Maraci et al. 2018; see Chap. 6), or to seek genomic regions showing signs of selection and domestication (e.g. Schreiber et al. 2019; see Chap. 6). Genes annotated near these regions of interest

M. T. Rabanus-Wallace (✉) · N. Stein
Leibniz Institute of Plant Genetics and Crop Plant Research (IPK), 06466 Seeland, Germany
e-mail: wallace@ipk-gatersleben.de

D. Wang · J. Yang · G. Li
College of Agronomy, State Key Laboratory of Wheat and Maize Crop Science, Center for Crop Genome Engineering, Henan Agricultural University, Longzi Lake Campus, Zhengzhou 450046, China

N. Stein
Center of Integrated Breeding Research (CiBreed), Department of Crop Sciences, Georg-August-University, Göttingen, Germany

© The Author(s), under exclusive license to Springer Nature Switzerland AG 2021

M. T. Rabanus-Wallace and N. Stein (eds.), *The Rye Genome*, Compendium of Plant Genomes,
https://doi.org/10.1007/978-3-030-83383-1_7

might then be shortlisted as candidates for further study, and functional geneticists can use the genome sequence to examine variation between gene orthologs/paralogs, which suggests possibilities about their function (e.g. Jung and Seo 2019; see Chap. 8). Access to intergenic sequence data yields not only enhanced possibilities for marker development, but a wealth of information about genome structure, including reconstruction of the repetitive element landscape, itself a primary driver of evolutionary change and genetic variation (e.g. Wicker et al. 2018; see Chap. 8). Comparative genomics provides further insight into structural variation, including cataloguing major structural variations that affect the evolution and expression of traits (e.g. Gabur et al. 2019). These uses and more are discussed elsewhere in this volume.

Genome assembly is challenging and the possibility of assembling a given genome is limited primarily by (i) the size and complexity of the genome, with repetitive DNA content being a major factor, (ii) the availability of technologies suitable to overcome the complexity and size of the genome, and (iii) the cost and time required to deploy the technology and integrate the data into an assembly. Triticeae genome assembly represents an important case study in the genome assembly sphere: The great economic importance of Triticeae cereals creates a substantial incentive to produce reference quality assemblies, while the immense complexity and size of Triticeae genomes creates an equally substantial technical challenge to overcome.

Cereal rye has secondary economic importance compared with bread wheat and barley, and a highly repetitive and long (around 8 Gbp, but varying among species; see Chaps. 3 and 8) genome. Further difficulties that once held back progress included cereal rye's outcrossing habit, meaning rare selfing-tolerant lines are required in order to create a reasonably homozygous genotype—though regions of residual heterozygosity are still expected to confound assembly efforts. Thus, rye received comparatively little early attention from the Triticeae genomics community.

7.1.1 Genome Assemblies

Rye entered the genomics era with the release of a virtual gene order (or 'genome zipper') (Haseneyer et al. 2011; Martis 2013), a collection of short *contigs* (fragments of *contig*uous DNA sequence, typically kilobases to megabases in length), enriched with transcriptome sequencing data, allowing a reasonably comprehensive representation of the gene-space, much of which (including 72% of the 31,008 annotated genes) was arranged into chromosome order by (i) anchoring the contigs to SNP-chip-based linkage maps from four mapping populations, and (ii) curating the order using conserved syntenic blocks shared between the rye genome and those of related grass species whose simpler genomes had already been assembled. This zipper also made use of *chromosome-sorted shotgun* (CSS) sequence data: short-read (typically one- to five-hundred base-pair) sequences made from preparations of individual, flow-sorted chromosomes (see Chap. 3). Because these reads are tagged with a probable chromosome of origin, they can be used to infer the chromosomal origin of a contig, by aligning the CSS reads to the contig. The chromosome whose CSS reads align most frequently to the contig is the likely chromosome of origin. The genome zipper was pivotal in establishing the syntenic relationships in genome structure between rye and the closely related cultivated Triticeae bread wheat and barley, showing that several major translocations had occurred along the rye lineage after its divergence from the ancestor of these groups (Martis 2013; Li et al. 2021).

A draft rye genome was subsequently produced by Bauer et al. (2017), composed of contigs assembled from short-read shotgun sequencing data, further joined into scaffolds by progressively adding data from *mate-paired* reads (pairs of reads separated by a known approximate distance, and oriented towards one another; the pairs can occasionally span two contigs allowing them to be joined) of increasing span, which had themselves been assigned to chromosomes using the CSS approach. The scaffolds were anchored to a high-density genetic map and annotated for gene features and transposable elements (TEs) to produce the draft genome, which achieved a length of approximately 2.8 Gbp representing around 35% of the expected genome sequence content, and likely capturing almost the entire gene-space. The genome was used as the basis for detecting genes under selection as a result of intensive breeding, and identified several candidates primarily affecting plant height.

Rye also lagged behind more commercially dominant crops in the transition from draft to reference quality genome. In 2017, barley became the first Triticeae crop to achieve a reference quality genome assembly, using a now-obsolete hierarchical BAC-by-BAC sequencing and assembly approach, coupled with high-throughput 3D conformation capture sequencing (Hi-C, more details below) (Mascher 2017). The bread wheat genome—a greater challenge owing to allohexaploidy—first achieved a reference quality assembly using a similar Hi-C-based methodology and was released in 2018 (IWGSC 2018). Even more complete genomes for both species, and many of their related wild and domesticated species, have since been released (Monat 2019; Alonge et al. 2020). True pangenomics projects involving the assembly of genomes of multiple varieties have also been achieved for both barley and wheat (Walkowiak 2020; Jayakodi 2020).

Reference quality rye genome assemblies have recently been completed by two independent groups. To simplify the assembly challenge, both groups worked with highly inbred, and hence largely homozygous, varieties. A short-read-and-Hi-C-based pseudo-haploid assembly of the breeding line 'Lo7' (a sixth generation self-pollinated individual) was assembled by the International Rye Genome Sequencing Consortium (IRGSC) (Rabanus-Wallace et al. 2021), a natural continuation of the previous work assembling the zipper and draft genomes that also made use of 'Lo7'. During the same period, a long-read-and-Hi-C-based assembly of the Chinese local variety 'Weining' (an eighteenth-generation self-pollinated individual) was produced by researchers at the Henan Agricultural University (Li et al. 2021). 'Weining' rye is particularly interesting to the crop science community owing to its broad-spectrum resistance to both powdery mildew and stripe rust, and its role as a translocation donor in 1BL/1RS wheat-rye translocation lines, that also possess this resistance.

The two assemblies are described in detail here, but to appreciate the major similarities and differences between them, we offer first a short primer on the overall process of Triticeae genome assembly at the time.

7.2 De Novo Assembly of Triticeae Genomes During the 2010s—A Broad Outline

The methods for assembling genomes de novo (or '*from scratch*') vary greatly and can be quite ad hoc, but a general overview of contemporary approaches during recent years is possible (Fig. 7.1).

High-throughput sequencing provides the raw, fragmentary sequence data. Traditionally, high-density random short-read sequencing (or *shotgun* sequencing) is used, since the error rate is low and sequencing errors can be corrected by consensus among the many reads. However, long-read sequencing has now advanced sufficiently in output and accuracy to be used as the primary source of sequence data, and this advance will likely continue until long-read sequencing comes to dominate the field and alleviate the need for complementary technologies (Amarasinghe 2020). Algorithms are applied to stitch the accumulated read data together into longer sequences. Typically, these algorithms work by representing sequence overlaps between reads, or between set-length sub-sequences (called *kmers*) represented frequently within the reads, as the edges of a graph. They then aim to parsimoniously resolve the graph into *contigs*—contiguous lengths of sequence that represent (as accurately as possible) sections of the original genome (Compeau et al. 2011).

Large genomes cannot typically be completely assembled in this way, owing to the

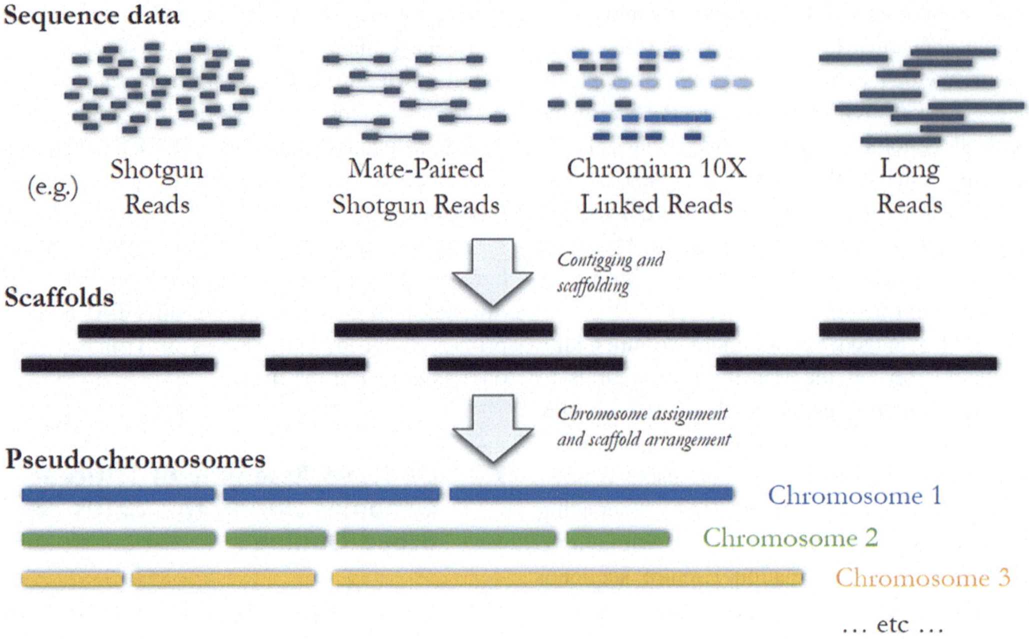

Fig. 7.1 Schematic overview of a generalised de novo assembly workflow, as used for Triticeae genomes in the 2010s—early 2020s. Details described in text

presence of repeated sequences occurring at multiple distinct places in the genome. If the length of the repeated unit exceeds the distance that a read can span, then it is impossible to know which of the interposed unique sequences should be adjacent—and so those unique sequences will be returned as individual contigs. For this reason, longer reads can yield longer contigs.

To continue the assembly further, information about the associations between unique sequence over longer ranges must be used. These longer-range technologies, some of which will be introduced later, have varied properties. For example, mate-paired reads provide accurate sequence association—but only at the 100 s-to-1000 s-of-bps scale. In contrast, Hi-C sequencing provides information on the Kbps-to-100 s-of-Mbps scale—but this information is noisy and not nearly as accurate at small scales as mate-pairs are. Molecule-linked reads, optical mapping, chromosome-sorted reads, and genetic map integration can all be conceptualised as means of associating sequence at a distance and will be discussed later in this chapter in the context of rye.

Integrating multiple longer-range technologies to make best use of their properties and attain a good overall result is a creative process, which differs greatly from assembly to assembly, but is usually implemented as a quasi-sequential workflow or *assembly pipeline*. To give a contrived example, molecule-linked reads and/or mate-paired reads might be first used to join adjacent contigs together (a step conventionally called *scaffolding*, and which produces *scaffolds*). Then, a genetic map might be used to assign as many scaffolds as possible to chromosome-associated linkage groups with an approximate order—a rudimentary genome structure. Finally, Hi-C may be used to link as many unassigned scaffolds as possible to the chromosomes, insert them into the rudimentary genome at a reasonable position, and then to optimise the order as a whole, producing a reference quality genome.

It is important to note that the results of hierarchical assembly steps differ so much between assemblies that words like *contig* and *scaffold* only have very general definitions (Hunt et al. 2014). The evolution of sequence throughout a pipeline can progress in many,

many steps, the results of which produce a new set of sequence entities by modifying the previous set. A plethora of terms and descriptors like *unitig*, *haplotig*, *super-/ultra-contig*, *super-scaffold, polished contig, seed read,* etc. are used to distinguish various such entities *within* a pipeline description, but are not used consistently *across* pipeline descriptions. As such, comparative statements like 'the average super-scaffold lengths differ by …' are misleading since the definition of a super-scaffold varies.

7.2.1 The IRGSC 'Lo7' Assembly

7.2.2 Methodology

The IRGSC 'Lo7' assembly (Rabanus-Wallace et al. 2021) relied on the effective integration of a broad suite of modern technologies (Figs. 7.2 and 7.3). Contigging and scaffolding (Fig. 7.3, step 1) were possible to greater effect than in previous efforts owing to (i) the development of PCR-free shotgun library construction(Aird 2011), which produces short-read data with a lower consensus error rate and a more even distribution over the original genome, and (ii) the availability of 10X Genomics Chromium™ molecule-linked read sequencing (www.10xgenomics.com (Ott et al. 2018). Molecule-linked reads (Fig. 7.2) are short reads derived from a collection of long DNA fragments, isolated and processed in such a way as to barcode the short reads with index sequences unique to their molecule of origin. Short reads bearing identical barcodes can be inferred to lie within a few 10 s or 100 s of Kbp from each other in the genome. They are therefore an effective source of medium-range information that can be affordably produced in high volume. This information was integrated

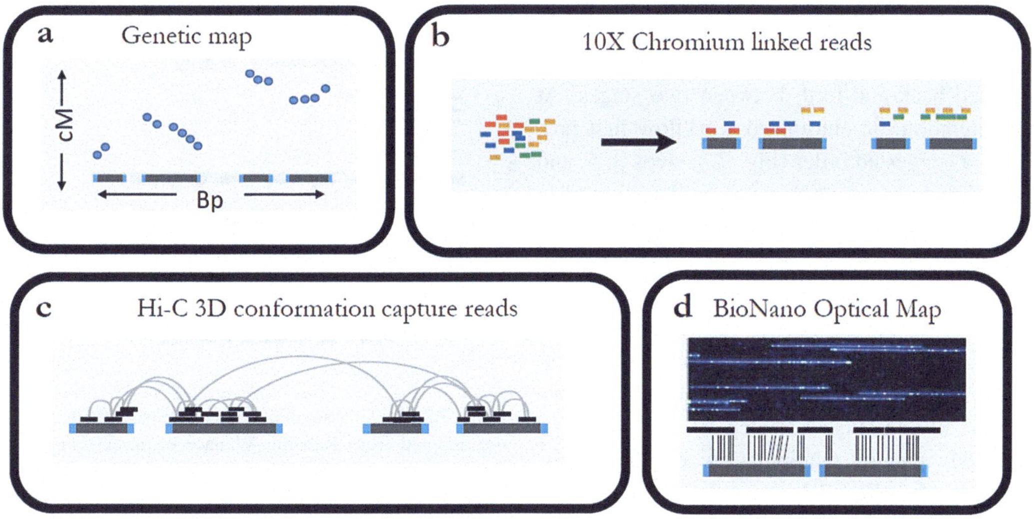

Fig. 7.2 Schematic representation of longer-range technologies used to inform scaffolding in the IRGSC 'Lo7' assembly. Grey bars represent scaffold sequences, which are typically ended by some irresolvable repeated sequence (blue tips). **a** The relative order and orientation of scaffolds containing map markers can be achieved by orienting scaffolds such that the genetic positions of the markers are continually increasing (or decreasing) across the pseudochromosomes. **b** Linked read groups (coloured) are expected to correspond to a sensible arrangement based on the span of their molecules of origin, and can hence be used by the scaffolding pipeline to infer order. **c** Hi-C links are more frequent between nearby loci, and can hence be used to order scaffolds over long ranges. **d** Optical maps (produced by aligning DNA molecules labelled at restriction sites—top inset) can be subsequently aligned to in silico labelled restriction sites on the scaffolds (bottom inset), allowing them to be ordered, and also confirming the internal correctness of the scaffolds

into a commercial scaffolding pipeline implemented by NRGene Inc. (www.nrgene.com), to produce longer scaffolds than those produced in previous assemblies.

A secondary advantage of molecule-linked reads is their usefulness in identifying *chimeric scaffolds*—scaffolds that are the result of a mis-join between two sequences that are not truly contiguous (Fig. 7.3, steps 2 & 4). Such scaffolds are inevitably present in most initial assemblies. Since molecule-linked reads can be used to establish an effective coverage of molecules over the genome based on the regions that they span, and since molecules are not expected to span the mis-joined breakpoints in a chimeric scaffold, a drastic drop in inferred molecule coverage can be used to infer—and to break—a chimera. In fact, several technologies (including linked reads, Hi-C, and optical maps) were all used to help spot and break chimeric scaffolds, but we defer methodological details until we explain how these technologies work below.

The arrangement of scaffolds into pseudochromosomes was achieved iteratively with a repeated back-and-forth between two stages. At each iteration, an automated workflow first produced a suggested order (Fig. 7.3, steps 3, 5, and 7), and this order was then manually adjusted in some way to progress towards an acceptable final optimum that best satisfied all the information available. The first set of manual adjustments involved breaking further chimeras (Fig. 7.3, step 4), the second involved joining unambiguously neighbouring scaffolds together to form super-scaffolds (Fig. 7.3, step 6), and the final involved manual manipulation of the super-scaffold order (Fig. 7.3, step 8.)

The automated order-suggestion steps made use of CSS data alongside a proxy genetic map (Fig. 7.2)—in fact simply the high-density-linkage-map-anchored contigs of the 2017 rye genome draft assembly (Bauer et al. 2017)—to associate the scaffolds with chromosomes and assign many of them an approximate relative position along those chromosomes. This 'backbone' was then used to constrain a Hi-C-based algorithm to produce a suggested scaffold arrangement for pseudochromosomes. Hi-C, briefly, is a sequencing library preparation method that provides long-range information about the relative distance between loci (Fig. 7.2). The data is generated by inducing

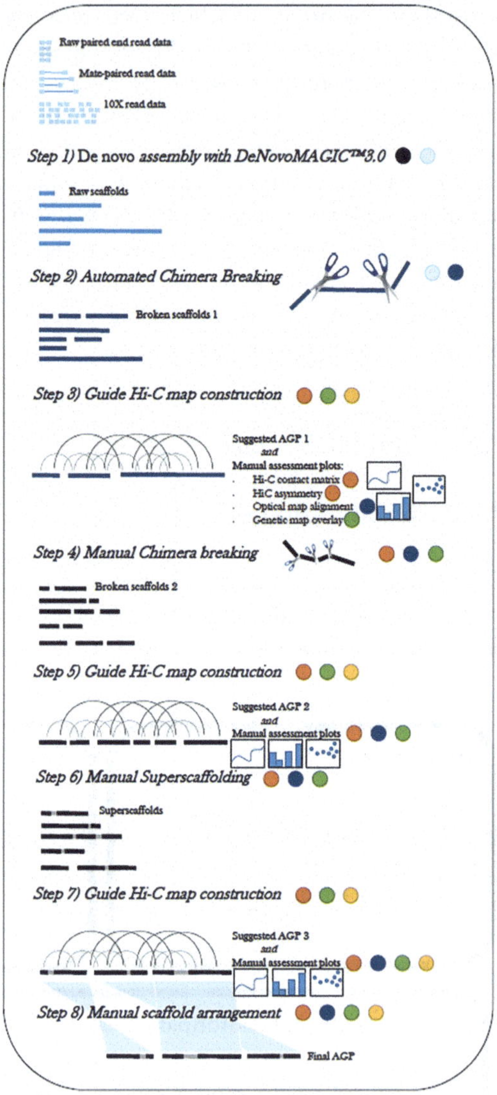

Fig. 7.3 Lo7 assembly procedure overview. Figure adapted from Rabanus-Wallace et al. (2021). Steps 1 to 8 are described in detail in the text. The term AGP ("A Golden Path") refers to a particular suggested scaffold order. Coloured circles represent data sources used in each of the steps: Black = Mate-paired and paired end read data; Pale blue = 10X Chromium linked reads; Orange = Hi-C data; Dark blue = BioNano optical map alignment; Green = Genetic map positions lifted from Bauer et al. (2017) assembly contigs. Yellow = CSS reads

chromatin links between nearby chromosomal regions, then fragmenting and re-joining the DNA, thus encouraging the formation of chimeric fragments consisting of two sequences from proximal loci joined by a recognisable linker sequence. The probability of two loci ending up in such a construct increases with proximity. Owing to this fact, the frequency of Hi-C read pairs that map to two respective loci becomes a proxy for their physical separation. An algorithm based on that published by Burton et al. (2013) and implemented in the public domain *TRITEX* pipeline(Monat 2019) represents the link frequencies between genomic bins as edge weights in a graph, and calculates a minimum spanning tree, the longest path through which becomes the basis for the suggested scaffold order. The influence of the genetic map is integrated into this procedure by removing graph edges that suggest disagreements with the map. Some more ad hoc methods are then applied to insert unincluded scaffolds into the path, and to optimise the final order by systematically trying local permutations of the order. Finally, the orientation of scaffolds containing multiple genomic bins is guessed by regressing the positions of the bins on the focal scaffold against the frequency of links between the bin and the bins of neighbouring scaffolds.

The manual scaffold order adjustment workflow played a defining role in the IRGSC rye genome assembly. The procedure is performed iteratively, beginning with the order suggested by the automated workflow described above. Each iteration consists firstly of producing a suite of custom visualisations that show the juxtaposition of several data sources against the suggested order from the previous iteration, such that they suggest where the assembly might be improved, and secondly making small adjustments to the order and orientation of problematic regions. Approximately twelve iterations were performed to attain an optimal result. A great deal of data is used to inform the adjustments (see Fig. 7.2), chiefly:

- The relative read depths of CSS reads, which flag instances in which the chromosome assignment may be suspect, or even where chimeric scaffolds composed of sequence from multiple chromosomes have been formed. These chimeras can be broken and their constituent parts re-placed in the pseudochromosomes.
- The juxtaposition of the genetic map marker genetic positions against their positions on the pseudochromosome. In a perfect assembly (and assuming a perfect map), these should be ordered monotonically.
- The alignment of optical maps against the pseudochromosomes (Fig. 7.4). Optical maps, such as produced for the 'Lo7' assembly on the BioNano Irys™ platform (bionanogenomics.com), are based on high-resolution photography of individual DNA molecules as they pass through a nanochannel array. The molecules are fluorescently labelled at restriction sites, and by aligning the molecules on this basis, Mbp-scale maps of the restriction site landscape can be assembled. Aligning these maps to the scaffolds can both confirm the contiguity of scaffold sequences, and demonstrate the arrangement of multiple scaffolds relative to each other (whenever a single optical map spans multiple scaffolds). Optical map alignments were also used to break possible chimeric scaffolds, which are frequently evident whenever the alignment suggests contradictory arrangements of sequence.
- Visualisations of the Hi-C link frequency data. Three such visualisations were used. A simple heatmap representing the matrix of link frequencies between bins along the chromosome is sufficient to identify large-scale misplacements or orientation errors. An asymmetry plot, which records differences in frequency between links extending left and right of each bin, is especially useful for identifying misoriented scaffolds: The ratio is expected to remain approximately constant near unity in a perfect assembly, but produces marked diagonals across any scaffold that is oriented the wrong way compared with its neighbours (Fig. 7.5). An intra-scaffold link count plot simply shows the number of links leading

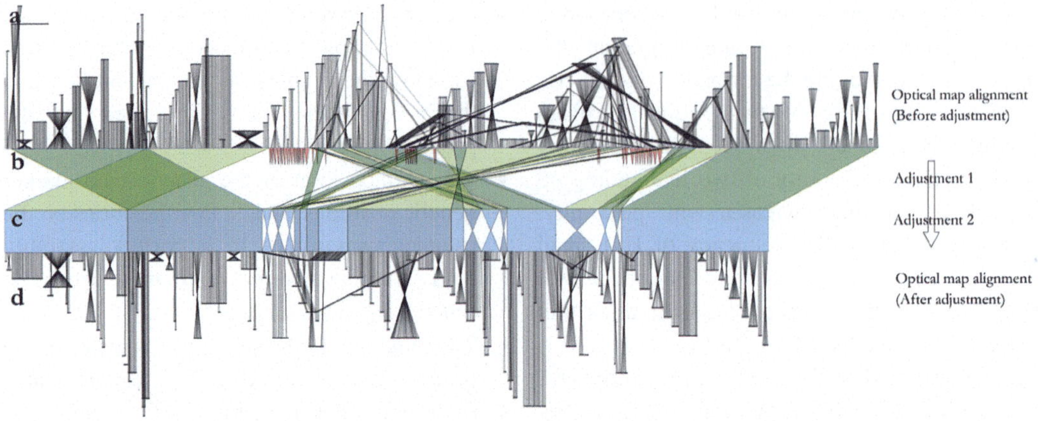

Fig. 7.4 A real example of iterative order optimisation during manual editing of the 'Lo7' alignment. The figure depicts the ordering of scaffolds in a region spanning approximately 70 Mbp around the centromere of chromosome 7R, demonstrating how the scaffold arrangements were changed during two rounds of manual editing (top to bottom), and how this is reflected in the optical map alignments. At the top of the figure **a** are drawn the optical map alignments to the scaffolds after automated ordering, but before manual adjustment; Optical maps are represented by horizontal bars, and alignments of restriction sites to the scaffolds are shown with grey lines. Improvements to the order will result in an increase in the number of optical maps that align along their entire lengths to a single contiguous stretch of the assembly. The order at the top of **b** represents an order suggested automatically by an algorithm, based on Hi-C and genetic map data. The middle level (joined by green blocks, **b**) shows the order after several rounds of manual editing, and the lower level (joined by blue blocks, **c**) shows the order after a final fine-tuning round. Red triangles correspond to scaffolds moved to regions outside of the one depicted. As can be seen in the lower part of the figure, many optical maps contain alignments to multiple non-contiguous parts of the assembly. After manual editing, only a few such instances remain. These may be caused by highly conserved repeated sequences, or they may be artefacts of residual assembly errors. The region was chosen because it represented the most difficult to resolve in the genome. Figure adapted from Rabanus-Wallace et al. (2021)

from each bin to any other bin on the same scaffold. Since Hi-C links are expected to occur between nearby loci on the same scaffold, a sudden drop in intra-scaffold Hi-C links suggests a chimeric breakpoint that should be broken.

7.2.3 Results and First Findings

The final IRGSC assembly contains 6.67 Gbp of sequence, representing approximately 85% of the estimated total genome size, of which about 6.25 Gbp was validated by alignment to optical maps, and 6.21 Gbp was arranged into pseudochromosomes. The genome release includes the suite of validation visualisations described above, which suggest a high degree of contiguity. Analysis of the expected-vs-realised kmer frequency profile in the assembly (where the expected value is calculated from the shotgun sequenced reads) suggests the assembly is very near complete and, in agreement with the observation that the chromosome-unassigned scaffolds are highly enriched in repetitive DNA, suggests the 'missing' portion is mostly accounted for in repeated sequences which are already represented in the assembly, but not at their true multiplicity. The genome was annotated for a range of features including genes, miRNAs and their targets, microsatellites, and transposable elements (Rabanus-Wallace et al. 2021, see also Chap. 8). The gene annotation allowed assessment of the assembly quality using the *BUSCO* (Benchmarking Using Single-Copy

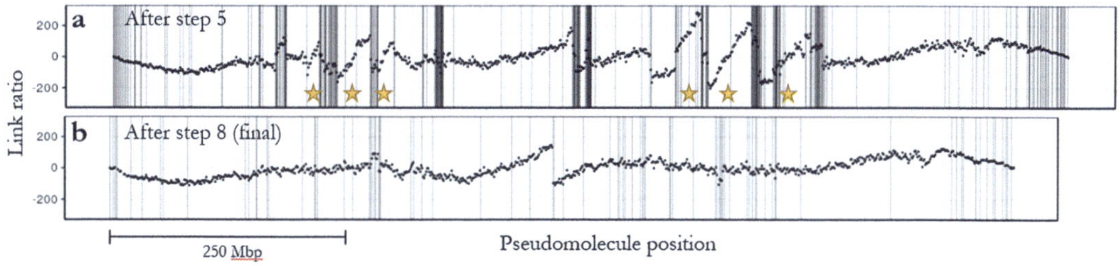

Fig. 7.5 A real example of iterative order optimisation during manual editing of the 'Lo7' alignment. The Hi-C asymmetry plots are shown for the whole of chromosome 7R at two different steps during manual editing. Vertical grey lines mark the scaffold boundaries. The upper stage **a** displays many discontinuities consistent with misplaced and misoriented scaffolds. Note in particular the distinct diagonals that indicate regions incorrectly oriented (examples highlighted with stars), and which are removed once the scaffold orientations are manually adjusted (lower; **b**). Figure adapted from Rabanus-Wallace et al. (2021)

Orthologs; busco.ezlab.org) method, which tests the completeness of the genome assembly by assaying for the presence (or absence) of a set of genes known to have single-copy orthologs in almost all species of a given group—*Viridiplantae* in this case, and achieved a score of 96.4%. The transposable element annotation allowed a comparative test of genome assembly quality: for highly complete Triticeae genomes, a strong linear relationship exists between the number of annotated full-length LTR retrotransposons, and the genome assembly size (Fig. 7.6). The 'Lo7' assembly meets this expectation.

The IRGSC rye assembly's release publication functions as a showcase of exploitation options, including as the basis for the statistical investigation of the agronomic effects of rye introgressions into wheat at unprecedented scales, the discovery of possible disease resistance genes, and analysis of the evolutionary relationships among rye species and varieties. Investigations using Hi-C identified genome structural variation between rye varieties, showing that the pericentromeric regions are prone to recombination-suppressing structural changes that may act to limit gene flow and initiate speciation. An analysis of the transposable element landscape in the 'Lo7' genome revealed fascinating intra-genomic niche partitioning by an active community of transposons (see Chap. 8).

7.2.4 The Henan Agricultural University 'Weining' Assembly

7.2.4.1 Methodology

The Henan Agricultural University 'Weining' assembly (Li et al. 2021) relied primarily on recent improvements in long-read sequence data generation, and—like the 'Lo7' assembly—on Hi-C scaffold arrangement technologies (Fig. 7.7).

Despite the comparatively high error rate (11–15%) of raw long-read sequence data compared to short-read sequencing alternatives, highly contiguous long reads are particularly useful for assembling complex genomes and are currently coming to dominate the field. The 'Weining' assembly used long reads with an average read length of nearly 8000 bp, generated using the PacBio Sequel I sequencing platform.[i] Since sequence errors are distributed randomly in the reads, errors can be corrected using consensus among the reads, provided sufficient sequencing coverage. This can be implemented at the read level (which was done in this case using the software tool Canu, https://github.com/marbl/canu; see Fig. 7.7), but the same principle can make even further corrections after the contigging/scaffolding steps, by aligning reads back to the assembled sequence and applying a similar consensus rule.

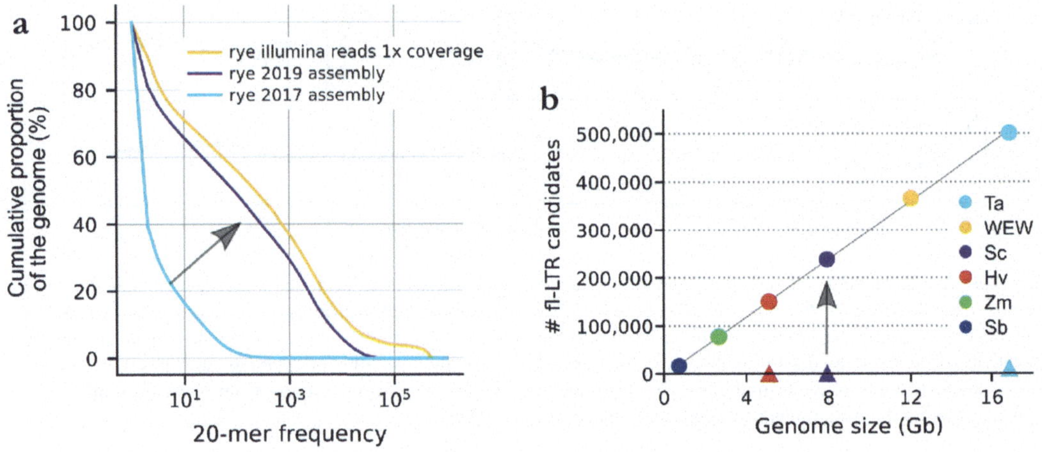

Fig. 7.6 Completeness assessment metrics for the 'Lo7' assembly. **a** A cumulative 20-mer frequency distribution is established from the reads (taken to be a near-random sample of the genome), is compared to the same distribution in the assembled sequence. The two are expected to become more identical as the assembled sequence becomes a more complete representation of the true genome. The 2021 IWGSC 'Lo7' assembly (Rabanus-Wallace et al. 2021) is compared to the 2017 draft assembly (Bauer et al. 2017). **b** More advanced Triticeae assemblies exhibit a linear correlation between the number of annotated candidate fl-LTR retrotransposons and the size of the assembly. The IRGSC 'Lo7' assembly falls directly in line with this relationship, indicating comparable quality to the most complete Triticeae genomes to date. Figure adapted from Rabanus-Wallace et al. (2021)

After read sequence correction, contig generation was performed using multiple public domain long-read-compatible assembly pipelines (WTDGB https://github.com/ruanjue/wtdbg; MECAT, https://github.com/xiaochuanle/MECAT; FALCON, https://github.com/PacificBiosciences/FALCON). While these pipelines all adhere to the generalised graph resolution methodology described previously in this chapter, differences between the algorithms and heuristics they employ lead to more or less subtle differences in the contigs they produce. To generate the most highly contiguous and integral possible set, an ensemble approach was implemented, wherein the contig sets of multiple assemblers were merged to gain a joint consensus contig set.

A secondary round of sequence correction was conducted on the contigs by aligning short-read data from the Illumina sequencing platform to the consensus contig set. Short-read data is desirable for this purpose since it is both highly accurate at the sequence level and can be generated in extreme quantities allowing high sequence coverage. At this stage, Hi-C-based detection and breaking of chimeric contigs were performed using similar methods to those described for 'Lo7'.

Contrary to the case with the 'Lo7' assembly, the resulting contig set was of sufficient length to allow arrangement of the scaffolds into seven full pseudochromosomes using the established pipeline LACHESIS, which employs a similar methodology to that described for 'Lo7' scaffold arrangement. Heatmap visualisation of the Hi-C link frequency was also used for detecting deficiencies in the suggested arrangement—though ultimately only orientation errors required manual correction. A genetic map was constructed specifically for the project from 295 F2 individuals derived from 'Weining' and 'Jingzhou' varieties. This map both provided validation for the arrangement of scaffolds and also allowed each of the seven pseudochromosomes to be assigned their correct chromosome numbers.

7.2.5 Results and First Findings

In total, the assembled genome sequence was 7.74 Gb in length, representing 98.47% of the

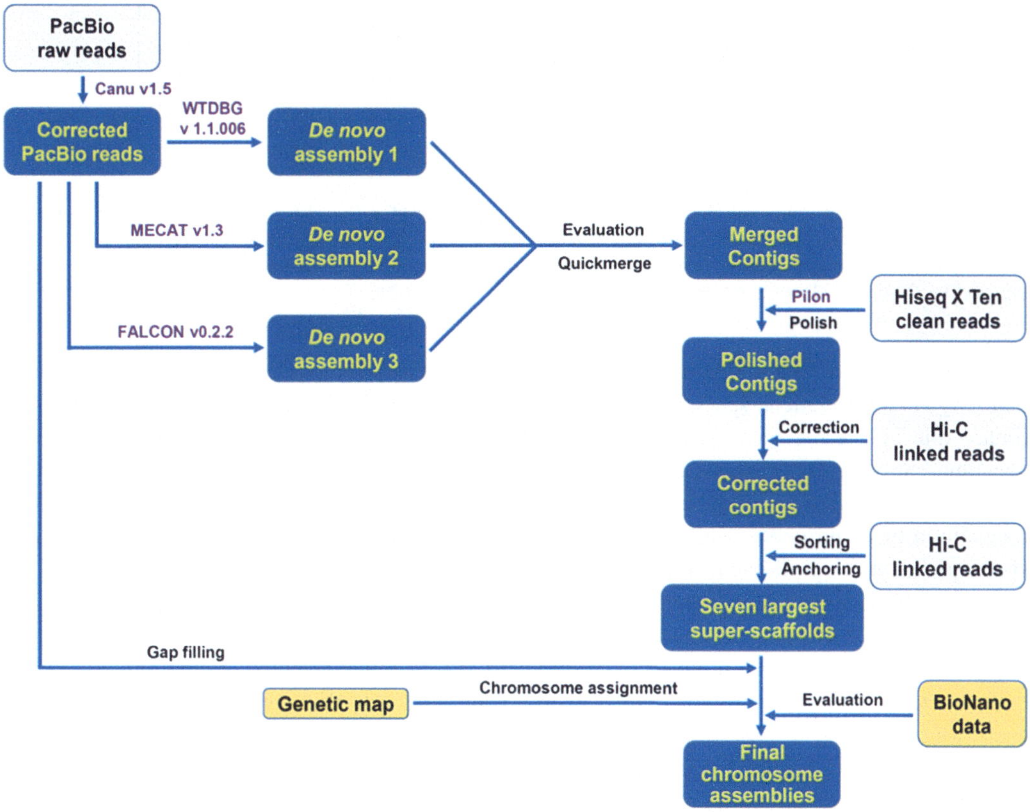

Fig. 7.7 Pipeline for constructing the seven pseudochromosomes of 'Weining' rye. See text. Figure sourced from Li et al. (2021)

estimated genome size of 'Weining' rye, of which 7.25 Gb (93.67% of the full genome) was anchored to seven pseudochromosomes. Several validation methods confirm the high quality of the assembly. A comparison of the pseudochromosome sequences to the linkage map used for the 'Lo7' assembly revealed exceptional correlation (Spearman's rank correlation coefficient 0.99). Mapping short reads back to the genome suggested a per-base sequence accuracy of 99.99%. The genome was annotated for gene features and transposable elements. BUSCO analysis revealed 98% presence of the Viridiplantae test gene set. As with the 'Lo7' assembly, a statistic based on the completeness of the transposable elements annotated in the genome was generated to allow comparative quality assessment with a 'gold standard' genome, the Japonica rice assembly MSUv7.0 (Kawahara 2013). The LTR Assembly Index (LAI) (Ou et al. 2018) of MSUv7.0 is 21.20 and 'Weining' rye achieved a comparable score of 18.42 (Fig. 7.8). The LAI statistic is unfortunately unstable when used on genome sequences with shorter scaffolds, and so was not applicable to 'Lo7'.

7.2.6 First Findings from the Weining Rye Reference Genome

The 'Weining' assembly was, at the time of publication, also used as the basis for a range of investigations. These included an analysis of the transposable element landscape which revealed, in agreement with the findings for 'Lo7', that different groups of elements specialise in distinctive genomic niches, and appear to have

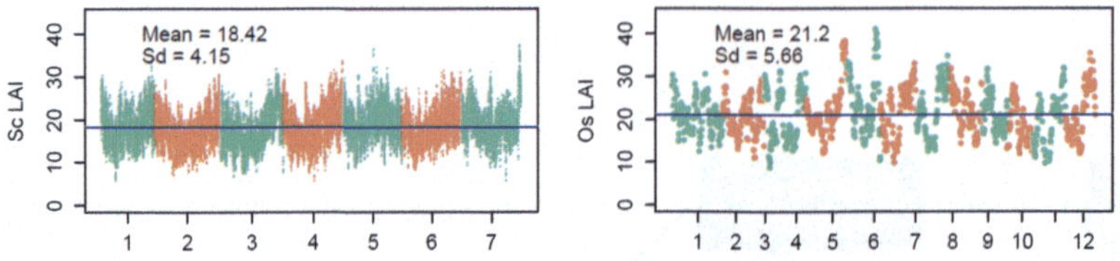

Fig. 7.8 Evaluation of genome assemblies by LTR Assembly Index (LAI). The *x*-axes show the chromosomes of each genome. LAI scores, represented by dots, were calculated using 3 Mb-sliding windows with 300-kb steps. The blue line indicates the whole genome average LAI score. Sc = Secale cereale ('Weining' rye); Os, O. sativa ssp. japonica ('Nipponbare'). Figure sourced from Li et al. (2021)

expanded at different times, suggesting TE population dynamics and niche specialisation play a primary role in altering gross genome structure over time. Specifically, Gypsy elements were shown to have undergone a recent expansion, especially in the centromere, while recent Copia expansion was more prominent in the interstitial regions. The release publication also investigated the associations between TEs and the genic space, showing that the short TEs SIEN and Harbinger tended to favour regions adjacent to genes, while the longer Gypsy and Copia elements displayed the opposite trend. Mariner and Gypsy/Copia elements in particular were shown to have a distinct affinity for short and long introns, respectively. The researchers also explored the tendency of particular functional classes of genes to undergo duplication, resulting in multiple functional paralogs. Notably susceptible gene families included those implicated in monooxygenase and oxidoreductase activities (functioning in antioxidant defence) and glucosidase/monosaccharidase activities (likely involved in the changes of cellular carbohydrates in responses to abiotic stresses such as temperature extremes).

7.3 Comparing Assemblies

Both the 'Lo7' and 'Weining' rye genome assemblies are of high quality, aptly passing standard quality checks for contemporary genome assemblies. Until these genomes are supplanted by even higher quality assemblies, the 'Weining' assembly, having benefited from long-read technology, should be preferred for analyses where assembly completeness is of paramount importance, but the true power of these two assemblies is to operate in tandem. Comparing both assemblies allows a first look at the genomic variability within rye, providing the first step into pan-genome analysis. We conducted two preliminary assays to allow us to compare the assemblies. Alignment of the genes in one genome to their best matches in the other allows us to see that the assemblies are highly collinear (Fig. 7.9), a further testimony to their accuracy, since the assemblies were entirely independently constructed. Curvature or breaks in the collinearity runs can indicate where a region in one assembly contains more or less sequence than the other, but the collinearity runs are more or less straight suggesting the length difference between the genomes are not especially frequent in any particular region. A few minor collinearity breaks around centromeres could equally represent assembly errors in one or both assemblies (since centromeric sequence is notoriously difficult to assemble), or structural rearrangements at the centromere (which is highly prone to local rearrangements). However, this confusion can be partially resolved by leveraging a Hi-C-based approach used in the 'Lo7' release publication, that allows structural variation between closely related varieties to be detected, by applying the Hi-C data generated from one variety to the genome assembly of another.[ii] The results suggest at a minimum that a large rearrangement near 300 Mb on chromosome 1R, and several large inversions surrounding the centromere of chromosome 2R may be genuine rearrangements and not assembly artefacts (Fig. 7.10). These are in fact expected in some degree, since the highly

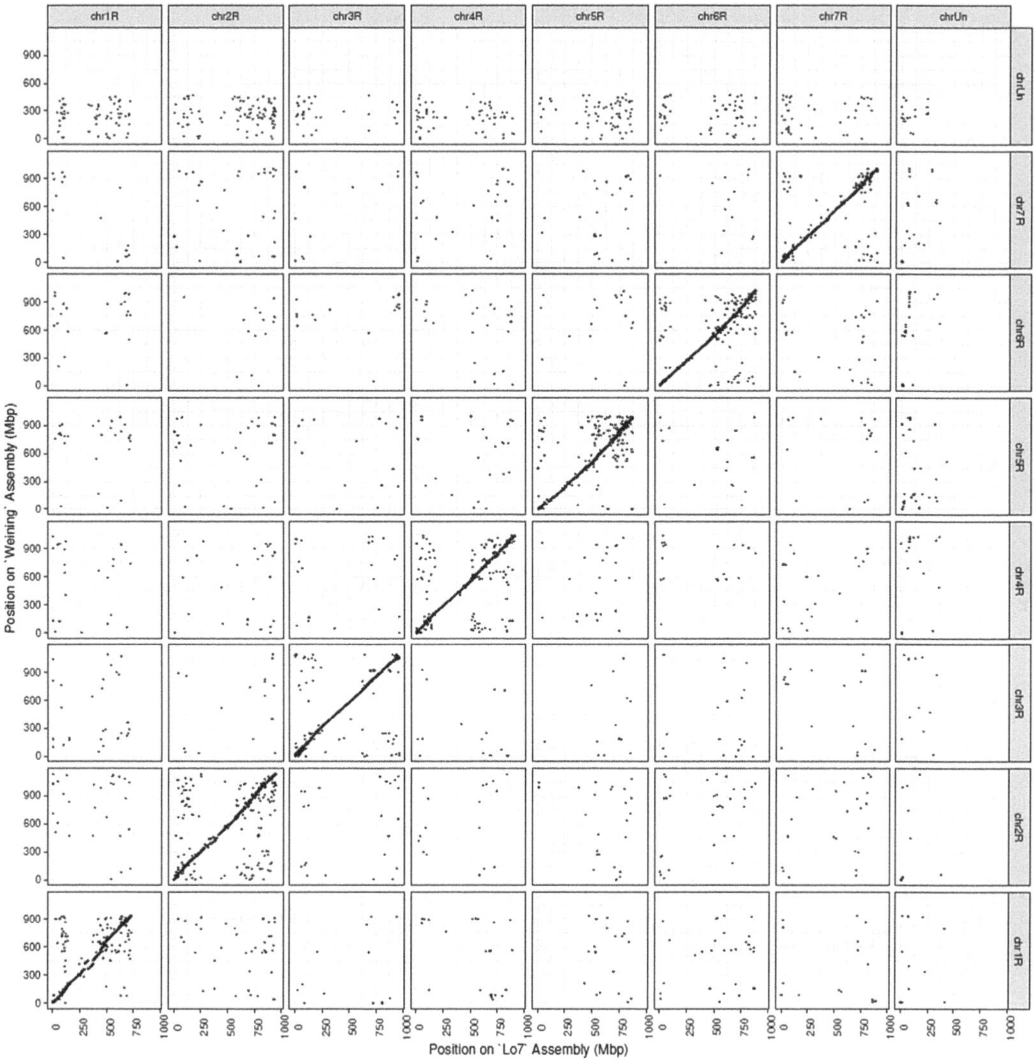

Fig. 7.9 Annotated gene-space collinearity between the 'Lo7' and 'Weining' rye assemblies. Points refer to annotated complementary DNA sequences (CDSs) from each assembly that reciprocally match each other better than any other candidate.[iii] 'ChrUn' represents the 'unknown chromosome', to which scaffolds which could not be placed in pseudochromosomes are relegated, in arbitrary order

reticulate evolutionary history of rye is thought to have contributed to a highly mosaicised genome structure, including frequent introgression and translocation events (Martis 2013; Rabanus-Wallace et al. 2021). Viewing genome collinearity at a finer, megabase-level scale (e.g. Figure 7.11) reveals the extensive intergenic rearrangements typical of Triticeae genomes, even at the intraspecific level. The effects of such rearrangements, especially on recombination among breeding lines, will be a fruitful area of future research.

7.4 Future Steps in Rye Genome Sequencing

The rate of technological improvement in the fields of DNA sequencing and genome assembly promises many more rye genome assemblies in

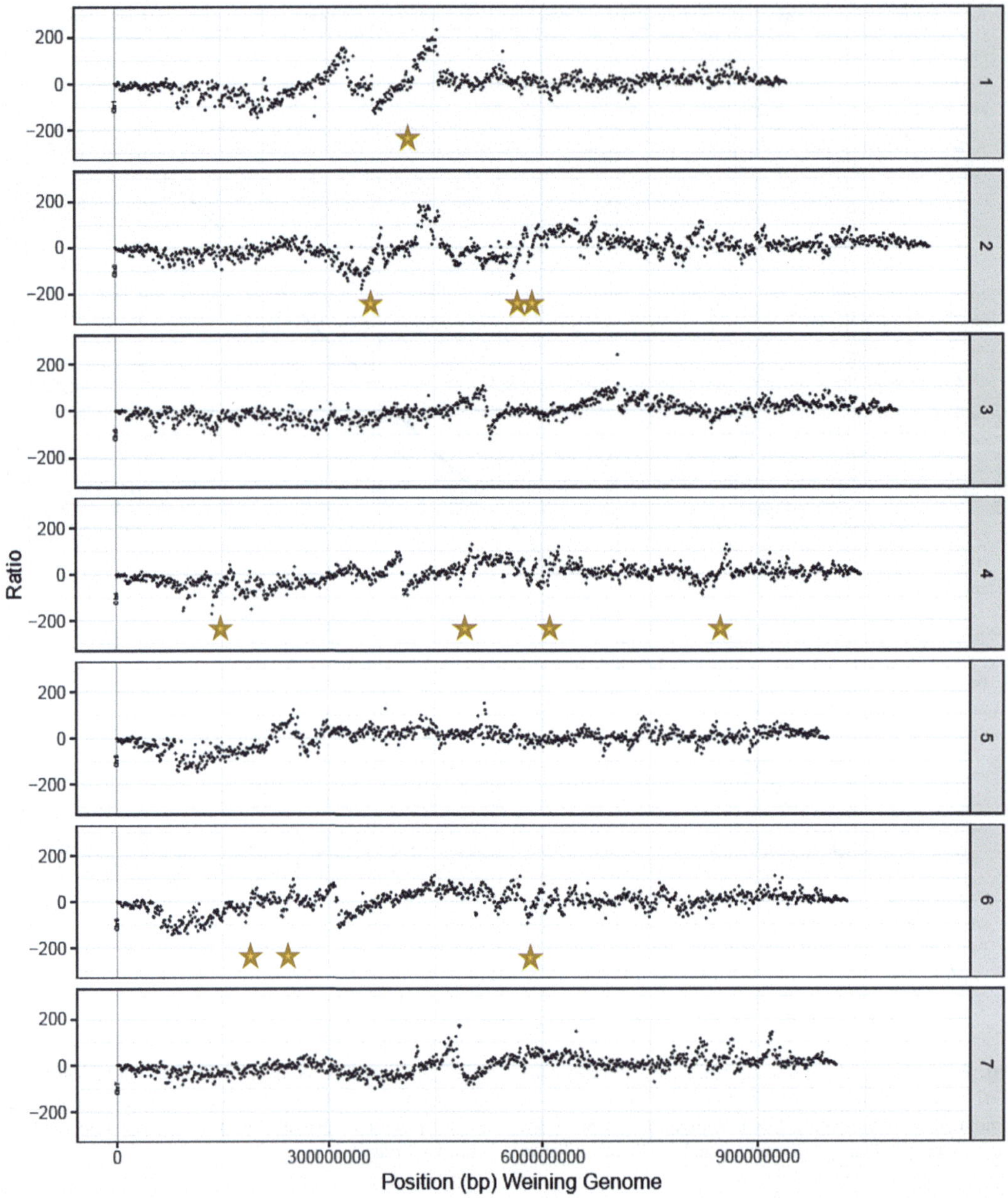

Fig. 7.10 Hi-C-based structural comparison of 'Lo7' and 'Weining' genomes, based on Hi-C link asymmetry (expressed as a left-to-right normalised link count ratio) calculated in 1 Mb bins across the 'Weining' genome sequence.[iv] Large inversions manifest as distinct local diagonals. Examples of candidate inversions are indicated with stars

the near future. Long-read sequencing and longer optical maps will likely be key among these near-term improvements, as has been demonstrated in the case of barley for which a genome assembled using high fidelity long reads has recently been published (Mascher et al. 2021). A natural next step for rye genome sequencing will be pangenomics: comparative genomics involving large numbers of fully assembled genomes (Walkowiak 2020; Jayakodi 2020; Golicz 2016), which

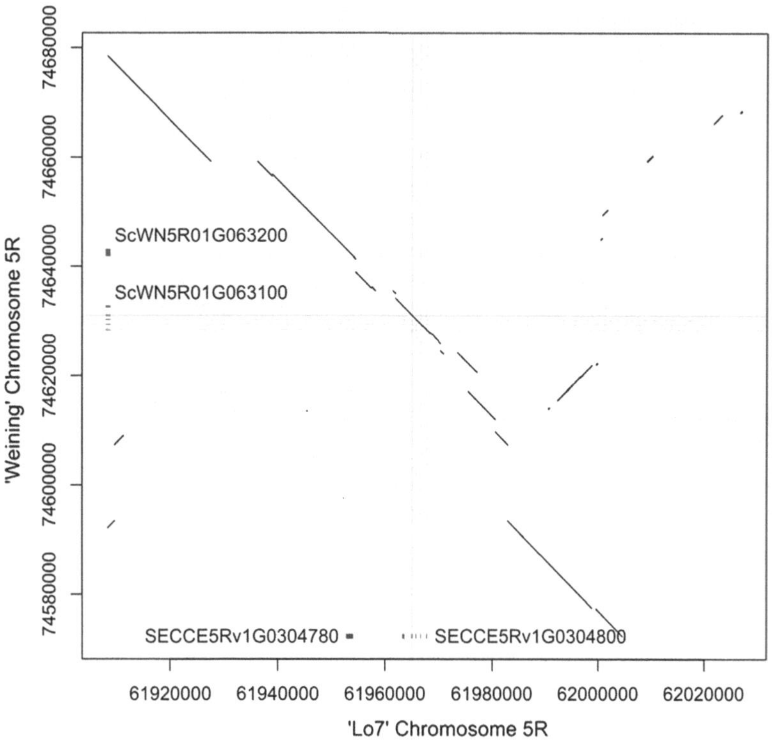

Fig. 7.11 Genomic alignment of homologous 5R regions on 'Lo7' and 'Weining' assemblies, showing extensive rearrangement of the intergenic space.[v] The region includes an annotated leucine zipper transcription factor (SECCE5Rv1G0304800/ ScWN5R01G063100), implicated in cold acclimation, and discussed in the 'Lo7' release publication (Rabanus-Wallace et al. 2021). Exons are shown with blue ('Lo7' and red 'Weining') bars

allows an even deeper understanding of how genomic variation is distributed within a species. Other directions will involve full sequencing of the B chromosomes that are present in many rye lines (see Chap. 4), and the challenge of distinguishing the B chromosomal sequence from the regular autosomes will likely depend upon proximity technologies including Hi-C.

Notes

i. Refers to Circular Consensus Sequence (CCS) subreads.

ii. *DpnII*-digested Hi-C data from the 'Lo7' genome release publication were aligned to the Henan 'Weining' assembly and processed using the run_hic_mapping.zsh wrapper script from the TRITEX pipeline (https://bitbucket.org/tritexassembly/tritexassembly.bitbucket.io/src/master/; accessed December 2020). Asymmetry plots were produced using the Hi-C R functions of the same pipeline.

iii. CDS transcript sequences from Li et al. (2021) and Rabanus-Wallace et al. (2021). Bi-directional alignments conducted using BLASTn (v2.9.0+), minimum match length 1000 bp, maximum reported e-value 0, with any secondary hits retained required to be at least 400 bp shorter than the primary match. Data processing and plotting using R (https://www.r-project.org/).

iv. The logic of the same-species Hi-C link asymmetry analysis (used for scaffold order optimisation), and cross-species Hi-C link asymmetry analysis (used for rearrangement detection) are very similar. In the same-species case, if the scaffold order is optimised, no notable asymmetry disruptions will be observed. If Hi-C data from a completely collinear sister species is mapped, the observation should be almost the same—no asymmetry disruptions. If the sister species' genome is not collinear, however, the results for link asymmetry will be similar to a scaffold misarrangement in the same-species case, i.e. it will cause the new introduction of marked asymmetry disruptions. For methodological details see Himmelbach et al. (2018).

v. Annotations and sequences extracted from Li et al. (2021; Chr5R: 74,572,362-74,698,872) and Rabanus-Wallace et al. (2021; Chr5R:61,908,294- 62,034,804) using bedtools (v2.26.0) getfasta. Alignment made with LASTZ (v1.04.03; arguments '–gfextend –chain –gapped –format = blastn –rdotplot = [output file]'). Data processing and plotting using R (https://www.r-project.org/).

References

Aird D et al (2011) Analyzing and minimizing PCR amplification bias in illumina sequencing Libraries. Genome Biol 12:1–14

Alonge M, Shumate A, Puiu D, Zimin AV, Salzberg SLJG (2020) Chromosome-scale assembly of the bread wheat genome reveals thousands of additional gene copies. Genetics 216:599–608

Amarasinghe SL et al (2020) Opportunities and challenges in long-read sequencing data analysis. Genome Biol 21:1–16

Bauer E et al (2017) Towards a whole-genome sequence for rye (*Secale cereale* L.). The Plant J 89:853–869

Braun E-M et al (2019) Gene expression profiling and fine mapping identifies a gibberellin 2-Oxidase gene co-segregating with the dominant dwarfing gene *Ddw1* in rye (*Secale cereale* L.). Front Plant Sci 10:857

Burton JN et al (2013) Chromosome-scale scaffolding of *de novo* genome assemblies based on chromatin interactions. Nat Biotechnol 31:1119

Compeau PE, Pevzner PA, Tesler GJ (2011) How to apply *de Bruijn* graphs to genome assembly. Nat Biotechnol 29:987–991

Gabur I, Chawla HS, Snowdon RJ, Parkin IAJT & genetics, a. Connecting genome structural variation with complex traits in crop plants. Theor Appl Genet 132:733–750

Gaikpa DS et al (2020) Genome-wide association mapping and genomic prediction of Fusarium head blight resistance, heading stage and plant height in winter rye (*Secale cereale*). Plant Breed 139:508–520

Golicz AA et al (2016) The pangenome of an agronomically important crop plant *Brassica oleracea*. Nat Commun 7:1–8

Haseneyer G et al (2011) From RNA-seq to large-scale genotyping-genomics resources for rye (*Secale cereale* L.). BMC Plant Biol 11:131

Himmelbach A et al (2018) Discovery of multi-megabase polymorphic inversions by chromosome conformation capture sequencing in large-genome plant species. Plant J 96(6):1309–1316

Hunt M, Newbold C, Berriman M, Otto TD (2014) A comprehensive evaluation of assembly scaffolding tools. Genome Biol 15:1–15

IWGSC (2018) Shifting the limits in wheat research and breeding using a fully annotated reference genome. Science 361:eaar7191

Jayakodi M et al (2020). The barley pan-genome reveals the hidden legacy of mutation breeding, pp 1–6

Jung WJ, Seo YWJG 920190 Identification of novel C-repeat binding factor (CBF) genes in rye (*Secale cereale* L.) and expression studies. Gene 684:82–94

Kawahara Y et al (2013) Improvement of the Oryza Sativa Nipponbare reference genome using next generation sequence and optical map data. Rice 6:4

Li G et al (2021) Analysis of rye genome sequence sheds new light on its genome expansion and agronomically important genes. Nat Genet 53:574-584

Maraci Ö, Özkan H, Bilgin RJ (2018) Phylogeny and genetic structure in the genus *Secale*. PLoS One 13: e0200825

Martis MM et al (2013) Reticulate evolution of the rye genome. Plant Cell 25:3685–3698

Mascher M et al (2017) A chromosome conformation capture ordered sequence of the barley genome. Nature 544:427–433

Mascher M et al (2021) Long-read sequence assembly: a technical evaluation in barley. Plant Cell 33:1888-1906

Monat C et al (2019) TRITEX: chromosome-scale sequence assembly of Triticeae genomes with open-source tools. Genome Biol 20:284

Ott A et al (2018) Linked read technology for assembling large complex and polyploid genomes. BMC Genomics 19:651

Ou S, Chen J, Jiang N (2018) Assessing genome assembly quality using the LTR Assembly Index (LAI). Nucleic Acids Res 46:e126–e126

Rabanus-Wallace M et al (2021) Chromosome-scale genome assembly provides insights into rye biology, evolution, and agronomic potential. Nat Genet 53:564–573

Schreiber M, Himmelbach A, Börner A, Mascher M (2019) Genetic diversity and relationship between domesticated rye and its wild relatives as revealed through genotyping-by-sequencing. Evol Appl 12:66–77

Vendelbo NM, Sarup P, Orabi J, Kristensen PS, Jahoor AJ (2020) Genetic structure of a germplasm for hybrid breeding in rye (*Secale cereale* L.). PloS one. 15:e0239541

Walkowiak S et al (2020) Multiple wheat genomes reveal global variation in modern breeding, pp 1–7

Wicker T, Gundlach H, Schulman AH (2018) The repetitive landscape of the Barley Genome. In: The Barley Genome. Springer, pp 123–138

8 The Gene and Repetitive Element Landscape of the Rye Genome

Alexander V. Vershinin, Thomas Lux, Heidrun Gundlach, Evgeny A. Elisafenko, Jens Keilwagen, Klaus F. X. Mayer, and Manuel Spannagl

Abstract

In the genomics era, the rye genome has been given as little attention by researchers as no other entity in the most renowned triad of cereal crops in the tribe Triticeae: wheat, barley, and rye. Some of the reasons behind this neglection are due to the large size of rye's genome and its abundance of various classes of repetitive DNA sequences. Despite substantial progress in sequencing and assembling methods, most of its genomic regions, which are made of alternating families of tandem repeats and transposable elements, lie ahead of the undertakers like uncharted dungeons. This chapter outlines the history of research into the molecular organization of the rye genome, from insights into the structure of the repetitive DNA sequences to the recently published annotated chromosome-scale genome assembly. In contrast to previous attempts (Martis et al, Plant Cell 25:3685–3698, 2013; Bauer et al, Plant J 89:853–869, 2017), which were focused on some parts of the rye genome, ours is trying to give a holistic understanding of the genome of this economically important cereal crop. To this end, we will consider the most complete set of classes of DNA sequences and see how their molecular structure and abundance are linked to their chromosomal locations.

8.1 Introduction

The genome of the cultivated rye (*Secale cereale*, $2n = 2x = 14$) is considered one of the largest cereals genomes, totaling about 8.0 Gb/1C (Leitch et al. 2019) and thus topping the average angiosperm genome (5.6 Gb) (Rabinowicz and Bennetzen 2006) and the genomes of its closest evolutionary neighbors, barley, *Hordeum vulgare*, (5.1–5.3 Gb) and diploid wheats (5.8–6.1 Gb). Whole-genome assemblies with the annotated genes of several species in rye's taxonomic neighborhood (*H. vulgare* and *Triticum aestivum*, to name a few) have allowed it to be seen that these plants have nearly an equal

A. V. Vershinin (✉) · E. A. Elisafenko
Institute of Molecular and Cellular Biology, SB RAS, Novosibirsk, Russia
e-mail: avershin@mcb.nsc.ru

A. V. Vershinin
Novosibirsk State University, Novosibirsk, Russia

T. Lux · H. Gundlach · K. F. X. Mayer · M. Spannagl (✉)
Helmholtz Zentrum München - German Research Center for Environmental Health, Plant Genome and Systems Biology (PGSB), Neuherberg, Germany
e-mail: manuel.spannagl@helmholtz-muenchen.de

E. A. Elisafenko
Institute of Cytology and Genetics, SB RAS, Novosibirsk, Russia

J. Keilwagen
Institute for Biosafety in Plant Biotechnology, Julius Kühn-Institute, Quedlinburg, Germany

© The Author(s), under exclusive license to Springer Nature Switzerland AG 2021

M. T. Rabanus-Wallace and N. Stein (eds.), *The Rye Genome*, Compendium of Plant Genomes,
https://doi.org/10.1007/978-3-030-83383-1_8

number of coding genes. What these facts suggest first of all is that the rye's genome is larger than that of other cereals' because of a higher content of diverse classes of repetitive DNA sequences. This implication receives support from the results of the first experiments on cereal DNA reassociation kinetics, suggested that the rye genome is comprised of more than 90% of repetitive DNA (Flavell et al. 1974).

The division of the whole lot of repetitive DNAs into two major classes according to their organization—(1) dispersed repetitive sequences or transposable elements (TEs) and (2) tandem arrays consisting of similar monomers monotonously following one another—was proposed in the 1980s (Flavell 1980) and still remains to be as clear as it was. As the scientists learned more and more about the primary structure and internal organization of repetitive DNAs, more detailed classifications of large groups of these elements were proposed, by Wicker et al. (2007) for TE and by Charlesworth et al. (1994) for tandem repeats. The molecular methods of the 1980s made tandemly arranged sequences the most accessible for identification and isolation with a view towards detailed analysis—because of their high copy number. That is why this class of repeated sequences in the rye genome attracted close attention from the scientific community and the papers describing the structural organization and chromosomal locations of tandem repeat families in rye have made their way to *Cell* and *Nature* (Bedbrook et al. 1980a, b). It was demonstrated that there are families, with their elements occurring in extremely high copy numbers, making up 8–12% of the rye genome (Bedbrook et al. 1980a; Jones and Flavell 1982a). These families were found in subtelomeric regions of chromosomes and gave strong signals following *in situ* hybridization (FISH) (see Chap. 3). These signals corresponded to C-bands obtained using differential chromosome staining with Giemsa dye, which reveals heterochromatic regions. It was thus determined that tandemly arranged DNAs exist in large copy numbers and comprise the bulk of the heterochromatic regions next to the telomeres. The presence of large heterochromatic blocks in subtelomeric regions is a distinctive feature of rye not shared by its most closely related grasses, wheat, or barley. At the same time, heterochromatic blocks are larger in the genome of cultivated rye than in the wild rye species' and tandem repeats contribute more to former than to the latter (Jones and Flavell 1982b; Cuadrado and Jouve 2002).

Tandem repeats are not the only repetitive DNA sequences in the heterochromatic regions of the rye chromosomes. Extended monomeric arrays are interspersed with TEs of different classes, primarily the Class I LTR retrotransposon superfamilies *gypsy* and *copia* (Evtushenko et al. 2016). TEs represent major contributors of repeated sequences, and have no restricted chromosomal regions (which tandem repeats do have), but are scattered throughout the rye genome.

BAC (Bacterial Artificial Chromosome) libraries and sequencing methods have enabled significant progress in the analysis of the molecular structure of tandemly organized repeats and transposable elements. The heterogeneous internal organization of the arrays of tandem repeat families and the presence of higher-order units of monomers in rye has been determined using these methods (Evtushenko et al. 2016). The rye genome, like others, has been found to have regions where TEs of different families were nested into each other (Bartos et al. 2008; Evtushenko et al. 2016), forming extended clusters of alternating, rearranged patchwork-like entities. Such regions are extremely difficult to sequence based on short-read technologies. This is one of the main reasons why the rye genome was until recently neither sequenced nor assembled with sufficient completeness (see this chapter). Admittedly, attempts were made and, as a result, the linear order was determined for 72% of rye genes, 17 conserved syntonic blocks shared by the rye and barley genomes were identified (Martis et al. 2013) and almost the entire low-copy number part of the genome containing coding regions was assembled and presented (Bauer et al. 2017). However, due to the underrepresentation of various classes of repetitive DNA, these drafts gave insights into only lower copy segments of the rye genome.

In recent years, "the third revolution in sequencing technology" has been proclaimed (van Dijk et al. 2018), which is another wording for long-read sequencing technologies that include the Pacific Biosciences and Oxford Nanopore platforms. These technologies are capable of generating ultra-long reads hundreds of kilobases and even a megabase in length. Such long reads make assembling easier to do and provide strong insight into the long-range organization of repetitive genomic regions. This "third-generation sequencing" is really worth resting our hope on, as it will bring us to a much better resolution as to why the rye genome evolved to be so large and so stuffed with repeats. Another promising approach is by marrying long-read sequencing with other approved sequencing and mapping techniques—to produce the so-called "hybrid assemblies" (Weissensteiner et al. 2017).

8.2 Tandem Repeats

Tandemly repeated DNA sequences occur as a monotonous succession of monomeric units which do not possess such complex internal structure as do transposable elements and for that reason are typically classified into microsatellite, minisatellite, and satellite DNA according to the length of monomers (Charlesworth et al. 1994). Microsatellites are shorter than 10 bp, minisatellites are longer than 10 bp but shorter than 100 bp and satellites are longer than 100 bp. According to the latest chromosome-scale assembly of the rye genome (Rabanus-Wallace et al. 2019), minisatellites occur at higher abundance in the centromeric regions of chromosomes, while microsatellites are distributed more or less evenly along chromosomes. However, there is one microsatellite DNA sequence that occurs at specific chromosomal locations, namely, telomere ends and plays an important role in chromosome structure.

8.2.1 Telomeric Repeats

Attempts at dissecting the nature of DNA telomeres revealed a tandem arrangement of short monomeric units of the telomeric repeat can be consensus sequence $(d[T/A\}_{1\text{-}4}dC_{1\text{-}8})n$ across a wide range of organisms (protozoa, fungi, insects, plants, and mammals). This range includes rye and its closest relatives, barley and wheat: the DNA at the ends of their chromosomes appears as long arrays of short sequences (TTTAGGG)n. The lengths of these arrays were estimated via processing chromosome ends with a specific exonuclease, *Bal*31, and the subsequent application of pulsed-field gel electrophoresis (PFGE) (Vershinin and Heslop-Harrison 1998). Rye telomere repeat arrays vary from 8 to 50 kb in length and are therefore much shorter than their counterparts in barley and especially wheat, in which the longest telomeres are longer than 100 kb. A substantial intraspecific range of the lengths of telomere repeat arrays suggests that telomeres are extremely variable in length, both across chromosomes and between chromosome arms, which is a shared feature of chromosomes in many species.

An intriguing question is how short conserved nucleotide sequences can be that are composed of monomeric units of the telomeric repeat, implicated in the most important telomere function—the maintenance of chromosome integrity. It has been established with telomeres in mammalian and some plant species' chromosomes that the DNA of the telomeric repeat is important for association with a six-member protein *shelterin* complex, which facilitates the formation of a lariat-like structure (the t-loop) to shield the exposed chromosome ends of telomeric DNA from the DNA damage machinery (de Lange 2005). The key feature of t-loops is that the end of any telomere is tucked in. In this way, telomeric DNA and telomeric proteins together preserve genome integrity. In addition, telomeric chromatin in rye chromosomes has a specific property: the internucleosomal spacing in it is 160 bp against 175–185 bp in bulk chromatin (Vershinin and Heslop-Harrison 1998).

8.2.2 Subtelomeric Repeats

Subtelomeres are chromosome regions, but they are especially difficult to define and delineate. In many organisms, telomere-specific features occur outside the telomeres. One such feature is the presence of telomere-adjacent heterochromatic regions with specific histone modifications. Other common features of subtelomeric heterochromatin are low genetic density and higher rates of evolutionary changes in subtelomeric DNA compared to bulk DNA (Young et al. 2020). It is hypothesized that in mammals telomere-adjacent DNA helps regulate telomere tract lengths because long noncoding telomeric repeat-containing RNA (TERRA) molecules are transcribed from subtelomeres into the tracts of telomeric repeats (Azzalin et al. 2007).

As was already mentioned, the ends of rye chromosomes feature large heterochromatic blocks—and it was there, where the story of research into the molecular structure of the rye genome began. Several families of highly repetitive DNA sequences have been identified in these regions (Appels et al. 1978; Bedbrook et al. 1980a, b). The molecular structure, copy number, and monomer lengths were furthermore determined for the three most abundant of them, pSc119.2, pSc200, and pSc250 (McIntyre et al. 1990; Vershinin et al. 1995). They are composed of monomeric units 118, 379, and 571 bp in length, respectively, with pSc200 contributing to ~2.5% of the genome, and pSc250 and pSc119.2 each contributing to ~1%. Fluorescence *in situ* hybridization (FISH) experiments have suggested that the pSc200 and pSc250 blocks coincide close to the telomere, while some pSc119.2 copies are located at interstitial sites. The pSc119.2 sequence is also represented in a number of other cereal genomes, but pSc200 and pSc250 are largely rye-specific.

Monomeric units of these three tandem repeat families are arranged into long arrays—longer than telomeric repeat arrays. The longest pSc250 arrays reach 600–700 kb in length, while pSc200 can even be longer (Alkhimova et al. 2004). These figures, however, may be inflated due to the introduction of sequences that are not tandem monomers, of which PFGE tells us nothing. Some families of tandemly repeated DNA sequences, such as human α-satellite DNA, are known to form higher-order repeat (HOR) units that may contain variable numbers of basic repeats (multimers) having highly similar monomer sequences. The tandem arrays pSc200 and pSc250 formed higher-order multimers, as shown by the ladder-like patterns seen in the Southern hybridization profiles of BAC clones (Evtushenko et al. 2016). The maximum lengths of these multimers appear to be ~3 kb for pSc200 (octamer) and ~3.5 kb for pSc250 (hexamer).

Despite a wealth of information regarding monomer length and sequence, knowledge about tandemly arranged repeats remains fragmentary and is usually limited to separate families. The long-range organization of monomers within arrays, the mutual arrangement of different families, and the molecular features of flanking regions between tandem arrays and neighboring non-tandem DNA remain poorly explored. The main reason for being poorly explored is their repetitive nature, which has set up a frustrating barrier to direct sequencing techniques. Therefore, it seems extremely exciting to learn how close to an answer the scrutiny of new assemblies from rye line 'Lo7' and 'Weining' genomes can bring us.

We determined the contribution of pSc119.2, pSc200, and pSc250 to the initial set of reads from the rye line 'Lo7'. The count of reads mapped onto the consensus monomer sequence in each family revealed the following figures: for pSc200, 1,125,291 copies or about 2.7% of the diploid rye genome; for pSc250, 228,121 copies or about 0.83%; and for pSc119.2, 282,256 copies or about 0.21%. Thus, we obtained good agreement with the hybridization-derived findings for the abundance of these three tandem repeat families in the rye genome.

The assemblies of the chromosome-scale genome sequence for the rye lines 'Lo7' and 'Weining' allowed for various classes of DNA sequences to be mapped, starting off at chromosomes telomeres. Localization of tandem repeats and transposable elements in the assemblies of

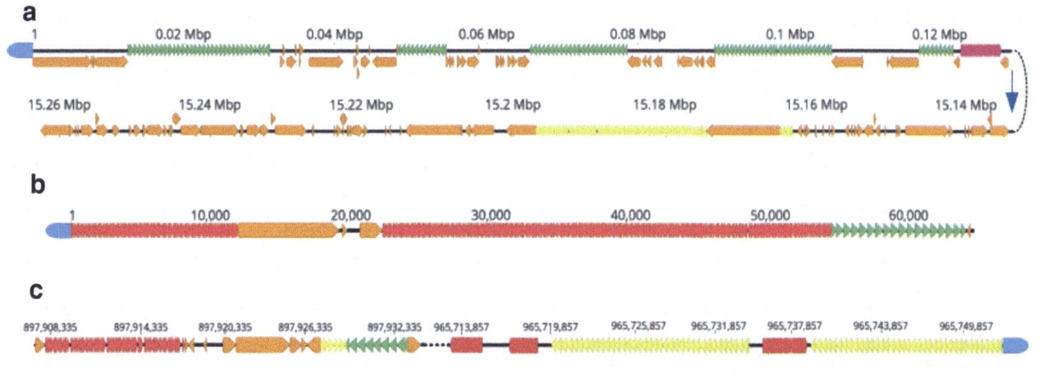

Fig. 8.1 Examples of the arrangement of tandemly repeated monomers pSc200 (yellow), pSc250 (green), and pSc119.2 (red) in the subtelomeric regions of rye chromosomes. **a** The region adjacent to the telomere (shown in blue) of chromosome 7R of the rye line 'Weining'. Fragments of transposable elements separating arrays of tandemly arranged monomers are shown in orange. **b** The region adjacent to the telomere of 'Weining' chromosome 3R has two pScl19.2 arrays separated by transposable elements and a pSc250 array adjacent to pScl19.2. **c** The subtelomeric region of one of the arms of 'Lo7' chromosome 3R, with all three tandem repeat families present. Transposable elements (orange finger-post arrows) and telomeres (blue rectangles each with a rounded side)

the rye 'Lo7' and 'Weining' genomes was carried out using the Viridiplantae RepeatMasker program and database (Smit et al. 2013) with reliance on the consensus monomer sequences. Visualization and further analysis were performed using Geneious 11.0.2 software (http://www.geneious.com) (Kearse et al. 2012). Figure 8.1 shows examples of the mutual arrangement of monomeric arrays of tandemly arranged repeat families with other classes of DNA sequences, most of which were TEs belonging to different families. Monomers of various tandem repeat families may lie immediately adjacent to telomeres, as does pSc119.2 on one arm of chromosome 3R of the line 'Weining' (Fig. 8.1b) or does pSc200 on the other arm of chromosome 3R of the line 'Lo7' (Fig. 8.1c). However, not in all chromosomes, the telomeric and subtelomeric repeat arrays lie immediately adjacent to each other. The telomere of 'Weining' chromosome 7R and the pSc250 array are separated by four fragments of the transposable elements *Xalax* and *Gypsy* 13_TA-I (Fig. 8.1a). Curiously, the pSc250 array is interspersed with various TEs. The tandem repeat arrays of all three families are present in the subtelomeric regions of both chromosomes, and their mutual arrangements are diverse. The arrays of different families can lie immediately adjacent to each other, be in various combinations, and be separated by tracts of varying length, with TE residing in. Considering the situation, we must confess that the general patterns in the formation of the molecular structure of subtelomeric, heterochromatic regions of rye chromosomes are desperately elusive.

The arrays of monomeric units shown in Fig. 8.1 are relatively short, not longer than 35 kb (pSc119.2, Fig. 8.1c). This length is not enough for HOR units to form inside the array. An example of the formation of such structures was found in unassigned scaffold s291 ('Lo7') (Fig. 8.2). Here, the pSc200 family cluster occupies 258.3 kb and consists of several long arrays separated by relatively small gaps populated by elements of the *copia*-like family WIS-2 and a short fragment of the *gypsy*-like family *Olivia*. From position 80 kb, Tandem Repeats Finder identifies not only monomers and dimers but also a succession of 13 units (mimicked in purple underneath), each consisting of five pSc200 monomers and being 1897 bp in length (Score 38,698, Percent Matches 90). As the new genome assemblies have helped revel this group is a HOR unit that resulted from unequal crossing-over, as most of the extended arrays of tandem repeats did.

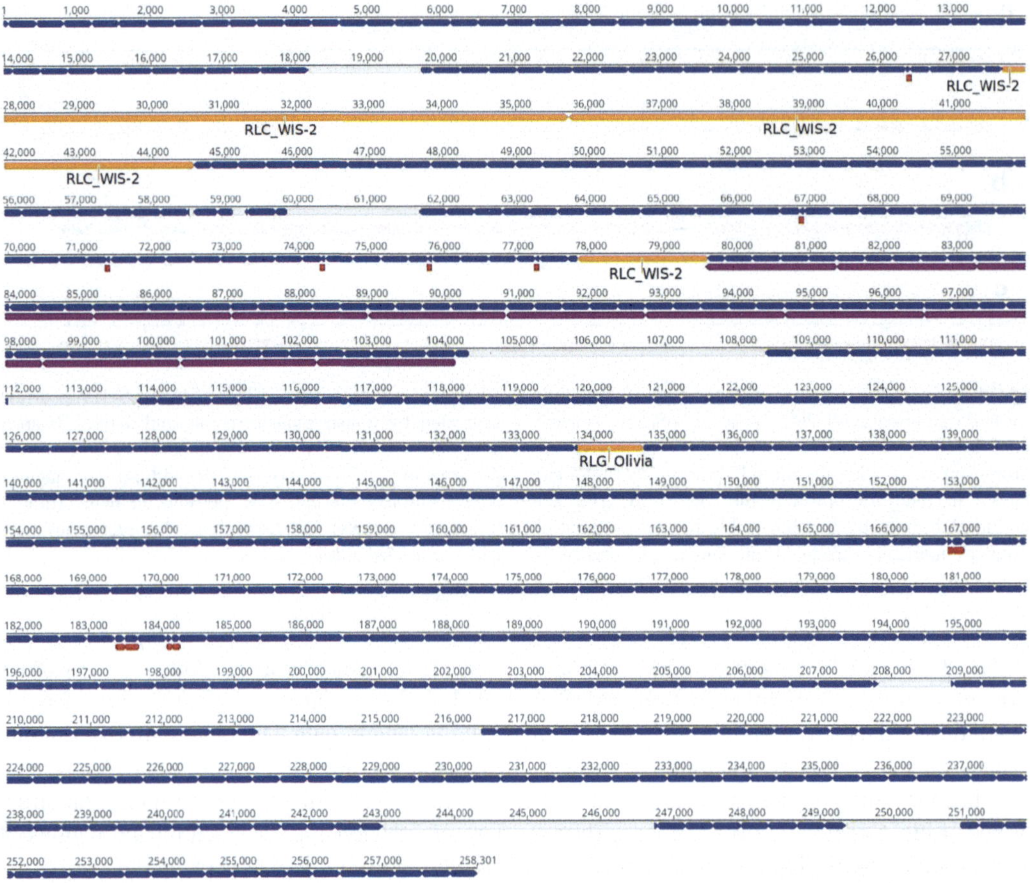

Fig. 8.2 Molecular organization of 'Lo7' scaffold s291. pSc200 monomers (dark-blue finger-post arrows) and the transposable elements separating them (orange finger-post arrows). The group of 13 HOR units, each containing five monomers, is mimicked in purple underneath. Truncated pSc200 monomers are mimicked in red underneath. Numbers above the monomers stand for length in kilobases

Rye's large genome contains not only the high copy number tandem repeat families pSc200, pSc250, and pSc119.2, but also other families that occur at much lower copy numbers and are not solely confined to subtelomeres. Some of these families deserve to be considered chromosome-specific. One of them, *Tai*I, occurs in some Triticeae species (Kishii and Tsujimoto 2002). In rye, it is found on two chromosomes, one being 1R and the other not yet identified (Vershinin and Evtushenko 2014). The interstitial regions of rye chromosomes reveal heterochromatic blocks that are much smaller than those in subtelomeres. For example, some rye cultivars have such blocks on the long arm of chromosome 2R (Nagaki et al. 1999). The DNA of these blocks is represented by the tandem repeat family JNK1 contributing about 4000 copies to the haploid genome. Monomeric elements of this family are 1192–1232 bp in size and none of its DNA sequences has homology to any other sequence in the rye genome.

8.3 Transposable Elements

In rye, it is not difficult to tell subtelomeric regions from the other parts of chromosomes—whether visually under a microscope or with reliance on the molecular composition of DNA sequences—while in many other species, it is. Where tandem repeats and heterochromatization decline, transposable elements—tiny islets of tandem arrays, the coding sequences of genes, and regulatory sequences—rise. There is one difficult point about sequencing genome regions rife with TEs: these elements tend to occur not only very close to each other but at times even within their own kind, making the region look like a mess of a mess. This train consisting of whole elements and their deformed pieces may be long enough. It is extremely difficult to parse it into its constituents, to be sure of where it begins and where it ends, or to assemble the linear structure that matches its counterpart in the chromosome. To be able to identify TEs, especially those put so close together, in large and highly repetitive genomes like rye's, one requires an efficient and fast matching tool. That is why it is deemed important to explain how such issues were dealt with while transposable elements in the genome assembly of the line Lo7 were being annotated.

8.3.1 Methods

Detection of All Transposon Classes in the Rye Genome

Transposons are divided by their replication intermediate into retrotransposons (Class I) and DNA transposons (Class II). The transposition mechanism of Class I is commonly called "copy-and-paste", and that of Class II, "cut-and-paste". Both main classes are subdivided into superfamilies and families. The long terminal repeat (LTR) retrotransposons contain two major Class I superfamilies which are termed *Gypsy* (the inner domain order RT-RH-INT) and *Copia* (INT-RT-RH), abbreviated *RLG* and *RLC*, respectively, in the notation from Wicker et al. (2007). *RLX* stands for LTR retrotransposons without further superfamilies classification.

To identify all classes of TEs in the rye reference genome sequence, a homology search against the *Triticeae* section of the PGSB transposon library (Spannagl et al. 2016) was performed. In order to have representative sequences for all full-length LTR retrotransposons (fl-LTR) families present in this library, a *de-novo* detection of LTR retrotransposons was done with a number of filtering steps following ("filter-2" criteria, details in Fig. 8.3). This resulted in a total of 644 template sequences that were added to the library and then the vmatch (http://www.vmatch.de) software was used.

Identification of Full-Length LTR Retrotransposons

To identify full-length LTR retrotransposons (fl-LTR) in the rye genome sequence the LTR harvest software was used (Ellinghaus et al. 2008). This resulted in a total of 238,543 sequence candidates from the LTR harvest output. These candidates were annotated for PfamA domains using hmmer3 (http://hmmer.org/). A critical step towards a high-confidence TE annotation is the filtering for false positive predictions and hybrid elements, which also has a potentially large effect on many downstream analyses. Two different filters were applied: as a less stringent approach, filter-1 criteria required the presence of at least one inner TE domain. Of the candidate TEs, 74% were retained after the filter-1 step. As a more stringent filter for high quality elements, filter-2 criteria required the presence of at least one typical retrotransposon domain, e.g., reverse transcriptase (RT), RNase H (RH), integrase (INT), protease (PR). Additional criteria included restrictions of the inner and LTR tandem repeat and gap content. Figure 8.3 provides an overview of the different filtering steps and criteria. Applying Filter-2 criteria resulted in a set of high-confidence fl-LTRs that retained about 20% of the initial candidates.

To estimate the insertion age of full-length LTR retrotransposons, the accumulated divergence between its 5' and 3' LTR was determined with a random mutation rate of 1.3×10^{-8} (SanMiguel et al. 1998). It is important to note that the type of downstream analysis and scope of the work mainly determines what TE dataset is most suitable and appropriate to use. Elements filtered by filter-2 criteria are for example well suited as templates for transposon libraries as they represent full length, high quality elements. Hybrid structures, like inconsistent TE pfam ordering or strand mismatches between inner parts, are the most frequent reasons for failing filter-2 (Fig. 8.3).

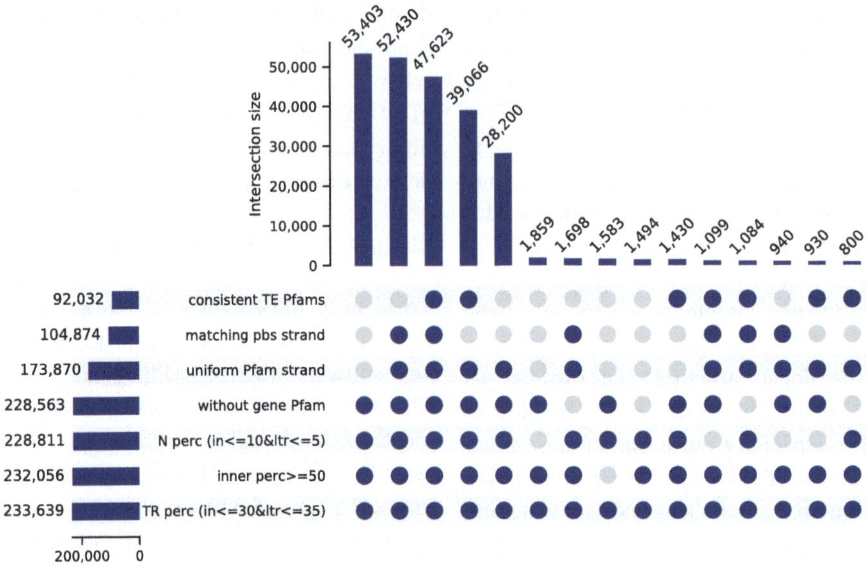

Fig. 8.3 Overview about the filter-2 criteria used for the selection of rye LTR retrotransposon candidate sequences

8.3.1.1 How Much TEs Contribute to the Rye Genome

All classes of transposable elements contribute to 82.4% of the genome sequences in the assembly of the line 'Lo7. Class I transposable elements account for 91.3% of all TEs and Class II transposable elements, 8.1% (Fig. 8.4). The most prevalent Class I elements are LTR-containing retrotransposons (91.0% of all TEs or 74.9% of the genome sequences in the assembly), while non-LTR elements are comparably rare. The prevalence of LTR retrotransposons is a shared

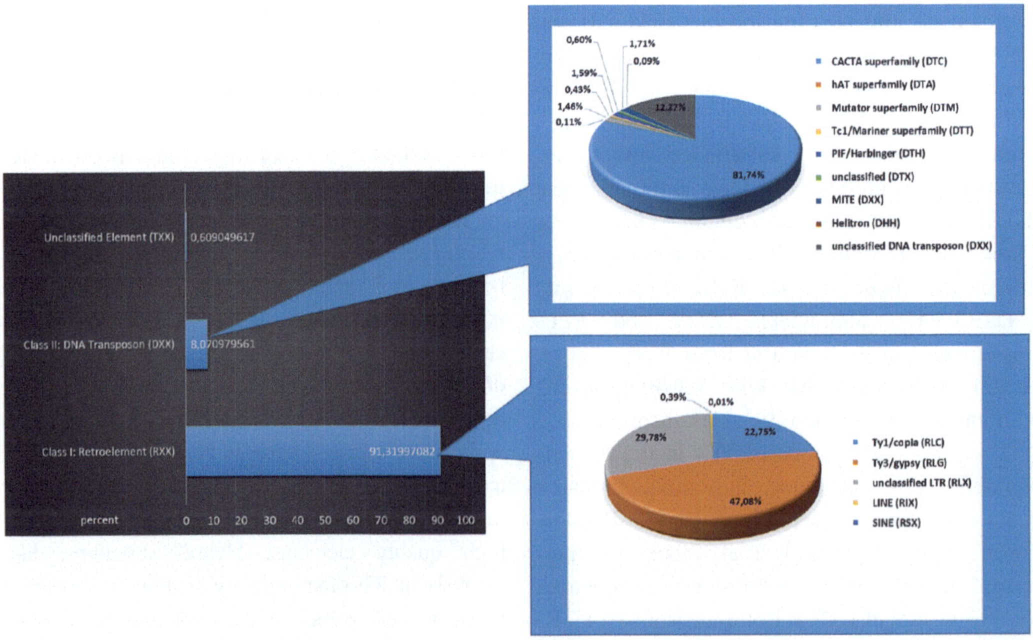

Fig. 8.4 Contribution of TE classes and superfamilies to the genome assembly of the rye line Lo7

Table 8.1 Contribution of the most abundant Class I and Class II families to the genomes of some Triticeae species

Species	***Gypsy*-like**			***Copia*-like**			***DNA transposons***		
	Family (*RLG*)**	% in super-family	% in genome	Family (*RLC*)**	% in super-family	% in genome	Family	% in superfamily *CACTA* (*DTC*)**	% in genome
S. cereale	*Sabrine*	18.23	8.75	*Angela*	37.38	5.37	*Jorge*	32.44	1.52
	Daniela	10.28	4.93	*WIS*	22.65	3.25			
	Erika	6.25	3.00	*Barbara*	14.41	2.07			
	Laura	6.02	2.89	*Inga*	5.15	0.74			
	Sabine	5.80	2.78	*Eugene*	2.79	0.40			
*H. vulgare**	*Sabrine*		9.14	*BARE1*		13.99	*Balduin*		1.59
	BAGY2		5.41	*Maximus*		2.49			
	WHAM		3.75	*Inga*		1.14			
	Surya		3.67						
	BAGY1		2.31						
T. urartu	*Fatima*	17.24	8.26	*Angela*	40.48	7.38	*Jorge*	49.62	4.19
	Sabrine	14.91	7.15	*WIS*	19.32	3.52			
	Erika	7.85	3.76	*Barbara*	11.61	2.12			
	WHAM	6.84	3.28	*Copia-2*	10.47	1.91			
	Romani	5.98	2.86	*Maximus*	3.73	0.68			
Ae. tauschii	*Sabrine*	14.31	5.54	*Angela*	45.03	7.22	*Jorge*	44.91	5.66
	Fatima	8.94	3.46	*WIS*	24.08	3.86			
	WHAM	6.28	2.43	*Maximus*	5.35	0.86			
	Romani	5.79	2.24	*Barbara*	5.25	0.84			
	Nusif	4.90	1.90	*Inga*	2.73	0.44			

* - from Wicker et al. (2018); ** - in brackets are codes in accordance with the classification Wicker et al. (2007)

feature of all plant species. For example, in Triticeae species other than rye, the percentage of LTR-containing retrotransposons is almost 81% of the genome sequences in *H. vulgare* (Mascher et al. 2017); 65.9% in *Aegilops tauschii* (Luo et al. 2017); and 71.8% in *Triticum urartu* (Ling et al. 2018). Two major superfamilies of LTR-containing retrotransposons, *gypsy* and *copia*, and the former contributes to the genome about twice as much as the latter. The most abundant members of these superfamilies are shown in Table 8.1. There is an interesting point about the contributions of separate families in these superfamilies. Three most abundant *copia* families *Angela*, *WIS* and *Barbara* contribute to nearly 75% of the entire superfamily, while five most prevalent *gypsy* families hardly reach a figure of 50%. The most prevalent Class II (transposon) superfamily is CACTA, which contributes to more than 81% of all transposon superfamilies (Fig. 8.4) and the family *Jorge* from this superfamily dominates in Class II (Table 8.1).

8.3.1.2 Evolutionary Dynamics of the Most Abundant LTR-Containing Retrotransposon Families in Triticeae Species

We have compared the contributions of the most abundant LTR-containing retrotransposon families to the rye genome and to the well-annotated genomes of closely related species in the tribe Triticeae: *T. urartu* (https://www.ncbi.nlm.nih.

gov/genome/), the putative donor of the A genome to the hexaploid wheat (*T. aestivum*); *Ae. tauschii* (https://www.ncbi.nlm.nih.gov/genome/), the putative donor of the D genome to the same wheat; and barley (*H. vulgare*) (data from Wicker et al. 2018). There is only one aspect in common between the most abundant TEs in rye and barley: the most prevalent *gypsy* family in both is *Sabrina*, with nearly equal contributions (Table 8.1). The most abundant *copia* family in the barley genome is *BARE1* (13.99%), while its contribution to the rye genome is 0.18%. All studied species show differences in the set of the most represented families, however, the closest sets of families and their percentages show *T. urartu* and *Ae. tauschii* genomes. A set of *copia* families in *T. urartu* and *Ae. tauschii* have more matches in rye than in barley. Estimating the insertion times of TEs for rye and barley (Fig. 8.5) showed that they are different. In barley, the process began 2 MYA and was long-lasting; while in rye it was a recent process, which started less than 1 MYA and did not last. A burst of TE amplification in the rye genome took place at the same time as those in *Ae. tauschii* (Luo et al. 2017) and *T. aestivum* (Ling et al. 2018). Noteworthy, in rye and in *T. urartu*, *copia* elements underwent amplification somewhat later than *gypsy* elements. A possible explanation of the differences in the times of colonization by TEs between barley and other Triticeae species may lie within differences in the split times and further divergence of these genera. The genus *Hordeum* and a common ancestor of rye and wheat split about 11.6 MYA, while the branches leading to the taxa *Secale*, *Triticum* and *Aegilops* split about 6.7 MYA (Chalupska et al. 2008). Note that the amplification of some TE families in rye was accompanied by a concurrent amplification of tandem repeat families, leading to heterochromatic subtelomeric chromosome regions to expand (Evtushenko et al. 2016). Neither *Triticum* nor *Aegilops* has experienced anything comparable, leaving us once again astonished about the diversity of ways, how genomes of different taxa were shaped during evolution by action of repetitive DNA.

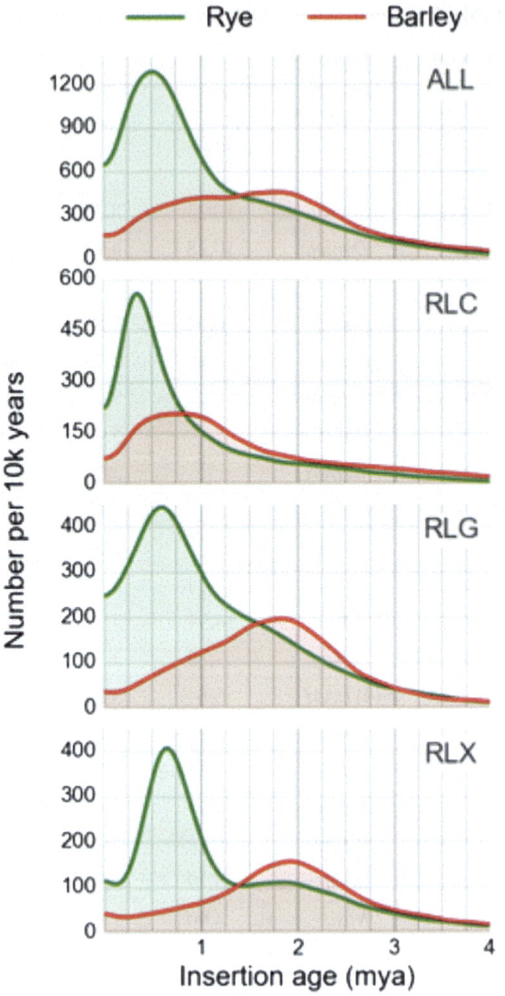

Fig. 8.5 Recent LTR retrotransposon expansion in rye in comparison to barley. Number of elements per 10,000 years with an insertion age estimation <1 MYA is clearly elevated in all rye LTR retrotransposon superfamilies as compared to barley

8.4 Gene Prediction in the Rye Genome

Gene prediction in plant reference genomes is a crucial task both for assessing the gene content in the first place and for use in various downstream analyses. The accuracy and completeness of the predicted gene calls can have a strong influence on conclusions and findings derived from these data and comparative analyses. However, due to intrinsic errors, missing or incomplete data, or conflicting evidences, gene predictions typically contain a relatively large number of gene structure errors and missing or incomplete gene models. This error rate can be reduced by using comprehensive and high quality data to support and inform gene models and structures. Gene prediction and annotation typically follow established routines with combinations of different gene prediction tools and software in so-called gene annotation pipelines. These gene prediction pipelines typically utilize genotype-specific transcriptomic data, protein homology to closely related reference genomes, and *de-novo* gene predictions (e.g., from *ab-initio* gene finders). This section describes the gene prediction procedure used for annotating the rye reference genome sequence and provides some information on the results and the basic evaluation of the gene calls.

8.4.1 Methods

Data Sets to Support Gene Identification in Rye

To assist the structural annotation, RNAseq data from 5 different tissues/developmental stages were obtained for 'Lo7': whole root (3 leaf stage), whole aerial organs (3 leaf stage/sampled at dusk), complete spike (complete heading), flag leaf (7 days post anthesis), whole grains (15 days post anthesis) and whole aerial organs (3 leaf stage), as well as IsoSeq data from the root and aerial organ tissue (3 leaf stage). IsoSeq nucleotide sequences were aligned to the rye genome sequence using GMAP (Wu and Watanabe 2005), while the RNASeq data were first mapped using Hisat2 (Kim et al. 2015) and subsequently assembled into transcript sequences using StringTie (Pertea et al. 2015). Transcripts from IsoSeq and RNASeq were combined using Cuffcompare (Ghosh and Chan 2016). To annotate gene models on the basis of sequence homology, all available *Triticeae* protein sequences utilized have been obtained from UniProt. These protein sequences were mapped to the rye pseudomolecules (on a nucleotide basis) using the splice-aware alignment software Genome Threader (Gremme et al. 2005). To further improve the structural gene annotation and identify gene models without support from sequence homology or gene expression data, an ab-initio annotation was performed with the Augustus software (Stanke et al. 2006).

The GeMoMa Gene Prediction Pipeline

In addition to the gene prediction pipeline outlined above, an additional, independent gene annotation was performed for the rye genome using the GeMoMa software (Keilwagen et al. 2016). The following plant species were used to give evidence for the homology-based gene prediction: *Arabidopsis thaliana* (167), *Brachypodium distachyon* (314), *Glycine max* (275), *Mimulus guttatus* (256_v2.0), *Oryza sativa* (323), *Prunus persica* (298), *Populus trichocarpa* (444), *Sorghum bicolor* (454), *Setaria italica* (312), *Solanum lycopersicum* (390), and *Theobroma cacao* (233). The numbers in brackets indicate the version of the respective data set in Phytozome. Initial homology search for coding exons was performed with mmseqs2 (Steinegger and Soding 2018). To identify splice sites, all mapped RNAseq data were used and combined with the other lines of evidence in GeMoMa. The resulting eleven gene annotation sets were combined and filtered using the module GAF in GeMoMa.

Combining and Classifying Gene Calls from Two or More Prediction Pipelines

To obtain a single, non-redundant set of candidate genes, the aforementioned individual gene predictions need to be consolidated, evaluated and any redundant calls removed/combined. All structural gene predictions resulting from the annotation efforts described above were combined using EVidenceModeller (Haas et al. 2008). A confidence classification protocol was applied to sort candidate gene models into the categories of complete genes, noncoding transcripts, pseudogenes, and transposable elements, respectively. All predicted protein sequences were searched against the following manually curated sequence databases using BLAST: (1) PTREP: this database contains (deduced) amino acid sequences of hypothetical proteins; (2) UniPoa, a database comprised of annotated *Poaceae* proteins; (3) UniMag, a database of validated magnoliophyta proteins. To classify genes into high-confidence (HC) and low-confidence (LC) protein sequences a number of filters were applied for each predicted protein to each of the three databases. High-confidence (HC) protein sequences are required to be complete (both start and stop-codon annotated) and have a significant hit in the UniMag database (HC1) or no blast hit in UniMag but in UniPoa and not TREP (HC2); low-confidence (LC) protein sequences can be incomplete (missing start- and/or stop-codon) and have a hit in the UniMag or UniPoa database but not in TREP (LC1), or no hit in UniMag and UniPoa and TREP but the protein sequence is complete.

The functional annotation (also termed as "human readable description") of all 'Lo7' predicted protein sequences was performed with the AHRD pipeline (https://github.com/groupschoof/AHRD).

Evaluation of the Gene Predictions

Completeness of the predicted gene space was assessed with the BUSCO software (version 3.02, orthodb9). BUSCO assesses the completeness of the expected gene content of a genome assembly or annotation. The resulting metric is complementary to other technical metrics like N50. Reported are complete and single-copy, complete and duplicated, fragmented or missing genes from a reference set of highly conserved genes.

8.4.2 Results

Table 8.2 provides an overview of the gene calls predicted on the 'Lo7' genome sequence. Gene annotation was validated by a search for 1419 BUSCO genes (Table 8.2) of which 1388 (96.4%) were correctly predicted among 57,222 genes. Thus, with the methodological approach that we used for assembling the genome of the line 'Lo7', we have substantially increased the number of genes annotated with high confidence: 34,441 against previous 22,426 (Martis et al. 2013) and 27,784 (Bauer et al. 2017). On the other hand, rye

Table 8.2 Gene prediction statistics for the rye genome

	Complete	HC	LC
Genes	57,222	34,441	22,781
Transcripts	57,222	34,441	22,781
Exons	193,073	152,231	40,842
CDs	193,073	152,231	40,842
Average gene length	2117	2892	946
Average CDS length	1025	1282	635
Average exon length	303	290	354
Average intron length	460	470	392
Exons per gene	3.37	4.42	1.79
Exons per transcript	3.37	4.42	1.79
Monoexonic genes	25,831	10,718	15,113
Complete transcripts	54,841	34,441	20,400
Start only	534	0	534
Stop only	739	0	739

HC high-confidence gene models; *LC* low-confidence gene models

Table 8.3 BUSCO gene space completeness report

	Complete %	HC	LC
Complete BUSCOs	98.5 (1419)	96.4 (1388)	5.8 (83)
Complete and single-copy BUSCOs	93.2 (1342)	91.7 (1321)	5.1 (73)
Complete and duplicated BUSCOs	5.3 (77)	4.7 (67)	0.7 (10)
Fragmented BUSCOs	0.8 (11)	0.6 (8)	2.2 (31)
Missing BUSCOs	0.7 (10)	3.0 (44)	92.0 (1326)

has somewhat fewer genes than other Triticeae species: *Ae. tauschii*, 38,775 (Luo et al. 2017), *T. urartu*, 37,516 (Ling et al. 2018) and *H. vulgare*, 39,734 (Mascher et al. 2017). Some figures regarding rye genes are shown in Table 8.2. As can be seen, the values fluctuate around their counterparts in the other Triticeae species mentioned, exhibiting a rather conserved averaged pattern of the exon-intron organization, exon sizes, intron sizes, and some more (Table 8.3).

Figure 8.6 provides an integrating view of the chromosomal architecture of the Lo7 pseudomolecule assembly. The background features stacked bar charts with the percent coverage of each major component per 4 Mb sliding window (0.8 Mb shift) along the chromosomes. The two curves depict the distribution of genes (green line) and basic repetitiveness in form of median 20mer frequencies (blue line). Centromere locations are marked by the dotted white lines and gray patches in the x-axes. Similar to other Triticeae genomes, LTR retrotransposons are dominating the scene with values between 40 and 82% per window, on average 75%. The LTR retrotransposons are more or less evenly distributed with slight decreases in the gene rich telomeric regions. Genes on the other hand have a highly biased distribution ranging from 0.6 genes per Mb in the proximal compartments up to 18 genes per Mb in the distal compartments. The overall coding sequence of genes representing, after all the main information content, averages to only 0.9% of the genome and is hardly visible in the bar chart. Basic repetitiveness partitions the chromosome into four zones: (1) low 20mer frequencies in the very small gene enriched distal compartments. (2) high 20mer values in the interstitial compartment which are caused by younger transposon insertions followed by (3) the distal compartment composed of older more degenerated transposons leading to decreased 20mer frequencies. The distinct high 20mer peaks at the direct centromeres (4) are a consequence of the highly repetitive centromere structure composed of tandem repeats and very young LTR retrotransposons. Besides their relatively even background amounts (~2%), prominent accumulations of satellite tandem repeats (up to 23% per window) are located at the distal short arms of chromosomes 3R, 4R, 5R, 6R, and the distal long arm of chromosome 3R. Two clusters of minisatellites are present on the long arm of 5R.

8.5 Conclusion

The wild and weedy rye species represent a lavish source of genetic diversity, and the rye genome has always been viewed as a pool of genes useful for improving other cultivated species, primarily wheat. From a consumer's point of view, the chromosome-scale genome assembly offers ample opportunities for breeders to benefit from insights into the genetic composition on their way towards improved breeding efficiency. However, rye is special in that it has features that neither wheat nor barley has: large heterochromatic regions, very distal on all chromosomes. These regions built up as rapidly as it took species in the genus *Secale* to shape, and especially so in the cultivated species *S. cereale*. In light of this fact, we must confess that neither structure nor evolutionary dynamics of the rye genome will be fully understood as a whole until the structure and evolutionary

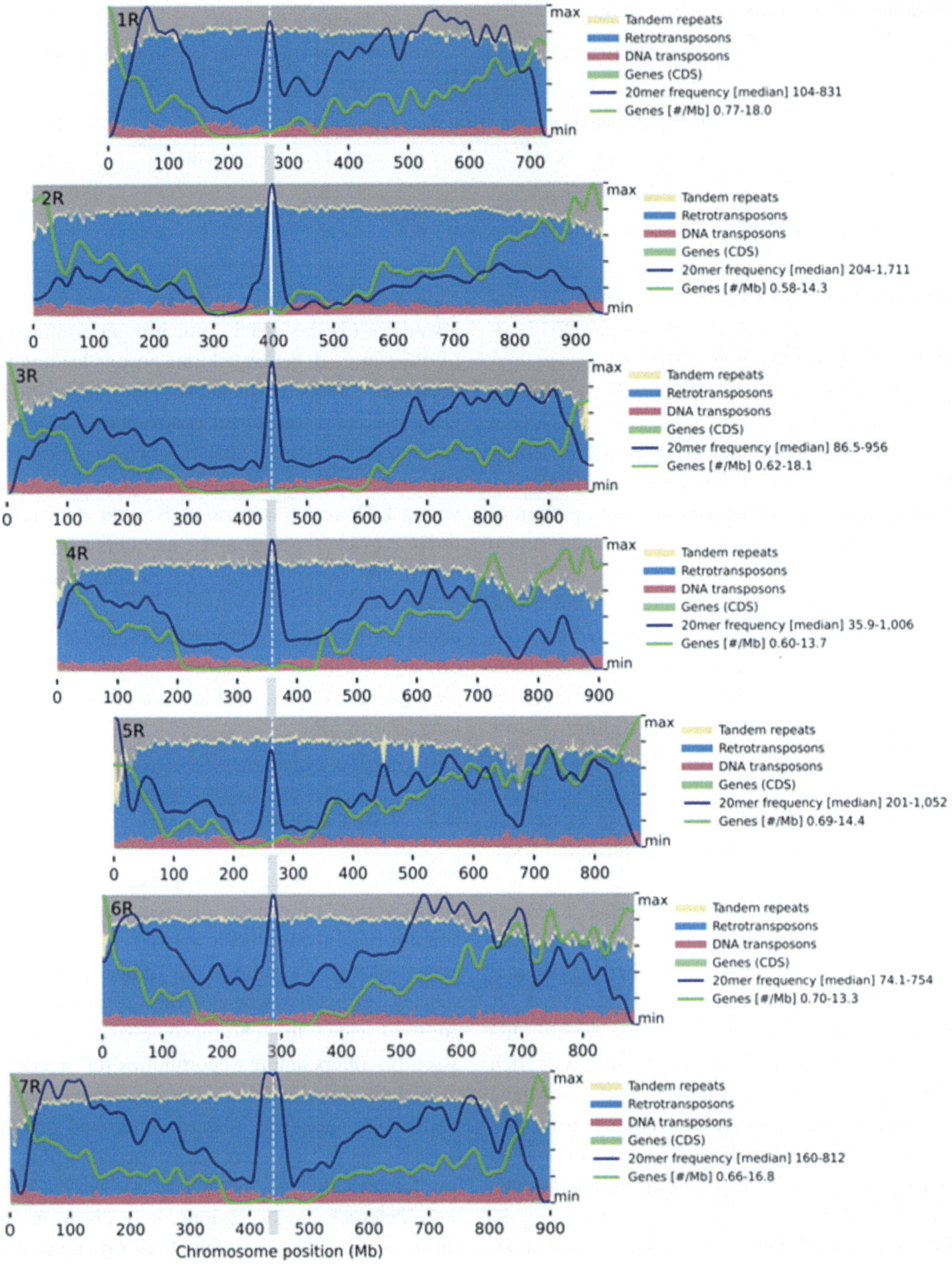

Fig. 8.6 Chromosomal architecture of the Lo7 pseudomolecule assembly

dynamics of these chromosomal regions are completely resolved.

Acknowledgments The authors thank Drs. G. Li and D. Wang, Henan Agricultural University, China, for access to the genomic assembly of 'Weining' rye. This work was supported by the German Ministry of Education and Research (BMBF) within the project de.NBI (FKZ 031A536B), by the Russian fundamental scientific research program (project No 23 0246-2021-0014); grants of Russian Foundation for Basic Research (project No 20-04-00699), and of the Russian Science Foundation (project No 19-14-00051).

References

Alkhimova OG, Mazurok NA, Potapova TA, Zakian SM, Heslop-Harrison JS, Vershinin AV (2004) Diverse patterns of the tandem repeats organization in rye chromosomes. Chromosoma 113:42–52

Appels R, Driscoll C, Peacock WJ (1978) Heterochromatin and highly repeated DNA sequences in Rye (*Secale cereale*). Chromosoma 70:67–89

Azzalin CM, Reichenbach P, Khoriauli L, Giulotto E, Lingner J (2007) Telomeric repeat containing RNA and RNA surveillance factors at mammalian chromosome ends. Science 318:798–801

Bartoš J, Paux E, Kofler R, Havránková M, Kopecký D, Suchánková P, Šafář J, Šimková H, Town CD, Lelley T, Feuillet C, Doležel J (2008) A first survey of the rye (*Secale cereale*) genome composition through BAC end sequencing of the short arm of chromosome 1R. BMC Plant Biol 8:95. https://doi.org/10.1186/1471-2229-8-95

Bauer E, Schmutzer T, Barilar I, Mascher M, Gundlach H, Martis MM, Twardziok SO, Hackauf B, Gordillo A, Wilde P, Schmidt M, Korzun V, Mayer KFX, Schmid K, Schon C-C, Scholz U (2017) Towards a whole-genome sequence for rye (*Secale cereale* L.). Plant J 89:853–869

Bedbrook JR, Jones J, O'Dell M, Tompson R, Flavell R (1980) A molecular description of telomeric heterochromatin in *Secale* species. Cell 19:545–560

Bedbrook JR, O'Dell M, Flavell RB (1980) Amplification of rearranged sequences in cereal plants. Nature 288:133–137

Chalupska D, Lee HY, Faris JD, Evrard A, Chalhoub B, Haselkorn R, Gornicki P (2008) *Acc* homoeoloci and the evolution of wheat genomes. Proc Natl Acad Sci USA 105:9691–9696

Charlesworth B, Sniegowski P, Stephan W (1994) The evolutionary dynamics of repetitive DNA in eukaryotes. Nature 371:215–220

Cuadrado A, Jouve N (2002) Evolutionary trends of different repetitive DNA sequences during speciation in the genus *Secale*. J Heredity 93:339–345

de Lange T (2005) Shelterin: the protein complex that shapes and safeguards human telomeres. Genes Dev 19:2100–2110

Ellinghaus D, Kurtz S, Willhoeft U (2008) LTR harvest, an efficient and flexible software for de novo detection of LTR retrotransposons. BMC Bioinform 9:18. https://doi.org/10.1186/1471-2105-9-18

Evtushenko EV, Levitsky VG, Elisafenko EA, Gunbin KV, Belousov AI, Šafář J, Doležel J, Vershinin AV (2016) The expansion of heterochromatin blocks in rye reflects the co-amplification of tandem repeats and adjacent transposable elements. BMC Genom 17(337):1–16. https://doi.org/10.1186/s12864-016-2667-5

Flavell RB, Bennett MD, Smith JB, Smith DB (1974) Genome size and the proportion of repeated nucleotide sequence DNA in plants. Biochem Genet 12:257–269

Flavell R (1980) The molecular characterization and organization of plant chromosomal DNA sequences. Annu Rev Plant Physiol 31:569–596

Ghosh S, Chan C-KK (2016) Analysis of RNA-Seq data using TopHat and cufflinks. Methods Mol Biol 1374:339–361

Gremme G, Brendel V, Sparks ME, Kurtz S (2005) Engineering a software tool for gene structure prediction in higher organisms. Inform Softw Technol 47:965–978

Haas BJ, Salzberg SL, Zhu W, Pertea M, Allen JE, Orvis J, White O, Buell CR, Wortman JR (2008) Automated eukaryotic gene structure annotation using EVidenceModeler and the program to assemble spliced alignments. Genome Biol 9:R7

Jones JDG, Flavell RB (1982) The mapping of highly-repeated DNA families and their relationship to C-bands in chromosomes of *Secale cereale*. Chromosoma 86:595–612

Jones JDG, Flavell RB (1982) The structure, amount and chromosomal localization of defined repeated DNA sequences in species of the genus *Secale*. Chromosoma 86:613–641

Kearse M, Moir R, Wilson A et al (2012) Geneious. Bioinformatics 28:1647–1649. http://bioinformatics.oxfordjournals.org/content/28/12/1647.short

Keilwagen J, Wenk M, Erickson JL, Schattat MH, Grau J, Hartung F (2016) Using intron position conservation for homology-based gene prediction. Nucl Acids Res 44:e89

Kim D, Langmead B, Salzberg SL (2015) HISAT: a fast spliced aligner with low memory requirements. Nat Methods 4:357–360

Kishii M, Tsujimoto H (2002) Genus-specific localization of the *TaiI* family of tandem-repetitive sequences in either the centromeric or subtelomeric regions in Triticeae species (Poaceae) and its evolution in wheat. Genome 45:946–955

Leitch IJ, Johnston E, Pellicer J, Hidalgo O, Bennett MD (2019) Plant DNA C-values database (release 7.1). https://cvalues.science.kew.org/

Ling H-Q, Ma B, Shi X, Liu H, Dong L, Sun H, Cao Y, Gao Q, Zheng S, Li Y, Yu Y, Du H, Qi M, Li Y,

Lu H, Yu H, Cui Y, Wang N, Chen C, Wu H, Zhao Y, Zhang J, Li Y, Zhou W, Zhang B, Hu W, van Eijk MJT, Tang J, Witsenboer HMA, Zhao S, Li Z, Zhang A, Wang D, Liang C (2018) Genome sequence of the progenitor of wheat A subgenome *Triticum urartu*. Nature 557:424–428

Luo M-C, Gu YQ, Puiu D, Wang H, Twardziok SO, Deal KR, Huo N, Zhu T, Wang L, Wang Y, McGuire PE, Liu S, Long H, Ramasamy RK, Rodriguez JC, Van SL, Yuan L, Wang Z, Xia Z, Xiao L, Anderson OD, Ouyang S, Liang Y, Zimin AV, Pertea G, Qi P, Bennetzen JL, Dai X, Dawson MW, Müller H-G, Kugler K, Rivarola-Duarte L, Spannagl M, Mayer KFX, Lu F-H, Bevan MW, Leroy P, Li P, You FM, Sun Q, Liu Z, Lyons E, Wicker T, Salzberg SL, Devos KM, Dvořák J (2017) Genome sequence of the progenitor of the wheat D genome *Aegilops tauschii*. Nature 551:498–502

Martis MM, Zhou R, Haseneyer G, Schmutzer T, Vrána J, Kubaláková M, Konig S, Kugler KG, Schiz U, Hackauf B, Korzun V, Schon C-C, Dolezel J, Bauer E, Mayer KFX, Stein N (2013) Reticulate evolution of the rye genome. Plant Cell 25:3685–3698

Mascher M, Gundlach H, Himmelbach A, Beier S, Twardziok SO, Wicker T, Radchuk V, Dockter C, Hedley PE, Russell J, Bayer M, Ramsay L, Liu H, Haberer G, Zhang XQ, Zhang Q, Barrero RA, Li L, Taudien S, Groth M, Felder M, Hastie A, Simkova H, Stankova H, Vrana J, Chan S, Munoz-Amatriain M, Ounit R, Wanamaker S, Bolser D, Colmsee C, Schmutzer T, Aliyeva-Schnorr L, Grasso S, Tanskanen J, Chailyan A, Sampath D, Heavens D, Clissold L, Cao S, Chapman B, Dai F, Han Y, Li H, Li X, Lin C, McCooke JK, Tan C, Wang P, Wang S, Yin S, Zhou G, Poland JA, Bellgard MI, Borisjuk L, Houben A, Dolezel J, Ayling S, Lonardi S, Kersey P, Langridge P, Muehlbauer GJ, Clark MD, Caccamo M, Schulman AH, Mayer KFX, Platzer M, Close TJ, Scholz U, Hansson M, Zhang G, Braumann I, Spannagl M, Li C, Waugh R, Stein N (2017) A chromosome conformation capture ordered sequence of the barley genome. Nature 544:427–433

McIntyre CL, Pereira S, Moran LB, Appels R (1990) New *Secale cereale* (rye) DNA derivatives for the detection of rye chromosome segments in wheat. Genome 33:317–323

Nagaki K, Tsujimoto H, Sasakuma T (1999) A novel repetitive sequence, termed the JNK repeat family, located on an extra heterochromatic region of chromosome 2R of Japanese rye. Chromosome Res 6:95–101

Pertea M, Pertea GM, Antonescu CM, Chang T-C, Mendell JT, Salzberg SL (2015) StringTie enables improved reconstruction of a transcriptome from RNA-seq reads. Nat Biotechnol 33:290–295

Rabanus-Wallace MT, Hackauf B, Mascher M, Lux T, Wicker T, Gundlach H, Báez M, Houben A, Mayer KFX, Guo L, Poland J, Pozniak CJ, Walkowiak S, Melonek J, Praz C, Schreiber M, Budak H, Heuberger M, Steuernagel B, Wulff B, Börner A, Byrns B, Čížková J, Fowler DB, Fritz A, Himmelbach A, Kaithakottil G, Keilwagen J, Keller B, Konkin D, Larsen J, Li Q, Myśków B, Padmarasu S, Rawat N, Sesiz U, Sezgi B, Sharpe A, Šimková H, Small I, Swarbreck D, Toegelová H, Tsvetkova N, Voylokov AV, Vrána J, Bauer E, Bolibok-Bragoszewska H, Doležel J, Hall A, Jia J, Korzun V, Laroche A, Ma X-F, Ordon F, Özkan H, Rakoczy-Trojanowska M, Scholz U, Schulman AH, Siekmann D, Stojałowski S, Tiwari V, Spannagl M, Stein N (2019) Chromosomal-scale genome assembly provides insights into rye biology, evolution, and agronomic potential. https://doi.org/10.1101/2019.12.11.869693

Rabinowicz PD, Bennetzen JL (2006) The maize genome as a model for efficient sequence analysis of large plant genomes. Curr Opin Plant Biol 9:149–156

SanMiguel P, Gaut BS, Tikhonov A, Nakajima Y, Bennetzen JL (1998) The paleontology of integene retrotransposons of maize. Nat Genet 20:43–45

Smit AFA, Hubley R, Green P (2013–2015) RepeatMasker Open-4.0. http://www.repeatmasker.org

Spannagl M, Nussbaumer T, Bader KC, Martis MM, Seidel M, Kugler KG, Gundlach H, Mayer KFX (2016) PGSB PlantsDB: updates to the database framework for comparative plant genome research. Nucleic Acids Res 44(D1):D1141–D1147

Stanke M, Schoffmann O, Morgenstern B, Waack S (2006) Gene prediction in eukaryotes with a generalized hidden Markov model that uses hints from external sources. BMC Bioinform 7:62

Steinegger M, Soding J (2018) Clustering huge protein sequence sets in linear time. Nature Commun 9:2542

van Dijk EL, Jaszczyszyn Y, Naquin D, Thermes C (2018) The third revolution in sequencing technology. Trends Genet 34:666–681

Vershinin AV, Heslop-Harrison JS (1998) Comparative analysis of the nucleosomal structure of rye, wheat and their relatives. Plant Mol Biol 36:149–161

Vershinin AV, Evtushenko EV (2014) What is the specificity of plant subtelomeres? In: Louis EJ, Becker MM (eds) "Subtelomeres". Springer, pp 195–209

Vershinin AV, Schwarzacher T, Heslop-Harrison JS (1995) The large-scale organization of repetitive DNA families at the telomeres of rye chromosomes. Plant Cell 7:1823–1833

Weissensteiner MH, Pang AWC, Bunikis I, Hoijer I, Vinnere-Petterso O, Suh A, Wolf JBW (2017) Combination of short-read, long-read, and optical mapping assemblies reveals large-scale tandem repeat arrays with population genetic implications. Genome Res 27:697–708

Wicker T, Sabot F, Hua-Van A, Bennetzen JL, Capy P, Chalhoub B, Flavell A, Leroy P, Morgante M,

Panaud O, Paux E, SanMiguel P, Schulman AH (2007) A unified classification system for eukaryotic transposable elements. Nature Rev Genet 8:973–982

Wicker T, Gundlach H, Schulman AH (2018) The repetitive landscape of the barley genome. In: Stein N, Muehlbauer GJ (eds). The barley genome Springer, pp 123–138

Wu T, Watanabe CK (2005) GMAP: a genomic mapping and alignment program for mRNA and EST sequences. Bioinformatics 21:1859–1875

Young E, Abid HZ, Kwok P-Y, Riethman H, Xiao M (2020) Comprehensive analysis of human subtelomeres by whole genome mapping. PLoS Genet 16(1): e1008347

Bridging the Genotype–Phenotype Gap for Precision Breeding in Rye

9

Bernd Hackauf, M. Timothy Rabanus-Wallace, and Viktor Korzun

Abstract

With release of two high-quality genome assemblies, rye has finally reached the genome era, enabling the integration and advancement of fundamental and applied breeding and research to understand how the genome builds, maintains, and operates rye. This chapter compiles a century of breeding research that aimed to describe and unravel the genetic diversity of rye. Systematic identification, management, and use of natural diversity became feasible in outbreeding rye with the establishment of hybrid breeding late in the twentieth century. Research conducted so far largely reflects target traits of rye improvement programs. We review progress achieved in the mapping of genes and QTL (quantitative trait loci) for agronomic traits, biotic and abiotic stress tolerance as well as grain quality. We describe how rye genome assemblies now enable association of the digital sequences of rye markers with locations in physical space, as an essential standard to conduct genome-based breeding and research in rye. Despite formidable achievements, major challenges in rye production remain, in particular concerning tailor-made grain qualities, to further advance rye from an all-rounder to an authentic high-performance crop with different and certified types of end-use. For this purpose, further progress in rye phenomics and functional genomics research is necessary to associate genome sequence information with phenotypes related to rye growth and development.

9.1 Introduction

9.1.1 Genomics in the Overlooked Cereal is Coming of Age

Rye (*Secale cereale* L.) belongs to the Triticeae tribe of the grasses and is a multipurpose cereal crop used for human consumption, animal feeding as well as agricultural bioenergy production. Rye, with its diploid genome consisting of seven chromosomes pairs ($2n = 2x = 14$), diverged from wheat approximately 3–4 million years ago (Middleton et al. 2014). As in barley and wheat (Sato 2020), the estimated rye genome size of 7–8 Gbp (Li et al. 2021; Rabanus-Wallace et al.

B. Hackauf (✉)
Institute for Breeding Research on Agricultural Crops, Julius Kühn-Institute, Rudolf-Schick-Platz 3a, 18190 Groß-Lüsewitz, Germany
e-mail: bernd.hackauf@julius-kuehn.de

M. T. Rabanus-Wallace
Leibniz Institute of Plant Genetics and Crop Plant Research (IPK), 06466 Seeland, Germany

V. Korzun
KWS SAAT & Co. KGaA, Grimsehlstr. 31, 37574 Einbeck, Germany

© The Author(s), under exclusive license to Springer Nature Switzerland AG 2021

M. T. Rabanus-Wallace and N. Stein (eds.), *The Rye Genome*, Compendium of Plant Genomes,
https://doi.org/10.1007/978-3-030-83383-1_9

2021) presents challenges to gene cloning based on high-resolution genetic and physical mapping. Rather, a low ratio of physical to genetic distance resulting from high levels of recombination and polymorphism as well as a low linkage disequilibrium (Li et al. 2011; Auinger et al. 2016) make rye a useful but so far underutilized resource for the functional characterization of genes in small-grain cereals. This applies in particular to genes involved in the control of agronomic traits including grain yield with a complex genetic architecture.

The rich diversity of rye is mirrored by more than 90 million single nucleotide variants (SNV) and short insertions/deletions, that were recently discovered by comparative analyses between a de novo assembly covering a total length of 2.8 Gbp of the inbred line 'Lo7', that has been obtained through whole-genome shotgun sequencing (WGS), and 10 resequenced rye inbred lines as well as one accession of the wild relative *S. vavilovii* (Bauer et al. 2017). These variants served to develop the Rye600k genotyping array, that enabled to establish a whole-genome draft sequence of rye based on a high-density genetic map (Bauer et al. 2017). The Rye600k genotyping array represents a state-of-the-art genomic resource to investigate genetic variation in rye DNA sequences and enables the development of single-plex SNP assays for high-throughput genotyping. SNP markers are transferable between mapping populations (Martis et al. 2013) and can serve as anchor markers linking genetic information from different experiments. The novel high-quality genome assemblies reporting the linear order of 45,596 and 34,441 high-confidence (HC) genes from two rye inbred lines (Li et al. 2021; Rabanus-Wallace et al. 2021) now enable to associate the digital sequences of rye SNP markers with locations in physical space as an essential standard to conduct genome-based breeding and research in this orphan crop. While we will mainly refer to the 'Lo7' genome sequence in this chapter, all given information should principally translate to the 'Weining' genome sequence (Li et al. 2021) as well, due to the almost perfect collinearity between both assemblies (see Rabanus-Wallace et al., Chap. 7 of this volume).

Previous studies enabled the genetic analysis and mapping of several traits and QTL in rye. The high-quality rye reference genome sequences enable to integrate the studied traits based on sequence information on linked or associated markers (Fig. 9.1). Subsequent to a short compilation of a century of breeding research that aimed to describe and unravel the genetic diversity of rye, progress achieved in the mapping of genes and QTL for agronomic traits, biotic and abiotic stress tolerance as well as grain quality in rye will be reported.

9.1.2 Cross-Pollination—Challenge and Opportunity

In contrast to other crop species (Scott et al. 2020), populations derived from experimental crosses mixing and recombining the genomes of multiple founders to dissect complex traits and support modern plant breeding have not yet been developed in rye. A strategy that combines the strengths of both linkage and association mapping, referred to as nested association mapping (NAM), has proven to identify functional markers for complex inherited traits in several plant species (Guo et al. 2010; Chen et al. 2019b; Scott et al. 2020). In rye, a NAM population for the Carsten pool would be extremely valuable due to the dominant role of Petkus as an ancestor of many elite open-pollinating rye varieties (OPVs) worldwide that reduces the probability of finding populations that are genetically diverse from the Petkus pool (Fischer et al. 2010). To establish the first rye NAM population, either of the two sequenced inbred lines can be used as a single reference to be crossed with 25 carefully selected founder inbred lines to derive multiple families of recombinant inbred lines (RILs). A rye NAM population will enable the melding of genome-wide association studies (GWAS) and expression QTL (eQTL) analysis: a promising approach for characterizing candidate genes that are involved in the genetic control of complex traits (Wang et al. 2018; Li et al. 2020) and adaptive responses to biotic and abiotic stress.

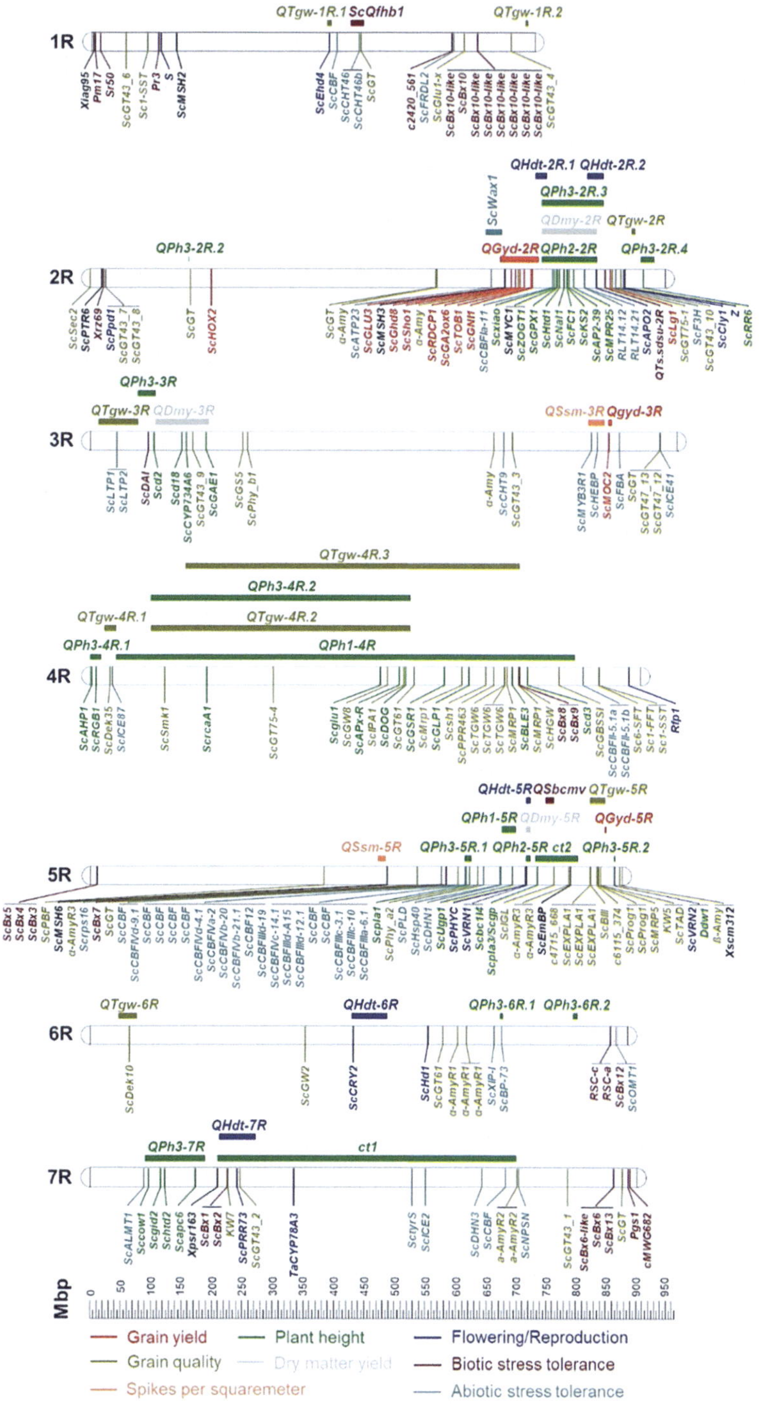

Fig. 9.1 Rye genes and QTL in the 'Lo7' physical map. The different classes of genes and QTL are shown in different colors. The symbols for agronomic traits including grain yield are explained in Sects. 9.2.1 and 9.3.1, for disease resistance in Sects. 9.2.2 and 9.3.2, for abiotic stress tolerance in Sects. 9.2.3 and 9.3.3, and for grain quality in Sects. 9.2.4 and 9.3.4. The QTL symbols indicate the position of the following quantitative traits: QGyd: grain yield, QPh: plant height, QHdt: heading date, QTgw: thousand-grain weight, QDmy: dry matter yield, QSsm: spikes per squaremeter, QFhb: Fusarium head blight, QSbcmv: Soil born cereal mosaic virus, QTs: Tan spot. The scale on the bottom indicates the distance in mega base pair (Mbp)

This ambitious goal requires common international efforts to receive financial support and scientific infrastructure on the long term. This necessary prerequisite to maintain and test a vast amount of recombinant inbred lines is obviously challenging to achieve for an orphan crop.

Species with functional self-incompatibility like rye (Lundqvist 1956) yield strong long-term evolutionary advantages and have been shown to diversify at a significantly higher rate than those without such genetic mechanisms promoting allogamy (Goldberg et al. 2010). The rich genetic diversity in randomly mating rye populations has been characterized using different DNA marker technologies (Bolibok-Brągoszewska et al. 2014; Hagenblad et al. 2016; Monteiro et al. 2016; Schreiber et al. 2018; Hawliczek et al. 2020; Targonska-Karasek et al. 2020), but the response to selection of favorable alleles in breeding open pollinating varieties (OPVs) is comparably low (Laidig et al. 2017). Notably, a heterotic pattern has been detected in diallel crosses between combinations of open-pollinating 'Petkus' and 'Carsten' varieties (Hepting 1978). These iconic cultivars have emerged as the main pillar of hybrid rye improvement programs. The phenomenon of heterosis or hybrid vigor was first detected and described in detail by the American botanist and plant geneticist George Harrison Shull (1908). In rye, the systematic use of heterosis by hybrid breeding started around 1970 and the first hybrid rye cultivars were released in 1984 (Geiger 1985). The natural genetic diversity in rye was the fundamental basis to achieve a series of technological advances, that facilitated to establish hybrid breeding. Meanwhile, whole-genome scans of single nucleotide polymorphisms (SNPs) revealed that highly heritable self-fertility mutations enabled the development of inbred lines invaluable to the effort of capturing and managing the rich genetic diversity in heterotic rye genepools (Bauer et al. 2017; Vendelbo et al. 2020). Nevertheless, the development of rye requires the meticulous selection of inbreds that need to be selectively mated in the final step of a breeding cycle, in order to reconstitute heterozygosity: an indispensable prerequisite to achieving the desired performance of rye in farmers' fields. A supporting pillar and game changing observation enabling selective mating of rye lines were identified by the German scientists Hartwig H. Geiger and Wolfgang Schnell about 50 years ago at the University of Hohenheim. They observed and analyzed plants from an Argentinian 'Pampa' rye, that failed to produce functional anthers, pollen, or male gametes (Geiger and Schnell 1970a). This so-called cytoplasmic male sterility (CMS) is the result of specific interactions between nuclear and mitochondrial genes (see Melonek et al., Chap. 10 of this volume). CMS creates an evolutionary advantage: empirical evidence has been reported that male-sterile genotypes produced more flowers, set more fruits and produced more seeds that were larger and germinated better than those of hermaphrodites from the same populations (Dufay and Billard 2012). Hybrid rye breeding resulted in more efficient responses to purifying selection as well as significant genetic gains in agronomic traits (Laidig et al. 2017), and contributed to keeping rye competitive in agricultural production systems. Hybrid breeding offers unique opportunities for cereal research aiming to (i.) understand the function of potentially all components of the genome, and (ii.) synthesize knowledge into an understanding of the phenotypic performance of rye and related species.

9.1.3 Unlocking Genetic Diversity in Rye

The strong self-incompatibility system of rye (see Melonek et al., Chap. 10 of this volume) constrains the development of purebred inbred lines with satisfying seed setting. This challenge is a long-lasting topic of breeding research and a reasonable number of rye inbred lines have been developed during the last century, for example, in Sweden (Heribert Nilsson 1916, 1953; Peterson 1934; Lundqvist 1960; Müntzing 1963), the USA (Davison et al. 1924), Germany (Ossent 1938; Mengersen 1951; Wricke 1969; Geiger and Schnell 1970b; Melz et al. 1990; Bauer et al. 2017; Vendelbo et al. 2020), the Netherlands (Sybenga 1958), Poland (Wolski 1970; Madej et al. 1990; Grochowski et al. 1995; Myśków et al. 2001), Canada (Musa et al. 1984; Ragaee et al. 2001a, 2001b), the Russian Federation (Voylokov et al. 2018; Goncharenko et al. 2019) and China (Li et al. 2021). The development of inbred lines uncovered the hidden variation of rye (Davison et al. 1924; Sybenga 1958; Müntzing and Bose 1969; Geiger and Schnell 1970b; Wricke 1973; Musa et al. 1984; Bauer et al. 2017; Voylokov et al. 2018; Vendelbo et al. 2020; Li et al. 2021) and contributed to describing more than 60 morphological traits (Melz et al. 1992; Schlegel et al. 1986, 1998). Several of these morphological markers were successfully deployed in rye improvement programs, such as the gene *hairy peduncle* (*hp*) (Chang 1975) in the cultivar 'Halo', the gibberellin insensitive dwarfing gene *ct1* (De Vries and Sybenga 1984) in 'Gülzower Kurzstrohroggen' (Dill 1983) or anthocyaninless (*an*) genes (De Vries and Sybenga 1984; Melz and Thiele 1990; Malyshev et al. 2001; Lykholay et al. 2014; Voylokov et al. 2015; Zykin et al. 2018; Braun et al. 2019) in 'Heines Hellkorn' and the hybrid varieties 'Hellvus' and 'Helltop' (Melz et al. 2003). Probably the most important impact of a single spontaneous mutant on rye breeding can be attributed to the dominant dwarfing gene *Ddw1* that originates from the germplasm collection preserved at the Vavilov Institute of Plant Industry in St. Petersburg (Kobylyansky, 1972) and that has been reported to improve the performance of many open-pollinating population varieties (Kobylyansky 1988; Torop et al. 2003; Kobylyansky and Solodukhina 2015). Likewise, high-quality whole-genome sequence information of an inbred line originating from the Chinese local rye variety 'Weining' (Li et al. 2021) as well as of the inbred line 'Lo7' from the Petkus pool (Rabanus-Wallace et al. 2021) provides strong momentum and invaluable template to study more than 45,000 high-confidence genes that will further promote harnessing the potential of the outbreeding rye and increase the significance of this overlooked cereal as a back-up for its close relative wheat.

Notably, the genetics of rye had initially not received much attention. Mapping genes and traits in rye have started late (Surikov 1971; Chang 1975; De Vries and Sybenga 1984; Smirnov and Sosnichina 1984). A straightforward approach based on biochemical markers provided early impetus to approach structural genes in rye. Almost 50 out of 122 chromosomally localized traits in rye represented gene products that function as enzymes (for a review see Schlegel et al. 1986). These so-called isozymes (Markert & Møller 1959) expanded the methods from merely phenotypically based linking of traits to certain chromosomes in aneuploid wheat/rye stocks or trisomics to genetic and linkage analysis, and enabled to establish an isozyme linkage map including all of the seven chromosomes corresponding to the haploid chromosome number of rye (Wehling 1985, 1991). Although isozymes served as markers for major genes in rye (Wricke and Wehling 1985; Gertz and Wricke 1989; Wricke et al. 1993; Benito et al. 1991; Fuong et al. 1993; Gallego and Benito 1997; Wricke 2002; Wehling et al. 2003; Roux et al. 2004; Voylokov and Priiatkina 2004; Konovalov et al. 2010), the limited number of genes encoding enzymes and available histochemical enzyme assays resulted in the development of linkage maps based on DNA markers. Genetic linkage maps dominated by restriction fragment length polymorphisms (RFLP), which consistently cover the seven rye chromosomes, have been developed in the last decade of the twentieth century using F2 populations (Devos et al. 1993; Korzun et al. 2001; Ma et al. 2001). In total, more than 980 marker loci including rye simple sequence repeat (SSR) markers (Saal and Wricke 1999; Hackauf and Wehling 2002; Chebotar et al. 2003) were genetically mapped in founding studies investigating genetic variation in rye DNA sequences (Devos et al. 1993; Philipp et al. 1994; Loarce et al. 1996; Senft and Wricke 1996; Korzun et al. 1998, 2001; Ma et al. 2001; Hackauf and Wehling 2003; Bednarek et al. 2003; Khlestkina et al., 2004, 2005; Hackauf et al. 2009). A further increase in the number of identified markers was achieved based on the microarray-based Diversity Arrays Technology (DArT^{TM}) (Jaccoud et al. 2001; Bolibok-Bragoszewska et al. 2009; Milczarski et al. 2011). While DArT technology can detect and type DNA variation at hundreds of genomic loci in parallel without relying on sequence information, Eva Bauer and coworkers contributed considerably to remedy the lack of genome sequence information in rye using next-generation sequencing (Haseneyer et al. 2011) and upgraded rye EST sequence resources, that had been obtained through Sanger sequencing (Milla et al. 2002; Zhang et al. 2004; Lazo et al. 2004). These EST contigs, later on complemented by a reference gene set of cDNA contigs assembled from rye 454-reads (Khalil et al. 2015), represented a basic step toward deciphering the rye genome and enabled to identify single nucleotide polymorphisms (SNPs) and the development of the Rye5K Infinium Bead Chip (Haseneyer et al. 2011). This high-throughput SNP genotyping array was used to establish an integrated high-density genetic map based on recombinant inbred lines (RILs) from four mapping populations. The subsequent integration of chromosome survey sequencing and conserved synteny information based on three sequenced model grass genomes led to the establishment of a virtual linear gene order model comprising 72% of 31,008 detected rye genes (Martis et al. 2013). Subsequently, this integrated map served as an invaluable reference to validate and establish an accurate genetic linkage map as the pivotal basis for the mapping of QTL and subsequent applications including comparative mapping, positional cloning, and particularly the transfer of these results in practical rye improvement programs (Hackauf et al. 2017a).

9.2 Mapped Major Genes—The Peak of an Iceberg

9.2.1 Agronomic Traits

Plant height is an important target trait in rye breeding and a major factor influencing lodging tolerance. Major genes reducing plant height are designated as dwarfing genes. In rye, several Gibberellin (GA)-sensitive and GA-insensitive dwarfing genes are known (Börner et al. 1996). In GA-sensitive mutants, a tall phenotype can be restored by the exogenous application of GA. In contrast, GA-insensitive mutants exhibit a reduced response or are completely insensitive to the application of GA. The GA-insensitive dwarfing gene *compactum1* (*ct1*) has been mapped within a 4 cM interval close to the centromere on chromosome 7R (Plaschke et al. 1995, Fig. 9.1). The *ct1* flanking markers *Xpsr163* and *α-Amy-R2* bracket a 489.5 Mbp interval on chromosome 7R (Fig. 9.1), indicating a high ratio of physical to genetic distance of 122.4 Mb/cM in this region. Genes located in the genetic centromeres of rye chromosomes, where meiotic recombination is severely suppressed (Rabanus-Wallace et al. 2021), are challenging to clone—but rarely recombining regions may be accessible to positional cloning by sequencing-based methods and the availability of a reference genome sequence (Mascher et al. 2014). More recently, the GA-insensitive dwarfing gene *dw9* has been mapped within a 0.6 cM interval on chromosome 6R (Grądzielewska et al. 2020). Another GA-insensitive dwarfing gene, *ct2*, is located on the distal end of the long arm of chromosome 5R (Plaschke et al. 1993; Braun et al. 2019; Fig. 9.1). Pleiotrophic effects have been reported for plants homozygous for *ct2* including erect leaves, nearly awnless ears, third flowers and short round kernels, resulting in an ear morphology similar to that of wheat (Braun et al. 2019). A marker-assisted introgression of recessive dwarfing genes like *ct1*, *ct2*, or *dw9* in elite inbred lines of both heterotic gene pools and hybrid breeding may serve to develop novel rye ideotypes with improved yield potential.

The most prominent among the dwarfing genes was described almost 50 years ago by the Russian scientist Vladimir D. Kobylyansky and was originally named *Humilus* (*Hl*) (Kobylyansky 1972). The *Hl* gene produced desirable pleiotropic effects on plant height, root system, tiller number, leaf surface area, and spike length as well as grain number, that lead to an increase in the yield potential of plants and substantially promoted Eastern European rye breeding programs (Kobylyansky and Solodukhina 2015). Renamed *Dominant dwarf 1* (*Ddw1*) by Melz (1989), this GA-sensitive gene (Börner and Melz 1988) was initially mapped to a 17.1 cM interval on the distal end of the long arm of chromosome 5R (Korzun et al. 1996). De novo transcriptome sequencing and differential gene expression analysis enabled the fine mapping of *Ddw1* to a 0.4 cM interval and revealed co-segregation with the C20-GA2-oxidase gene *GA2ox12* (*syn. SECCE5Rv1G0374310,* Fig. 9.1), that is up-regulated in culms of *Ddw1* genotypes (Braun et al. 2019). As GA2-oxidases inactivate endogenous GA to maintain the dynamic homeostasis of GA for plant growth (Thomas et al. 1999), *Ddw1* likely affects and reduces the concentration of bioactive GA in semi-dwarf rye by transcriptional activation of *GA2ox12*. Thus, *Ddw1* offers a genetic solution to reduce endogenous GA concentrations as an ensurance that the full yield potential of rye cultivars can be achieved and maintained until harvest. In this way *Ddw1* alleviates the necessity of fine-tuning rye growth by farmers, who are generally currently responsible for regulating the plant's hormonal status by treatment with chemical inhibitors of GA biosynthesis (Rademacher 2016), the effects of which are subject to a range of unpredictable environmental factors. For this purpose, the novel markers were used for the introgression of *Ddw1* in elite seed parent lines of the Petkus pool. Homozygous semi-dwarf seed parent genotypes in the male sterility inducing Pampa (P) cytoplasm are now available and enable the development of semi-dwarf P-type hybrids in rye improvement programs (Fig. 9.2), to extend the portfolio of vigorous hybrid varieties and change the renowned image of rye for being a tall crop (Fig. 9.3). *Ddw1* is an example of a SMART breeding strategy to offset linkage drag effects of effective *Restorer-of-fertility* genes (*Rf*) on plant height in P-type CMS hybrids (Miedaner et al. 2017). The genetic mapping of further GA-sensitive dwarfing genes like *Ddw3* on chromosome 1R (Stojałowski et al. 2015), *Ddw4* on chromosome 3R (Kantarek et al. 2018) or other dominant dwarfing genes identified in Estonian (Kukk and Tupits 1996) and Ukrainian (Skoryk et al. 2010) rye breeding programs may provide alternative options to breeders for improving lodging resistance, and increasing the harvest index and yield potential of rye.

The genetic adaptation of rye to a changing climate is of particular importance for grain production, as rye is mainly cultivated on light soils with low fertility and water holding capacity. An average drought induced grain yield reduction of 23.8% has been reported for hybrid rye under a non-irrigated regime, compared with an irrigated control regime (Hübner et al. 2013). Up to 57% grain yield reduction was observed in controlled environments under different drought regimes (Kottmann et al. 2016). Noteworthy in this context is the growing body of evidence that the GA class of plant hormones has pivotal relevance to the response of plants to abiotic stresses (Colebrook et al. 2014). Positive effects on grain yield, lodging tolerance as well as drought tolerance in tef [*Eragrostis tef* (Zucc.) Trotter] and finger millet (*Eleusine coracana* Gaertn) have been reported as a result of chemically induced GA deficiency (Plaza-Wüthrich et al. 2016). In rice, induced mutants of the GA deactivation gene *GA2-oxidase 6* (*GA2ox6*) moderately reduced GA concentration and reprogrammed transcriptional networks, leading to reduced plant height, more productive tillers, an expanded root system, higher water use efficiency and photosynthesis rate, and elevated abiotic and biotic stress tolerance (Lo et al. 2017). In accordance with these observations, the analysis of the *Arabidopsis* NAC-like GIBBERELLIN SUPPRESSING FACTOR (GSF) have been shown to perform a novel function in the regulation of gibberellin biosynthesis (Chen et al. 2019a). The ectopic expression of GSF lacking a transmembrane domain (GSF-TM) caused a dwarf phenotype, which was correlated with the upregulation of *GA2ox2/6* and increased drought tolerance compared to the wild-type plants. The described progress in our understanding of the importance of GA-homeostasis for drought stress tolerance, thus, implies that the established semi-dwarf seed parent lines most likely confer both lodging and drought tolerance in rye, as the *Ddw1* mutant alters the GA content in a favorable manner. Interestingly, a root-derived precursor of bioactive gibberellins has recently been described to mediate thermo-responsive shoot growth in *Arabidopsis* suggesting that root-to-shoot translocation of GA12 enables flexible growth

Fig. 9.2 In 2020, seeds of semi-dwarf P-type experimental hybrids have been produced in rye for the first time. The GA-sensitive dwarfing gene *Ddw1* results in a natural reduction of plant height by ca. 30% as compared to their near-isogenic tall full-sibs in the background

responses to ambient temperature changes (Camut et al. 2019). According to this research, it can be assumed, that the root signal cannot trigger stem growth in GA-sensitive semi-dwarf rye, at least to the same extent as in wild-type plants. As a consequence, dry matter in semi-dwarf rye is allocated to the grain rather than to the stem, even under elevated temperature. In addition to genes governing the biosynthesis of GA, the high-quality reference genome assemblies (Li et al. 2021; Rabanus-Wallace et al. 2021) allow researchers to comprehensively approach further important players in plant responses to drought stress like genes controlling abscisic acid biosynthesis and signaling (Seiler et al. 2014; Vishwakarma et al. 2017).

The genetic modification of cell walls to enhance plant mechanical strength has been suggested as another approach for the improvement of lodging resistance (Dill 1983; Liu et al. 2018). Common components of cell wall include cellulose, hemicelluloses, pectins, lignins, and pectic polysaccharides, of which cellulose is the main constituent of cell wall in most plant species. Brittleness mutants commonly show variation in cellulose content (Davison et al. 1924; Rao et al. 2013). In rye, brittleness of straw has been observed in 21% of inbred lines developed from a random mating rye population (Davison et al. 1924) and one brittle stem associated mutation has been genetically mapped on chromosome 5R (Konovalov et al. 2010).

Leaf angle (LA) is a critical agronomic trait, with more upright leaves allowing higher planting density, leading to efficient light capture and increased yield (Cao et al. 2020). In a recessive rye *liguless* (*lg*) mutant (De Vries and Sybenga 1984) the absence of ligular tissue results in erect leaves and variation in LA. This rye *lg* mutant has been mapped on the long arm of chromosome 2R (Benito et al. 1991; Korzun et al. 1997; Dobrovolskaya et al. 2009). The rye ortholog of *lg1* from maize (Moreno et al. 1997), *SECCE2Rv1G0127610,* maps proximal to the self-incompatibility locus *Z* on chromosome 2R (Fig. 9.1). Two SNP markers from the Rye600k array (Bauer et al. 2017) serve as promising tools for introgression of the rye *lg* allele in elite germplasm and SMART breeding of new rye varieties with improved plant architectures. *SECCE2Rv1G0127610* encodes a SQUAMOSA-PROMOTER BINDING PROTEIN-LIKE (*SPL*) transcription factor (Rabanus-Wallace et al. 2021). The *SPL* family of transcription factors is functionally diverse, controlling a number of fundamental aspects of plant growth and

Fig. 9.3 Ferdinand III. von Lochow within elite germplasm of his Petkuser winter rye that was taller than a man in 1912. The outstanding international success of his rye breeding program distinguishes Ferdinand III. von Lochow as a pioneer of systematic plant breeding. The high productivity of Petkuser winter rye served to establish the seed parent pool in modern hybrid rye breeding programs. Carrying the initials of his surname, the 'Lo7' sequence assembly (Rabanus-Wallace et al. 2021) recognizes the visions of Ferdinand III. von Lochow to increase efficiency in the genetic improvement of rye

development, including vegetative phase change, flowering time, branching, and leaf initiation rate (see Preston and Hileman 2013 and Liu et al. 2016 for review). As variation in branching and developmental rate impact biomass and yield, allele mining for the 24 predicted 'Lo7' *SPL* genes (Rabanus-Wallace et al. 2021) merits further research.

The control of flowering is central to reproductive success in plants and has a major impact on grain yield in crop species (Cockram et al. 2007). In wheat and barley, the closely linked genes *VRN-1* and *VRN-2* are primarily responsible for the induction of flowering subsequent to long periods of cold exposure (Tranquilli and Dubcovsky 2000; Danyluk et al. 2003). The alleles for winter growth habit are the ancestral forms of these genes in the Triticeae, and independent mutations in both genes have resulted in the recurrent generation of spring forms in the temperate cereals (Cockram et al. 2007). *VRN-1* is a meristem identity gene and dominant for spring growth habit (Yan et al. 2003). Likewise, the vernalization gene *VRN-2* is a dominant repressor of flowering that is down-regulated by vernalization (Yan et al. 2004). Notably, the outstanding performance of hybrids has not yet been successfully transferred to spring rye. The spring growth habit gene in rye is located on the long arm of chromosome 5R (De Vries and Sybenga 1984; Melz 1989; Plaschke et al. 1993; Philipp et al. 1994). The rye orthologs of *VRN-1* and *VRN-2* in the 'Lo7' assembly (Fig. 9.1), *SECCE5Rv1G0353290* and *SECCE5Rv1G0373910*, respectively, are separated by 145.8 Mbp and a high-throughput SNP marker for *VRN1* (Erath et al. 2017) as well as the three *VRN2*-SNPs (Bauer et al. 2017) enable an efficient introgression of the mutant alleles in elite germplasm by precision breeding. For this purpose, the Rye600k genotyping array (Bauer et al. 2017) and both high-quality rye reference genome sequences (Li et al. 2021; Rabanus-Wallace et al. 2021) represent genomic resources that enable marker-assisted development of near-isogenic lines for *VRN-1* and *VRN-2,* and to estimate the tolerated freezing temperature of the allele for winter and spring growth habit with unprescedented precision, as a prerequisite to identifying target environments of spring type rye hybrids.

Mutations in the gene *Heading date 1* (*Hd1*) offer a further option to sustain rye productivity in a changing climate by increasing the flexibility in varietal flowering time. *Hd1* is a key regulator of photoperiodic flowering in plants of short-day zones (Cockram et al. 2007) and has been mapped to chromosome 6R (Święcka et al. 2014). *Hd1* is encoded by *SECCE6Rv1G0407180* in rye (Fig. 9.1) and represented by 5 SNP markers on the rye 600 k SNP array (Bauer et al. 2017). The effects of *ScHd1* observed in lines *per se* included enhanced performance of heading date as well as spike length, number of spikelets per spike, and thousand-grain weight (Święcka et al. 2014). As demonstrated in a case study based on the 'Weining' rye genome assembly (Li et al. 2021), the rye reference genome sequences facilitate comprehensive exploitation of knowledge on the genetics and genes involved in cereal flowering pathways (Cockram et al. 2007) for the development of varieties with different developmental timing, which functions as a mitigation option (Sheehan and Bentley 2020) for high-performance rye cultivars in a changing climate. A promising, yet under-considered target in rye breeding programs appears to be the pseudo-response regulator gene *Photoperiod-1* (*Ppd-1*), residing on the short arm of chromosome 2R (Fig. 9.1), which controls photoperiod-dependent floral induction and can be used to form an inflorescence with increased number of grain producing spikelets in wheat (Boden et al. 2015).

The leaf cuticle serves as the major barrier preventing nonstomatal water loss and contributes to protect plant surfaces from pathogens and ultraviolet radiation (Ji and Jetter 2008). Waxes are integral components of plant cuticles and are found either deposited within the cutin matrix (intracuticular wax) or accumulated upon the leaf surface as epicuticular wax (Yeats and Rose 2013). In rye, the gene *wa1* is associated with waxless cuticles and has been mapped to chromosome 7R (Korzun et al. 1997). Further progress in identifying genes involved in cuticle biosynthesis of rye has been achieved by mapping the gene *wax1* on chromosome 2R (Góralska et al. 2020). The 'Lo7' genome assembly will allow to further narrow down the 27.1 Mbp segment on chromosome 2R (Fig. 9.1), to ultimately identify the *wax1* gene and improve our understanding of cuticle biosynthesis at the molecular level.

9.2.2 Abiotic Stress Tolerance

Among small-grain cereals, rye possesses an exceptionally high level of tolerance to abiotic stress conditions. For example, rye has an excellent ability to withstand primary stress caused by below-freezing temperatures (Limin and Fowler 1991). Winter hardiness is important for establishing highly productive rye varieties in marginal high latitude production regions with continental climates and severe winters which require high levels of frost tolerance (FT). Recently, new insights on winter hardiness of rye in a Canadian cropping region have been reported. Winter field survival of 96 random mating populations and a few inbred lines revealed that desired and indispensable alleles for winter hardiness are available in the Petkus pool, and may exist at low frequency in the Carstens pool as well (Bahrani et al. 2019). The broad genetic diversity available in both heterotic genepools may, therefore, be useful to identify valuable alleles and mechanisms of winter hardiness in rye. However, winter hardiness is a complex trait influenced by diverse environmental factors such as the presence of snow cover or ice encasement, soil fertility, and biotic factors like disease pressure (Limin and Fowler 1991). The *Fr2* locus on chromosome 5R in rye has previously been reported to improve the recovery of plants after winter (Li et al. 2011b; Erath et al. 2017). Notably, none of the analyzed *C-repeat Binding Factor* (*Cbf*) genes residing at the *Fr2* locus (Fig. 9.1) affected post-winter survival. This result indicates that factors beyond the *Fr2* locus might control the outstanding winter hardiness of rye. This assumption is supported by the observation that the low-temperature tolerance (LTT) of the wheat cultivar 'Norstar' was not significantly altered by a translocation of the long arm of chromosome 5R and the replacement of the *Cbf* and *Vrn1* cluster with the orthologous rye locus originating from the rye cultivar 'Puma', which exhibits exceptional LTT (Rabanus-Wallace et al. 2021). In fact, a functional genomics approach identified rye candidate genes involved in LTT and provided empirical evidence for a MYB-like transcription factor, *MYB3R1* (Cao et al. 2016), corresponding to *SECCE3Rv1G0197840,* which resides in the vicinity of two SNP markers from the Rye600k array (Bauer et al. 2017) as a novel target gene for the improvement of survival rates after winter. Likewise, the O-methytransferase *OMT1* is up-regulated by low temperatures (NDong et al. 2003), and requires further research. *OMT1* corresponds to *SECCE6Rv1G0450300* in the 'Lo7' genome assembly and maps to the distal end of the long arm of chromosome 6R (Fig. 9.1). *OMT1* is represented by 4 SNP markers originating from the Rye600k genotyping array (Bauer et al. 2017) that will allow researchers to systematically approach the genetic variation of this cold-responsive gene in rye germplasm collections. This applies to the low-temperature responsive genes *RLT14.12* and *RLT14.21* (Zhang et al. 1993), and two chitinase genes *CHT9* and *CHT46* (Yeh et al. 2000) as well. The expression of *RLT14.12* and *RLT14.21* is higher in the more frost hardy Canadian rye cultivar 'Puma' as compared with the less frost tolerant rye variety 'Rhayader' (Zhang et al. 1993). Both genes belong to a cluster of eight genes residing on chromosome 2R of the 'Lo7' genome assembly (Rabanus-Wallace et al. 2021; Fig. 9.1). *CHT9* and *CHT46* accumulated in rye leaves following only five weeks of cold acclimation, and this may be related to a developmental process that occurs at a later phase of cold acclimation, possibly at a time when plants in the field are more prone to be exposed to subzero temperatures and/or low-temperature pathogens (Yeh et al. 2000). The chitinase genes *SECCE1Rv1G0031140* and *SECCE1Rv1G0031150* in the 'Lo7' genome assembly confirmed two predicted copies at the *CHT46* locus on chromosome 1R (Yeh et al. 2000). Moreover, the 'Lo7' genome sequence (Fig. 9.1) closed a knowledge gap by mapping *CHT9* on chromosome 3R based on the corresponding gene *SECCE3Rv1G0185830*. A total of 6 SNP markers represent an efficient means to approach both *CHT46* and *CHT9* in dedicated experiments. Previously analyzed genes of the frost response network in rye like *Inducer of Cbf Expression* (*ICE*), that are basic helix-loop-helix transcription factors and bind to promoters of the *Cbf* gene family, or dehydrins, also known as *Late Embryogenesis Abundant II* (LEA II) (Li et al. 2011a), map on chromosomes 3R, 4R, 5R and 7R (Fig. 9.1). Antifreeze proteins (AFPs) encoding genes like *LTP1* and *LTP2* from rye (Doxey et al. 2006) confer the ability to survive at subzero temperatures and are essential for the survival of organisms in cold environments (Naing and Kim 2019). *LTP1* and *LTP2* belong to a gene family of 28 genes residing on all rye chromosomes except of 6R (Rabanus-Wallace et al. 2021) and map to a cluster of 10 genes within a 405 kb segment on chromosome 3R (Fig. 9.1).

Genome-wide approaches led to the discovery of further candidates. Signatures of positive selection were significantly stronger in ten low-temperature induced (LTI) gene trees (Fig. 9.1; Table 9.1), the selection signals having accumulated after the rice and Pooideae split but before the *Brachypodium* divergence (Viegland et al. 2013). The orthologs of these genes exemplify that the rye reference genome assemblies (Li et al. 2021; Rabanus-Wallace et al. 2021) facilitate with an unprecedented precision an in-depth characterization of candidate genes beyond the *Cbf* regulatory hub (Campoli et al. 2009; Thomashow 2010) residing at the *Fr2* locus and offer a novel opportunity to gain insights into the genetic architecture of winter hardiness in rye. A total of 30 SNP markers originating from the Rye600k genotyping array provide a starting point to assess the genetic diversity contained within eight of the LTI genes (Table 9.1). The establishment of experimental hybrids with a contrasting genetic make-up would enable to study effects of individual LTI genes on winter field survival by pairwise comparisons between alternative genotypes. The rich genetic diversity of rye in the target trait (Bahrani et al. 2019) promises an increase in the power of this approach, since genotypes with very low winter field survival can be leveraged as CMS-testers to identify germplasm carrying superior alleles increasing FT of the elite gene pool.

Genes involved in the anthocyanin biosynthesis pathway count among further targets that might improve winter field survival, as the most cold-hardy rye populations exhibit an overall higher abundance and diversity of anthocyanins compared to the less hardy cultivars (Bahrani et al. 2019). Compared to *Arabidopsis* (Shi and Xie 2014), maize (Petroni et al. 2014), and rice (Zheng et al. 2019), the regulation of anthocyanin biosynthesis is less well-described in rye. However, genetic diversity in the anthocyanin pigmentation of rye coleoptiles served to identify the six recessive *viridis* (lat.: green) genes *vi1* to *vi6* (Voylokov et al. 2015). A novel 20 k custom wheat/rye/triticale SNP array, that contains 5 k SNPs selected from the rye 600 k SNP array (Bauer et al. 2017), and bulked segregant analysis (Michelmore et al. 1991) enabled the genetic mapping of genes on chromosomes 2R, 3R, 4R and 7R (A. V. Voylokov, personal communication); a first step toward unraveling the anthocyanin pathway in rye.

Aluminum (Al) toxicity is one of the major constraints to agricultural production on acid soils (Zhang et al. 2019). As the main rye cultivation area in the cold and temperate climate of Northern latitudes coincides with one of two global belts with acid soils (Von Uexküll and Mutert 1995), rye has evolved the most effective tolerance of Al of cultivated Triticeae species (Collins et al. 2008). Two loci controlling Al tolerance, *Alt1* and *Alt4*, have been mapped on the short arms of chromosome 6R (Gallego et al. 1998) and 7R (Matos et al. 2005; Fontecha et al. 2007; Collins et al. 2008; Shi et al. 2009), respectively. Recently, a comparative genetics approach identified *MATE3* as candidate gene residing at the *Alt1* locus (Santos et al. 2020). *MATE3* corresponds to *SECCEUnv1G0528200* facilitating the assignment of this 'Lo7' gene model to the short arm of chromosome 6R. Further markers need to be developed and integrated in the rye reference genome sequence (Rabanus-Wallace et al. 2021) to exploit this genomic resource for fine mapping and cloning of the *Alt1* gene. In contrast, the *Alt4* locus could be perfectly integrated into the 'Lo7' genome sequence based on the available comprehensive sequence information (Collins et al. 2008). The B1-B4 interval (Collins et al. 2008) covers 5.12 Mbp in the 'Lo7' assembly delimited by *SECCE7Rv1G0467250* and *SECCE7Rv1G0467750*. While the *Alt4* locus in the Al-tolerant rye genotype contains a cluster of 5 genes homologous to the single-copy Al-activated malate transporter (*ALMT1*) Al-tolerance gene of wheat (Collins et al. 2008), the *ALMT1* ortholog *SECCE7Rv1G0467600* is a single-copy gene residing in the 5.12 Mbp interval of the 'Lo7' reference genome sequence (Fig. 9.1). This observation is in line with copy number variation between Al-tolerant and Al-intolerant rye genotypes that has been reported for the *ALMT1* locus (Collins et al. 2008). The 'Lo7' genome sequence was established for an inbred line whose Al tolerance is uncharacterized, and thus, the assembly may well not contain the target gene. Drawbacks like this are well known from previous studies where disruption of collinearity, as well as non-conserved gene content, were observed between related species or even within the same species. Non-conserved gene content can disrupt the final step of map-based cloning projects for a genomic library of a genotype that contains the gene of interest (Isidore et al. 2005), as has been discussed for *Alt4* (Shi et al. 2009). The multidrug and toxic compound extrusion (MATE) protein encoding gene *FRDL2* represents an alternative gene that may be involved in Al detoxification in rye (Yokosho et al. 2010). *FRDL2* corresponds to *SECCE1Rv1G0043400* and maps to chromosome 1R in the 'Lo7' genome assembly (Fig. 9.1). Three SNP markers representing *SECCE1Rv1G0043400* will create novel momentum and support the unambiguous indexing of further major Al-tolerance genes like *Alt3*, that remain to be mapped in rye (Gallego and Benito 1997).

9.2.3 Disease Resistance

Resistance breeding in rye is presently focused on leaf (*Puccinia recondita* f.sp. *secalis*) and stem rust (*P. graminis* f.sp. *secalis*), ergot (*Claviceps purpurea*) and Fusarium diseases (Geiger and Miedaner 2009).

Leaf rust is the most important windspread pathogen in rye and is endemic in all rye growing regions (Wehling et al. 2003). Under natural infection in continental climate conditions, yield losses of up to 40% have been reported

Table 9.1 Rye orthologs of low-temperature-induced (LTI) genes under positive selection

LTI-Gene[a]	Gene ID	'Lo7' gene ID[b]	Annotation[b]	Putative function(s)	N° of SNPs[c]
BRADI2G38290.1	ATP23	SECCE2Rv1G0108050	Mitochondrial inner membrane protease ATP23	Mitochondrial protein processing	1
BRADI5G25050.1	F3H	SECCE2Rv1G0128670	Flavanone 3-hydroxylase	Flavanoid biosynthesis	4
BRADI2G55070.1	HEBP	SECCE3Rv1G0198580	Soul heme-binding family protein	Red/far-red light signaling	1
BRADI2G58050.1	FBA	SECCE3Rv1G0202100	Fructose-bisphosphate aldolase	Glycolysis	6
BRADI4G34170.1	rps16	SECCE5Rv1G0335530	30S ribosomal protein S16-like	Translation	
BRADI4G36800.1	PLD	SECCE5Rv1G0344160	Phospholipase D	Cell membrane lipid hydrolysis/signaling	4
BRADI1G13640.1	Hsp40	SECCE5Rv1G0347900	Chaperone protein dnaJ, putative	Co-chaperone activity	9
BRADI4G09430.1	XIP-I	SECCE6Rv1G0418540	Xylanase inhibitor protein 1	Disease response	4
BRADI3G17200.1	tyrS	SECCE7Rv1G0491530	Tyrosine–tRNA ligase	Translation	
BRADI1G35200.2	NPSN	SECCE7Rv1G0502960	Novel plant snare, putative	Membrane receptor/protein transport	1

[a]According to Viegland et al. (2013)
[b]According to Rabanus-Wallace et al. (2021)
[c]According to Bauer et al. (2017)

(Kobylyansky and Solodukhina 1983). A comprehensive survey of 2,500 rye populations revealed genotypes resistant to the pathogen in only 2% of the studied germplasms and resulted in the identification of 6 major genes (Kobylyansky and Solodukhina 2015). A genetic analysis indicated that these race-specific genes reside on distinct loci in the rye genome (Solodukhina 2002). Due to extensive genetic analysis, *Lr-a* and *Lr-b* (Solodukhina 2002) were renamed *Pr1* and *Pr2,* and mapped to the proximal part of rye chromosome 6RL and the distal part of chromosome 7RL, respectively (Wehling et al. 2003). Likewise, the three dominant and race-specific resistance genes *Pr3*, *Pr4* and *Pr5* have been mapped to the centromeric region of rye chromosome 1R (Roux et al. 2004). Notably, *Pr3* maps to the short arm of chromosome 1R closely linked to the *S* locus in rye. Two genes encoding NBS-LRR disease resistance like proteins, *SECCE1Rv1G0014190* and *SECCE1Rv1G0014220*, are residing in the *S* locus genomic regions 132 kbp and 10 kbp distal from a candidate for the *S* gene, *SECCE1Rv1G0014240* (Fig. 9.1). The position of *Pr3* relative to *Xiag95* and the *Sec-1* locus indicates that the *Pr3* gene is not identical with *Lr26* which confers resistance to leaf rust in T1BL·1RS translocation wheat lines (Singh et al. 1990) because the latter is reported to be closely linked to *Xiag95* and to the seed-storage protein gene *Sec-1* located near to the end of the chromosome arm 1RS (Hsam et al. 2000).

Table 9.2 Rye orthologs of the benzoxazinones (BX) biosynthesis pathway

Maize gene model[a]	Gene symbol	'Lo7' gene ID[b]	Annotation[b]	N° of SNPs[c]
CAA54131.1	Bx1	SECCE7Rv1G0477710	Tryptophan synthase alpha chain	1
GRMZM2G085661	Bx2	SECCE7Rv1G0477720	Cytochrome P450	4
GRMZM2G167549	Bx3	SECCE5Rv1G0298510	Cytochrome P450	
GRMZM2G172491	Bx4	SECCE5Rv1G0298500	Cytochrome P450	5
GRMZM2G063756	Bx5	SECCE5Rv1G0298490	Cytochrome P450	3
GRMZM6G617209, AC148152.3	Bx6, Bx13	SECCE7Rv1G0518770	2-oxoglutarate (2OG) and Fe(II)-dependent oxygenase superfamily protein, putative	
		SECCE7Rv1G0518760	2-oxoglutarate (2OG) and Fe(II)-dependent oxygenase superfamily protein, putative	
GRMZM2G441753	Bx7	SECCE5Rv1G0336260	O-methyltransferase-like protein	
GRMZM2G085054, GRMZM2G161335	Bx8, Bx9	SECCE4Rv1G0261920	UDP-glycosyltransferase	5
		SECCE1Rv1G0047950	O-methyltransferase-like protein	5
GRMZM2G311036, GRMZM2G336824, GRMZM2G127418	Bx10, Bx11, Bx14	SECCE1Rv1G0047970	O-methyltransferase-like protein	5
		SECCE1Rv1G0047980	O-methyltransferase	5
		SECCE1Rv1G0047990	O-methyltransferase-like protein	
		SECCE1Rv1G0048000	O-methyltransferase-like protein	
		SECCE1Rv1G0048010	O-methyltransferase-like protein	
		SECCE1Rv1G0048020	O-methyltransferase-like protein	
		SECCE1Rv1G0048030	O-methyltransferase-like protein	4
GRMZM2G023325	Bx12	SECCE6Rv1G0450290	O-methyltransferase-like protein	

[a]According to Portwood et al. (2019)
[b]According to Rabanus-Wallace et al. (2021)
[c]According to Bauer et al. (2017)

Four additional genes designated *Pr-d, Pr-e, Pr-f* and *Pr-n* have been mapped on chromosomes 1R, 2R and 6R (Roux et al. 2007).

Rye is known to reduce weed growth in organic farming regimes, and has gained importance as a cover crop or mulch for allelopathic weed control, e.g., in maize, cotton, and soybean fields (Schulz et al. 2013). The main secondary metabolites that function as bioherbicides are glucosylated benzoxazinones (BX) (Niculaes et al. 2018). BXs were first discovered and characterized in rye (Virtanen and Hietala 1955a, b) and serve as part of defense mechanisms directed toward insects and microbial pathogens in economically important crops such as maize, wheat, and rye, while neither rice, oat, sorghum, nor cultivated barley produce BXs (Frey et al. 1997; Niemeyer 2009). BXs were found to be present in mature cereal grains, being abundant in rye, and attracted studies on putative pharmacological and health-protecting properties of these natural chemical compounds (Adhikari et al. 2015; Tanwir et al. 2017). The BXs biosynthetic pathway has been most extensively studied using maize as a model (Niculaes et al. 2018). In rye, the core genes *Bx1* to *Bx5* were mapped to chromosomes 5R and 7R (Nomura et al. 2003), while the modifying genes encoding *Bx glucosyltransferase* (*GT*) and *Bx-glucoside glucosidase* (*Glu*) were located on chromosomes

4R and 2R (Sue et al. 2011). Sequence information has been generated for the *Bx1* to *Bx7* genes in rye (Bakera et al. 2015; Tanwir et al. 2017; Bakera and Rakoczy-Trojanowska 2020) and used to identify marker-trait associations (Milczarski et al. 2017; Rakoczy-Trojanowska et al. 2017). Based on the fully elucidated BX biosynthesis in maize (Frey et al. 2009; Portwood et al. 2019), the 'Lo7' genome sequence enabled the completion of a list of genes encoding the enzymes of the pathway in this small-grain cereal (Fig. 9.1; Table 9.2). Thus, the rye genome assemblies (Li et al. 2021; Rabanus-Wallace et al. 2021) extensively address the well-appreciated need for further investigation into the specific genes that are involved in the biotransformations of BXs in rye (Ahikari et al. 2015), and will support further research toward a better understanding of BX biosynthesis in this cereal crop. There is accumulating evidence that BXs may have important functions in regulating a broad range of defense responses, flowering time, auxin metabolism, iron uptake and perhaps aluminum tolerance in maize (Zhou et al. 2018). Recent genomic advances in rye offer targeted investigations of natural variation leading to the discovery of yet unknown genes and functions of BX metabolism.

A reduction of post-harvest crop losses in the course of storing grain in warehouses is an important factor in enhancing food safety (Schmidt et al. 2018; Hamel et al. 2020). A rye seed chitinase-a (RSC-a) has been shown to inhibit fungal growth more effectively than rye seed chitinase-c (RSC-c; Taira et al. 2002). The cloned cDNAs (Ohnuma et al. 2002, 2004) are represented by *SECCE6Rv1G0448630* and *SECCE6Rv1G0448610* and map within a common 72 kb segment on chromosome 6R (Fig. 9.1). Five nearby SNPs on the Rye600k array (Bauer et al. 2017) demonstrate diversity among rye inbred lines in the coding sequence of RSC-a *syn. SECCE6Rv1G0448630,* and enable MAS of this gene in rye improvement programs. To conclude, the novel rye genome assemblies facilitate the systematic identification of chitinase genes on a genome-wide scale, and further study on their impact in host–pathogen interactions (Langner and Göhre 2016) in rye.

A functional α-amylase inhibitor (BIII) from rye seeds, that reduced the activity of α-amylases of larvae of coleopteran pests (Dias et al. 2005), maps to chromosome 5R in the 'Lo7' genome assembly (Fig. 9.1). Likewise, the dimeric α-amylase inhibitor encoding gene *DAI* (Wang et al. 2010) maps to chromosome 3R (Fig. 9.1).

9.2.4 Grain Quality

Rye grain-storage proteins have only secondary relevance to rye bread production; however, they are important for the typical flavor and taste of rye sourdough bread (Deleu et al. 2020). As a consequence, neither of the three separate seed-storage protein loci gained serious attention in rye improvement programs: the ω- and 40 K γ-secalins at the *Sec-1* locus on the short arm of chromosome 1R (Shewry et al. 1983, 1984; Carillo et al. 1992; Clarke et al. 1996; Shimizu et al. 1997; Chai et al. 2005; Li et al. 2016a, 2016b, 2021), the high molecular weight (HMW) secalins at *Sec-3* on the long arm of chromosome 1R (Lawrence and Shepherd 1981; Singh and Shepherd 1984; De Bustos et al. 2001; De Bustos and Jouve 2003) and the 75 K γ-secalins encoded at *Sec-2* locus on the short arm of chromosome 2 (Malyshev et al. 1998). Instead, a high grain weight and resistance to pre-harvest sprouting have historically been the main target traits to genetically improve the baking quality of rye (Geiger and Miedaner 2009) although water-extractable arabinoxylans (WEAX) are also an important focus of rye dough improvement efforts (Cyran and Cygankiewicz 2004; Hansen et al. 2004; Buksa et al. 2010; Stępniewska et al. 2019; Deleu et al. 2020).

Thousand-grain weight (TGW) can be used as an effective proxy for starch content during selective breeding programs (Geiger and Miedaner 2009). Assessed in a period of 26 years, TGW increased by a slightly higher albeit not significant rate in hybrids than in population varieties which, at 38.6 g and 39.0 g respectively, reached about the same level (Laidig et al. 2017). These results indicate that population breeding is not inferior to hybrid breeding for the genetic improvement of this yield component. Indeed, high heritability estimates and a preponderance of additive inheritance for TGW have been reported for random mating rye populations (Wolski et al. 1972). In fact, a modified recurrent mass selection scheme enabled a simple, yet highly effective strategy to substantially improve TGW from initially 19 g up to 57 g in a random mating rye population within two decades of breeding (Dill 1983, 1989). This impressive selection gain suggests the influence of major genes controlling grain phenotypes in rye. Subsequent research integrated SSR markers and aimed to test the hypothesis of complementary action between two major genes, designated *Kernel weight 5* (*KW5*) and *KW7,* residing on chromosomes 5R and 7R, respectively (Wricke 2002). Both genes appear to be responsible for a substantial part of the heritable variance of grain weight in rye. Likewise, an independent study reported regulation of grain size in rye by two complementarily acting major genes *lg* (large grain) and *tg* (thick grain; Skoryk et al. 2010). The *KW5* marker *WMS6* maps between *SECCE5Rv1G0370050* and *SECCE5Rv1G0370060*, while the *KW7* marker *SCM40* maps between *SECCE7Rv1G0477730* and *SECCE7Rv1G0477740* (Table 9.3). In silico mapping of the DArT marker loci *XrPt-400590* and *XrPt-507373* in the 'Lo7' genome assembly confirmed the hypothesis that the QTL *QTgw-5R* includes the *KW5* marker *WMS6* (Hackauf et al. 2017a) and emphazises the relevance of this subgenomic region in terms of the genetic control of grain weight in rye. Additionally, the rye genome sequences enable the comparison of *QTgw-5R* with genes that have recently been reported to control grain size and weight in barley (Wang et al. 2019). Among the 395 genes residing in the 25.3 Mbp *QTgw-5R* interval of the 'Lo7' genome assembly, six genes are prime candidates to control grain weight in rye, according to knowledge gained in rice and barley (Fig. 9.1; Table 9.3). The gene *SECCE5Rv1G0370090*, mapping less than 365 kb distal from *WMS6*, appears to be particularly promising. *SECCE5Rv1G0370090* is predicted to encode a WD40-repeat-like protein (Rabanus-Wallace et al. 2021) and shares 91% identity at the amino acid level to the rice gene *Os03g0123300* (Table 9.3). *Os03g0123300* is, in turn, involved in maintaining seed size formation by mediating the exit from mitotic cell division to enter the endoreduplication cycles in rice endosperm (Su'udi et al. 2012). The endoreduplication event is the final step for seed size regulation during endosperm development prior to undergoing programmed cell death (Su'udi et al. 2012), and could explain the observed effect on grain weight by the predicted major gene *KW5* in rye (Wricke 2002), which may owe to close linkage between the marker *WMS6* and *SECCE5Rv1G0370090* (Table 9.3). However, *SECCE5Rv1G0367670* residing proximal to *WMS6* appears to represent an interesting candidate gene for *QTgw-5R* as well. *SECCE5Rv1G0367670* maps to an 11.6 Mbp chromosome segment defined by markers *c4715_668* and *c6115_374*, respectively, that have previously been associated with a significant increase in the performance of TGW in an inbred line (Mahone et al. 2015) originating from a rye introgression line library (Falke et al. 2008). *SECCE5Rv1G0367670* is predicted to encode an Expansin-like protein (Rabanus-Wallace et al. 2021) and shares 84% identity at the amino acid level to *Os03g0132200* that increases cell growth and seed development in rice, and which is controlled by *Grain Width 2* (*GW2*), a RING-type E3 ubiquitin ligase-encoding gene (Choi et al. 2018). In contrast to *KW5*, none of the QTL for TGW identified in rye hybrids so far (Miedaner et al. 2012; Mahone et al. 2015; Hackauf et al. 2017a) could be associated with the *KW7* marker locus *SCM40*, which hampers delineation of the target segment in the 'Lo7' genome sequence. Two genes residing proximal and distal of *SCM40* appear to be interesting due to their known effects on grain size and weight in rice and barley (Wang et al. 2019).

TGW is significantly affected by *Restorer-of fertility* (*Rfp*) genes which explain, at least in part, the yield penalty observed in male fertile P-type CMS hybrids (Miedaner et al. 2017). Knowledge of major genes like *KW5* and *KW7* may, thus, offer a chance to develop genome-based precision breeding strategies to counterbalance the linkage drag effects of effective *Rfp* genes and simultaneously minimize the costs of restoration in terms of grain yield.

Since starch is by far the major component of the mature grain, it has been assumed that variation in the capacity for starch synthesis during grain filling can influence final grain weight (Fahy et al. 2018; Qu et al. 2018). In the inbred line 'Lo7', *SECCE4Rv1G0272640* encodes the granule-bound starch synthase (*GBSS*; Fig. 9.1), responsible for the synthesis of amylose and the extra-long-chain fraction of amylopectin. *GBSS* was the first gene of the starch biosynthesis pathway that has been genetically mapped (Korzun et al. 1997) and comprehensively characterized at the sequence level in rye (Xu et al. 2009). Subsequent research identified the key catalytic sites of the gene and a significant negative correlation between enzyme activity and amylose content, as well as amylopectin/amylose ratio (Meng et al. 2014). The 'Weining' rye genome assembly served to conduct an in-depth analysis of the starch biosynthesis pathway in rye (Li et al. 2021), and offers new opportunities to understand genes and pathways that control grain size and weight in rye. Dedicated research may be supplemented by recent progress concerning the genetic control of grain weight (Brinton and Uauy 2019; Wang et al. 2019; Olsen 2020). Next to genome-based prediction of breeding values for TGW (Auinger et al. 2016), this knowledge makes up a fundamental basis upon which to further improve the rye grain as a source supplying calories and food as well as industrial raw materials to mankind. At this point, the strong negative correlation between protein and starch content of the rye grain (Miedaner et al. 2012; Kunkulberga et al. 2017) indicates that the comprehensive knowledge on the rye endosperm storage protein loci (Li et al. 2016a, 2021) holds potential for genome-based precision breeding in rye. For instance, phenotypic differences in grain weight may be affected by the *prolamin-box binding factor1* (*PBF1*) on chromosome 5R (Fig. 9.1), a *cis* regulatory element that controls the expression of seed-storage protein genes (Ravel et al. 2006; Lang et al. 2014).

9.3 QTL Mapping in Rye Hybrids—The Gene Discovery Tool

9.3.1 Agronomic Traits

A Scopus database search using the terms *Oryza* & QTL, *Triticum* & QTL, *Hordeum* & QTL as well as *Secale* & QTL in January 2021 returned 2,381, 2,241, 919 and 82 results, respectively, and demonstrates, that knowledge on QTL in the 8 Gb genome of rye is still comparably scarce. The pioneering QTL studies on line *per se* performance of agronomic traits have been conducted in two bi-parental populations once RFLP maps covering the entire rye genome were established (Börner et al. 2000; Milczarski and Masojć 2003). In both studies the influence of major developmental loci biased the QTL identification, indicating a need for further research and approaches to discover non-developmentally related QTL in rye. Indeed, inroads to this goal were made in a subsequent study on *per se* performance in plant height and yield parameters using two high-density genetic maps and single plant precision phenotyping (Myśków et al. 2014).

A sophisticated strategy for QTL discovery from unadapted germplasm in the genetic background of elite breeding lines has been applied in rye (Falke et al. 2008), and enabled the identification of segments in the rye genome with positive effects on agronomic traits including grain yield (Falke et al. 2009). Subsequently, linear model analysis has been applied extensively to the localization of QTL responsible for phenotypic variation among lines carrying multiple large donor chromosome segments (Mahone et al. 2013). As demonstrated for two donor segments carrying favorable QTL for TGW, genome-wide prediction methods based on DNA profiles established with the Rye5k array (Haseneyer et al. 2011) enabled to describe introgressed QTL more precisely (Mahone et al. 2015). The 'Lo7' genome assembly facilitates to further specifiy the position of the QTL on chromosome 5R and revealed that this QTL is linked to *KW5*, while the second QTL with

Table 9.3 The *KW5* and *KW7* genomic segments controlling grain weight in rye

Chr	QTL[a]	Marker/gene ID	Position (bp)[b]	Annotation[b]	Barley ortholg[c]	Rice ortholog[c]	Gene name[c]	Regulated traits[c]
5R	*QTgw-5R*	*rPt-400590*	721,960,899					
5R	*QTgw-5R*	*TNAC1394*	821,974,903					
5R		*tcos3066, tcos1010*	822,615,159					
5R		*c4715_668*	822,615,159					
5R		*SECCE5Rv1G0367670*	823,104,927	Expansin-like protein	HORVU4Hr1G081990	Os03g0132200	EXPLA1	Gw, Tgw
5R		*SECCE5Rv1G0367690*	823,288,365	Expansin-like protein				
5R		*SECCE5Rv1G0367700*	823,333,046	Expansin-like protein				
5R		*c6115_374*	834,295,498					
5R		*SECCE5Rv1G0369580*	836,229,455	Zinc finger protein, putative	HORVU4Hr1G084080	Os07g0153600	prog1	Gn
5R		*SECCE5Rv1G0369590*	836,332,477	Zinc finger protein, putative	HORVU4Hr1G084140	Os07g0153600	prog1	Gn
5R		*SECCE5Rv1G0369730*	837,418,410	Multidrug resistance protein	HORVU4Hr1G084290	Os03g0142800	OsMRP5	Tgw
5R	*KW5*	*WMS6*	839,551,942					
5R		*SECCE5Rv1G0370090*	839,916,856	WD-repeat protein, putative	HORVU4Hr1G084710	Os03g0123300	OsCCS52A/TAD/TE	Gw, Gt
5R	*QTgw-5R*	*rPt-507373*	847,292,398					
7R		*SECCE7Rv1G0470770*	123,899,651	Alpha/Beta Hydrolase	HORVU5Hr1G089150	Os03g0203200	HTD2/D88/D14	Small grain
7R		*SECCE7Rv1G0477730*	226,980,162					
7R	*KW7*	*SCM40*	227,032,174					
7R		*SECCE7Rv1G0477740*	227,144,857					
7R		*SECCE7Rv1G0483100*	317,567,940	Otubain-Like Protease	HORVU7Hr1G055280	Os08g0537800	WTG1/OsOTUB1	Gn, Tgw

[a]According to Hackauf et al. (2017a), [b]According to Rabanus-Wallace et al. (2021), [c]According to Wang et al. (2019)

favorable effects maps 41.4 Mbp proximal from *KW7* on chromosome 7R.

In contrast to wheat and barley, target genotypes in outbreeding rye are highly heterozygous. This results in decreasing genotypic correlations between line *per se* and test-cross performance with increasing complexity of the trait (Sprague and Tatum 1942; Miedaner et al. 2014). CMS-based hybrids and progress achieved in DNA profiling enabled the accommodation of critical aspects of QTL mapping in rye. Two studies dissecting the genetic architecture of complex agronomic traits including grain yield by linkage mapping and assessing the test-cross performance in the non-restorer (Miedaner et al. 2012) and restorer (Hackauf et al. 2017a; Miedaner et al. 2018) gene pools have been conducted. Between one and ten QTL per trait have been detected, including several large-effect QTL for plant height, heading date, TGW and grain yield, demonstrating that the genetic diversity of elite rye germplasm allows the assessment of QTL segregating for quantitative traits, even in bi-parental populations. As demonstrated for *QTgw-5R*, the rye genome assemblies (Li et al. 2021; Rabanus-Wallace et al. 2021) facilitate the integration of sequence information of DNA markers mapped in rye and related species. A comprehensive genome-based index will efficiently complement inevitable validation studies (Melchinger et al. 1998) to qualify the importance of reported rye QTL for practical breeding programs. For example, the flanking markers of *QHdt-2R.1*, *QHdt-2R.2*, *QHdt-5R*, *QHdt-6R* and *QHdt-7R* for heading date can be indexed in the 'Lo7' genome assembly (Fig. 9.1; Table 9.4), to further narrow down previously suggested candidate genes of the flowering time pathway as genetic factors controlling heading date in rye (Hackauf et al. 2017a). Besides the five-fold cross-validation for QTL detection, localization, and estimation of genetic effects (Hackauf et al. 2017a), the localization of know genes controlling the flowering time pathway in temperate cereals (Cockram et al. 2007) within defined 'Lo7' intervals further suggest that the superimposed QTL are biologically relevant.

Plant height in elite rye germplasm is controlled by an array of QTL (Fig. 9.1) illustrating the challenge of achieving progress in the genetic improvement of this trait by simultaneously selecting favorable alleles at individual QTL. In total, 35 rye orthologs of cloned rice QTL (see Yonemaru et al. 2010 for further informations) reside within the confidence intervals and serve as prime target genes for further dissection of individual rye QTL. Rather, a single dominant genetic factor like *Ddw1* offers an option of developing ideal shoot architectures with an optimized height, to fully exploit the yield potential of rye. Previous studies discussed whether plant height QTL identified in elite germplasm coincide with *Ddw1* and represent lower effect alleles of the same gene (Miedaner et al. 2012, 2018). Indeed, the position of the EST-derived SSR marker *Xscm312* on the distal end of the long arm of chromosome 5R in the 'Lo7' genome assembly (Fig. 9.1) supports the conclusion (Braun et al. 2019) that QTL #8 (Miedaner et al. 2012) is located neither in the telomeric region of chromosome 5RL, nor is allelic to *Ddw1*—since the map position of this QTL and *Ddw1* do not correspond. Moreover, the 'Lo7' genome sequence (Rabanus-Wallace et al. 2021) closed a knowledge gap by mapping *Ddw1* within the 0.413 Mbp interval of *QPh3-5R.2* (Fig. 9.1).

Plant height is an important component of biomass yield in rye. The overlapping QTL *QPh2-2R* and *QPh2-5R* with *QDmy-2R* and *QDmy-5R* (Fig. 9.1) are in line with a suggested selection strategy to improve biomass yield in rye improvement programs (Miedaner et al. 2018). Indirect selection on the QTL alleles of *QPh2-2R* and *QPh2-5R* by marker-assisted selection offers a higher indirect selection gain and increased efficiency than the direct selection on dry matter yield for breeding high biomass yielding cultivars (Miedaner et al. 2018). The 'Lo7' genome assembly (Rabanus-Wallace et al. 2021) as well as next-generation sequencing-based approaches like mapping-by-sequencing (James et al. 2013) now offer efficient means to approaching these QTL, in order to validate the

Table 9.4 Integration of rye heading date QTL in the 'Lo7' genome assembly

QTL[a]	Marker/Gene ID[a]	Chr	Position (bp)[b]	Rye gene model[c]	Annotation[b]
QHdt-2R.1	*rPt-507619*	2R	730,731,884		
QHdt-2R.1	*rPt-508957*	2R	749,356,857		
QHdt-2R.2	*rPt-509592*	2R	817,051,299		
	ScAPO2	2R	831,645,819	*SECCE2Rv1G0124300*	LEAFY-like protein
QHdt-2R.2	*rPt-402599*	2R	842,166,921		
QHdt-5R	*tcos1359*	5R	714,458,198		
	ScPHYC	5R	715,135,833	*SECCE5Rv1G0353240*	Phytochrome
	ScVRN1	5R	715,810,716	*SECCE5Rv1G0353290*	MADS box transcription factor
QHdt-5R	*rPt-400590*	5R	721,960,899		
QHdt-6R	*Xtnac1727*	6R	429,009,008		
	ScCRY2	6R	431,907,364	*SECCE6Rv1G0399150*	Cryptochrome
QHdt-6R	*rPt-5403*	6R	486,901,940		
	Xtnac1419[3]	7R	212,708,584		
QHdt-7R	*rPt-402149*	7R	*n.a*		
	ScPRR73	7R	242,013,209	*SECCE7Rv1G0478830*	Pseudo-response regulator
QHdt-7R	*rPt-399686*	7R	272,245,168		

[a]According to Hackauf et al. (2017a)
[b]According to Rabanus-Wallace et al. (2021)
[c]Maps 0.3 cM proximal of *rPt-402149*

assumption that both traits are encoded by the same genes.

Juxtaposition of QTL 1-LOD confidence intervals (Hackauf et al. 2017a; Miedaner et al. 2018) against the 'Lo7' physical map reveal low physical distances for QTL on all 'Lo7' pseudomolecules, except for *QPh1-4R*, *QPh3-4R.2*, *QTgw-4R.2* and *QTgw-4R.3* (Fig. 9.1). It is currently unclear whether these four QTL reside in a rarely recombining region on chromosome 4R, or whether the observation simply owes to the many inversions and large structural rearrangements that have been observed among non- 'Lo7' *Secale* genotypes relative to 'Lo7' (Rabanus-Wallace et al. 2021).

The dissection of molecular mechanisms controlling yield traits is of particular importance to supporting the development of rye cultivars with high yield potential. Sequence information on flanking markers allows the superimposition of three grain yield QTL—*QGyd-2R*, *QGyd-3R* and *QGyd-5R* (Hackauf et al. 2017a)—against the 'Lo7' genome assembly, and confirms the predicted localization of a *Gibberellin (GA) 2-beta-dioxygenase* gene in the 64.6 Mbp *QGyd-2R* interval on chromosome 2R (Fig. 9.1; Table 9.5). Among the remaining five orthologs of cloned rice QTL (Yonemaru et al. 2010) that reside at *QGyd-2R* and that control root cell elongation and division, shoot apical meristem formation, drought tolerance, or spikelets morphology in rice, *SECCE2Rv1G0110420* appears to be particularly interesting, since it may explain the large observed effect of *QGyd-2R* on grain yield. *SECCE2Rv1G0110420* shares 58.3% identity at the amino acid level with *Os08g0174500*, a major QTL that coordinates grain productivity, plant height, and heading date in rice (Yan et al. 2011). Most interestingly, this QTL up-regulated *MOC1*, a key gene controlling tillering and branching, that increases the number of tillers in rice, resulting in 50% more grain per plant (Yan et al. 2011), and which has been identified as a major QTL for yield heterosis in rice (Li et al. 2016c). Considering that the number of productive tillers is the key determining factor for the yield progress gained in hybrid rye (Laidig et al. 2017), this QTL (designated *Ghd8* in rice) alongside *SECCE3Rv1G0200400* (which resides within the 5.3 Mbp *QGyd-3R* interval) underscores the biological relevance of both QTL to genetic control of the most important target trait in rye breeding. *SECCE3Rv1G0200400* shares 89.4% identity at the amino acid level with *Os01g0866400*, a QTL that controls photosynthetic rate (Lee et al. 2008) and tiller number (Koumoto et al. 2013) in rice. Another promising gene residing at *QGyd-2R* is *SECCE2Rv1G0113870*, the ortholog of the wheat gene *GNI1*. *GNI1* encodes a homeodomain leucine zipper class I (HD-Zip I) transcription factor and increases grain production in wheat by controlling floret fertility and grain number (Sakuma et al. 2019). The illustrated in-depth knowledge on the gene space of grain yield QTL in rye as well as progress in our knowledge on the genetic architecture of grain yield from related cereal species (Abbai et al. 2019, 2020; Radchuk et al. 2019) motivates the development of solutions to use natural genetic diversity and counterbalance the yield penalty observed in male fertile P-type CMS hybrids (Miedaner et al. 2017).

9.3.2 Disease Resistance

In a study based on bi-parental mapping populations, major QTL conferring resistance toward stem rust have been identified on chromosomes 1R, 2R, 6R, as well as on the distal end of the long arm of chromosome 7R (Gruner et al. 2020). A large-effect QTL on 7R has been designated *Pgs1*, presuming a major gene residing at this QTL, although the inheritance pattern of *Pgs1* based on measurable numerical relationships of entities from cross-breeding experiments has not yet been reported (Gruner et al. 2020). A KASP-marker developed for *Pgs1* maps to Lo7_v2_ contig_127744 (Gruner et al. 2020), that is represented in the 'Lo7' genome assembly by *SECCE7Rv1G0524340* residing 3.47 Mbp proximal of the *Pr2* marker *cMWG682* (Fig. 9.1). In total, 36 genes encoding disease resistance proteins predominated by the NBS-LLR class have been annotated in the 'Lo7' genome sequence, distal from *SECCE7Rv1G0524340* (Rabanus-Wallace et al. 2021). The co-localization of two functional resistance (*R*) genes with a cluster of *R* gene analogs qualifies this subgenomic segment as a probable 'hotspot of resistance' in the rye genome and provides an excellent template for the development of functional markers for *Pr2* and *Pgs1*. This consideration is supported by a recent report on a T7BS.7RL translocation line conferring high levels of resistance against stripe rust, powdery mildew, and *Fusarium* head blight (FHB) to common wheat (Ren et al. 2020). The major stem rust resistance QTL residing on chromosome 1R is tagged by the SNP marker *c2420_561* (Gruner et al. 2020) and represented by *SECCE1Rv1G0043120* in the 'Lo7' assembly (Fig. 9.1). *SECCE1Rv1G0043120* is, in turn, located 573.9 Mbp distal from the stem rust resistance gene *Sr50* cloned in wheat, the latter of which maps to the short arm of chromosome 1R (Mago et al. 2015). Thus, the 'Lo7' genome sequence (Rabanus-Wallace et al. 2021) enables the assertion that the stem rust resistance QTL is, in fact, not encoded by *Sr50,* and enables use of the extensive haplotype diversity at the rye *Sr50* locus for mining stem rust resistance genes in rye germplasm collections that are, for example, effective against the broadly virulent Ug99 race lineage (Mago et al. 2015).

Ergot is a disease of cereals and grasses caused by fungi in the genus *Claviceps,* and numbers among the most economically important diseases in rye (Miedaner and Geiger 2015). At anthesis the open, non-fertilized florets of rye enable ergot spores to access the stigma and mimic pollination. Restorer genes originating from adapted European germplasm result in P-type CMS-based hybrids with an unsatisfactory restoration level. These hybrids hence reveal reduced pollen shedding and are notably

Table 9.5 Integration of rye grain yield QTL in the 'Lo7' genome assembly

Rye QTL[a]/Rice gene[b]	Marker/Gene ID[a]	Chr	Position (bp)[c]	Rye gene model[c]	Annotation[c]
Qgyd-2R	*Xscm188*	2R	670,800,088	SECCE2Rv1G0108940	
Os04g0497200	***ScGLU3***	2R	678,773,739	SECCE2Rv1G0109590	Endoglucanase
Os08g0174500	***ScGhd8***	2R	689,781,500	SECCE2Rv1G0110420	Nuclear transcription factor
Os04g0509300	***ScSho1***	2R	696,743,151	SECCE2Rv1G0110930	Dicer-like 3
Os04g0530500	***ScRDCP1***	2R	704,463,613	SECCE2Rv1G0111800	E3 ubiquitin-protein ligase
Os04g0522500	***ScGA2ox6***	2R	711,956,774	SECCE2Rv1G0112640	GA2-beta-dioxygenase
Os04g0536300	***ScTOB1***	2R	723,615,868	SECCE2Rv1G0113560	YABBY transcription factor
Qgyd-2R	*rPt-508470*	2R	735,374,696		
Qgyd-3R	*rPt-401113*	3R	851,718,435		
Os01g0866400	***ScMOC2***	3R	852,981,330	SECCE3Rv1G0200400	Fructose-1,6-bisphosphatase
Qgyd-3R	*rPt-398525*	3R	856,966,480		
Qgyd-5R	*Xtnac1388*	5R	846,612,344	SECCE5Rv1G0371300	
Qgyd-5R	*Xtcos3096*	5R	847,717,152	SECCE5Rv1G0371550	

[a]According to Hackauf et al. (2017a)
[b]According to Yonemaru et al. (2010)
[c]According to Rabanus-Wallace et al. (2021)

susceptible to ergot as the fungal spores have no competitors during the infection of the stigmatic tissue (Miedaner and Geiger 2015). Though several *Claviceps purpurea* virulence factors were identified, RNAseq analysis of *in planta* expressed fungal genes identified more than 400 highly expressed transcripts including an elevated frequency of genes encoding putative effectors that might be involved in repelling the fungal attack or interfering with host-defense interactions (Oeser et al. 2017). Although genetic variation for ergot resistance has been reported (Miedaner et al. 2010), specific plant defense reactions are unknown and the complexity of the system make identification of major effects triggered by single genes unlikely (Oeser et al. 2017). As a consequence, *Rf* genes are of central importance for hybrid rye breeding, both for achieving maximum seed setting as well as for minimizing ergot infestation (Miedaner and Geiger 2015). QTL mapping has identified major *Rf* genes for P-type CMS (*Rfp*) that originate from unadpated genetic resources and that reside on chromosomes 1R, 4R and 6R, while minor QTL have been identified on chromosomes 3R, 4R and 5R (Miedaner et al. 2000). Substantial progress has been achieved in marker-assisted stacking of the dominant restorer gene *Rfp1* on the long arm of chromosome 4R (Fig. 9.1) in hybrids (Stracke et al. 2003; Hackauf et al. 2012, 2017b; Wilde et al. 2017), which is currently the most outstanding example of a gene for improving the efficiency and precision of rye breeding via marker-assisted selection (MAS). However, hybrids carrying an effective *Rfp* gene suffer from a significant grain yield reduction (Miedaner et al. 2017). As a consequence of this high yield penalty, a restricted integration of *Rfp* genes from weedy rye in the pollinator gene pool gaining a restorer index of ~ 50% is considered as a feasible practice (Miedaner et al. 2017). However, the risk of ergot contamination in rye can be particularly high in years with cool and rainy weather during flowering, since wet pollen agglutinates and is dispersed over short distances only. Thus, a restorer index of ~50% may result in insufficient quantities of pollen to combat the fungus adequately. In order to comprehensively reduce the risk of ergot infection in hybrid rye, varieties must be developed with a restoration index of close to 100%, i.e., male fertility restoration is realized for nearly every plant. Therefore, further research is necessary to develop strategies that counterbalance the linkage drag effects of effective *Rfp* genes and simultaneously minimize the costs of restoration in terms of grain yield.

Unlike other cereal crops affected by FHB, knowledge about the genetic architecture of FHB resistance in winter rye has just recently been elucidated. A genome-wide association analysis identified 15 QTL for FHB resistance that jointly explained 74% of the genotypic variance (Gaikpa et al. 2020). Two major QTL on chromosome 1R and 5R explaining 33% and 14% of the genotypic variance are also particularly interesting for further investigations. The SNP marker *contig1930* associated with the major QTL on chromosome 1R maps at position 67.47 in the high-density map of rye (Bauer et al. 2017). In total, 43 gene models are located within a 2 cM segment flanking this position and enable the mapping of this QTL to a 21.4 Mbp target interval in the 'Lo7' genome sequence (Fig. 9.1), defined by *SECCE1Rv1G0030080* and *SECCE1Rv1G0031670*, respectively. The gene space in this interval will serve as a blueprint to fine map and clone this major QTL responsible for the genetic control of FHB resistance in rye. Searching chromosomal survey sequences generated from flow-sorted and amplified rye chromosomes (Martis et al. 2013) with *SECCE1Rv1G0030080* and *SECCE1Rv1G0031670* as query reveals that the FHB QTL fall on the long arm of rye chromosome 1R. This observation prevents the use of this valuable gene variant for the genetic improvement of FHB resistance in wheat via 1RS wheat-rye translocation lines.

Rye serves as an alternative host of *Pyrenophora tritici-repentis* (PTR), and may influence the evolution of PTR races as the cause of tan spot disease in wheat (Abdullah et al. 2017). A comprehensive evaluation of 178 geographically diverse accessions of four *Secale* sp. for response to tan spot caused by PTR race 5 revealed resistance or moderate resistance in approximately 59% of accessions (Sidhu et al. 2019). A genome-wide association study (GWAS) performed on random mating *S. cereale* subsp. *cereale* accessions identified two QTL, *QTs.sdsu-2R* and *QTs.sdsu-5R* on chromosomes 2R and 5R conferring PTR resistances (Sidhu et al. 2019). In wheat the PTR resistance gene *Tsc2* has been mapped on the short arm of chromosome 2B, distal from the marker *RZ69* (Friesen and Faris 2004). The high-quality genome assemblies (Li et al. 2021; Rabanus-Wallace et al. 2021) help to close a knowledge gap; *RZ69* is represented by *SECCE2Rv1G0068050* in the 'Lo7' genome sequence and maps 824.9 Mbp proximal from the *QTs.sdsu-2R* marker on chromosome 2R (Fig. 9.1) demonstrating that *QTs.sdsu-2R* is not orthologous to the wheat gene *Tsc2*. Thus, the developed SNP markers (Sidhu et al. 2019) represent efficient means to introgress rye chromatin carrying *QTs.sdsu-2R* or *QTs.sdsu-5R* in T2BS.2RL (Lee et al. 2009) and T5A.5RL (Owuoche et al. 1996) or T4BL.5RL (Schlegel and Korzun 2008) wheat-rye translocation lines, which will bolster the available wheat germplasm resources with improved resistance to PTR (Crespo-Herrera et al. 2017). The *QTs.sdsu-5R* marker (Sidhu et al. 2019) represents the 'Lo7' gene *SECCEUnv1G0564640* (Rabanus-Wallace et al. 2021), which enables assignment of this coding sequence to chromosome 5R.

Soil-borne viral diseases caused by the *Soil-borne cereal mosaic virus* (SBCMV) and the *Wheat spindle streak mosaic virus* (WSSMV) are gaining recognition among European rye cultivation communities (Huth 2002). QTL mapping identified a major QTL for SBCMV resistance on chromosome 5R, while for WSSMV resistance, a major QTL was detected on chromosome 7R (Erath et al. 2016). The desired allele conferring SBCMV resistance originates from a genetic resource of rye, which unfortunately may carry gene variants in the neighborhood of the resistance gene which show negative impact on the performance of a recurrent parent. The SBCMV QTL is defined by *SECCE5Rv1G0357520* and

SECCE5Rv1G0359200 and covers a 13.2 Mbp segment in the 'Lo7' assembly (Fig. 9.1), which may now serve as a blueprint for the development of additional SNP markers to increase the efficiency of introgression breeding and optimize the recovery of the recurrent parent genome, free of undesirable linkage drag. This consideration applies in particular for a major QTL conferring SBCMV resistance, that has been identified on the long arm of chromosome 2R, and which originates from the wild species *S. montanum* (Schlegel et al. 2020). Notably, the favorable allele for WSSMV resistance was discovered among inbred lines of a commercial hybrid breeding program, demonstrating that elite germplasm provides substantial diversity for WSSMV resistance in rye.

9.3.3 Abiotic Stress Tolerance

The main driver enabling rye cultivation on light soils with low fertility and low water capacity is its highly developed root system, which facilitates efficient uptake of water and nutrients (Starzycki 1976). The root system of a single rye plant consisted of 13,815,672 branches, with a total length of 622 km, a surface area of 401 m^2, and a total root hair length of 11,000 km (Dittmer 1937; Ryser 2006). In an experiment comparing the yield reduction effects of dehydration among small-grain Triticeae, rye outperformed wheat, barley, and even triticale (Schittenhelm et al. 2015). As a further example demonstrating the great potential of its root system, rye has been shown to reduce nitrate leaching at the field scale by up to 93%, greater than barley (87%) and wheat (57%) (Yeo et al. 2014). In fact, the increased Nitrogen Use Efficiency (NUE) of rye has been mainly attributed to its higher specific root length (Paponov et al. 1999). The high NUE of winter rye enables grain production with the lowest carbon footprint as compared to wheat, triticale, barley and oats (Wojcik-Gront and Bloch-Michalik 2016). However, no mapping data on NUE are currently available for rye.

In wheat, the introduction of GA-insensitive dwarfing genes caused a reduction of root mass and root length, even though crop productivity was significantly increased (Waines and Ehdaie 2007; Subira et al. 2016; Voss-Fels et al. 2017). It is currently under discussion whether the reduction in plant height does also reduce the water uptake capacity of cereal crops. Zhang et al. (2009) suggested that recent semi-dwarf wheat cultivars are even more efficient with regard to water use, because the decrease in total root length was mainly at the expense of root length in the top soil, and led to a higher proportion of roots in deeper soil layers. On the other hand, Waines and Ehdaie (2007) assume that the root system sizes of modern, GA-insensitive wheat cultivars are sufficient under optimal growth conditions, but insufficient when growth conditions are unfavorable. Notably, there is currently no information available about the diversity in rooting patterns of rye and possible changes as a result of breeding activities during the last decades. A distinct impression about the power of the 'hidden half' of rye results from studies on the root system in aneuploid wheat/rye stocks (Sharma et al. 2009, 2010, 2011; Howell et a. 2019; Gabay et al. 2021), which have seen an overall increase in root biomass and improved tolerance to environmental stress (Ehdaie et al. 2003).

Root phenotypic plasticity has been proposed as a target trait in crop improvement programs due to beneficial effects in low-input systems (Schneider and Lynch 2020), which predominate among target environments of rye production. Integration of novel root phenotyping technologies (Tracy et al. 2020) in rye improvement programs promise substantial progress in germplasm enhancement by completing phenotypic datasets to accurately predict the genetic value of selection candidates.

9.3.4 Grain Quality

The grain serves as the basic propagation unit and represents a particular sensitive part within a cereal's life cycle. Pre-harvest sprouting (PHS) is a phenomenon by which grains germinate on the plant before harvest. This is a major problem in rye production as it leads—via α-amylase activity—to degradation of the starch and a considerable reduction in baking quality, which results in economic losses just as in wheat and barley (Nakamura 2018). The Hagberg falling number enables indirect selection for low α-amylase (α-Amy) activity in the grain, without negatively influencing yield and other agronomic traits (Geiger and Miedaner 2009). As predicted by RFLP analyses (Masojć and Gale 1991), the 'Lo7'genome assembly (Fig. 9.1) contains three *α-Amy1* structural genes on chromosome 6R, two *α-Amy2* genes on chromosome 7R as well as the three *α-Amy3* genes on chromosome 5R. Two further *α-Amy* structural genes are annotated in the 'Lo7' genome sequence on chromosome 2R and one gene on chromosome 3R, that shed new light on mapping QTL for α-amylase activity and PHS of rye grains (Masojć and Milczarski 2005; Masojć et al. 2007; Tenhola-Roininen et al. 2011). Overlapping α-amylase activity and PHS QTL (Masojć and Milczarski 2009; Myśków et al. 2012) are in line with the selection strategy to improve PHS resistance in rye improvement programs and support the assumption that—at least in some cases—both traits are indeed encoded by the same gene. The high-quality rye reference genome sequences offer advancement in dedicated experiments (Myśków et al. 2010; Bienias et al. 2020) designed to identify novel inroads to improve rye grain quality (Gao and Ayele 2014; Nakamura 2018; Tai et al. 2021).

A stronger focus on grain quality parameters in rye breeding appears to be particularly rewarding, since rye has a higher content of the essential amino acid lysine compared to wheat (Villegas et al. 1968; Riley and Ewart 1970; Deleu et al. 2020) and represents, thus, as a healthy choice and valuable alternative protein supply of adequate quality in cereal-based diets. For this purpose, the natural genetic diversity of rye offers an attractive yet underutilized means to developing rye varieties with health-promoting effects (Zykin et al. 2018). The outstandingly high enzymatic activity of rye phytases (Nielsen et al. 2007) is important in terms of the bioavailability of phytate, the major storage form of phosphorus in plant seeds and the single most important antinutritional compound in the grain (Madsen et al. 2013). The purple acid phosphatase phytase (*PAPhy*) genes *PAPhy_b1* and *PAPhy-a2* in rye (Madsen et al. 2013) map to chromosome 3R and 5R, respectively (Fig. 9.1) and can be monitored in rye improvement programs by six and five SNPs, respectively (Bauer et al. 2017). The rye reference genome sequences permit breeders to approach genes controlling WEAX in rye. Arabinoxylans (AX) are non-starch polysaccharides and the predominant components within the endosperm cell walls in rye and, to a lesser degree, in wheat (Buksa et al. 2016; Freeman et al. 2016; Oest et al. 2020). High AX content increases the falling number, dough yield, bread volume, and bread shelf-life (Weipert, 1995, 1997; Buksa et al. 2010; Oest et al 2020). Current methods of rye breeding and severe drought conditions in a changing climate are thought to negatively influence bread qualities, which demands improved understanding of the mechanisms by which proteins, starch, and AX—the most prominent hemicelluloses—might interact (Oest et al. 2020). With respect to AX, chromosomes 2R, 5R and 6R have been associated with significantly increased amounts of WEAX in aneuploid wheat/rye stocks, while chromosomes 1R, 3R, 4R and 7R conferred significantly lower levels of WEAX (Cyran et al. 1996; Boros et al. 2002). Meanwhile, substantial progress has been achieved in identifying genes responsible for synthesis of xylan in general and AX in cereal endosperm—in particular based on knowledge from fundamental research in *Arabidopsis* as well as steadily increasing sequence information on plant genomes (Mitchell et al. 2007; Zeng et al. 2010; Anders et al. 2012; Pellny et al. 2012; Lovegrove et al. 2013; Freeman et al. 2015; Pellny et al. 2020). The rye genome assemblies (Li et al. 2021; Rabanus-Wallace et al. 2021) now enable researchers to capitalize upon this knowledge for rye breeding. A subset among the 425 glycosyltransferases encoding rye genes (Fig. 9.1; Table 9.6) are orthologs with known phenotypic effects on WEAX in wheat (Lovegrove et al. 2013; Freeman et al. 2015; Pellny et al. 2020). These protein coding sequences enable the systematic assessment of the rich natural genetic diversity of rye for the content of WEAX (Madej et al. 1990; Scoles et al. 1993; Weipert 1996; Nilsson et al. 1997; Ragaee et al. 2001a, 2001b; Cyran and Cygankiewicz 2004; Hansen et al. 2004; Goncharenko et al. 2008, 2017; Goncharenko and Timoshchenko 2010; Jürgens et al. 2012; Cyran and Dynkowska 2014; Ponomareva et al. 2017; Kobylyansky et al. 2019) at the molecular level to select haplotypes with desired effects on quality traits. Grain quality traits differ somewhat depending on the end-use of rye. To increase the value of rye as livestock feed, protein content should be maximized and WEAX content minimized (Fernadez et al. 1973; Antoniou et al. 1981; Ward and Marquardt 1987; Fengler and Marquardt 1988; Scoles et al. 1993; Gan et al. 1998; Thacker et al. 2002; Lazaro et al. 2003, 2004; McGhee and Stein 2018, 2020; Smit et al. 2019; Muszyński et al. 2020), in stark contrast to the optimal needs for breadmaking. However, the evaluation of end-use quality parameters like WEAX requires precision laboratory protocols for selecting both parents and progeny (Reynolds et al. 2020) that are currently not suitable for large-scale phenotyping of WEAX in rye improvement programs.

The WEAX content of cereals is related to the content of fructan, and the amounts of both decrease in the order rye > wheat > barley > oat > maize (MacLeod and Preece 1954). In humans, fructans are prebiotics that may promote growth of healthy gut bacteria, aid in immune support, reduce colon cancer incidence, and support bone health (Veenstra et al. 2017). However, fructans represent the prevailing group of short-chain carbohydrates called FODMAPS, that escape absorption in the small intestine and are rapidly fermented in the upper section of the colon and may cause unpleasant ailments of the gastrointestinal system and should be mitigated by the ingenstion of cereal products rich in dietary fiber, but with low fructan content (Pejcz et al. 2020). Similarly high concentrations of these antinutritive factors limits the use of rye grain as feed material in poultry nutrition (Bederska-Łojewska et al. 2017). As outlined for WEAX, the reference genome assemblies (Li et al. 2021; Rabanus-Wallace et al. 2021) enhance integration of knowledge on genes encoding the fructan biosynthesizing enzymes *sucrose:sucrose 1-fructosyltransferase* (*1-SST*), *fructan:fructan 1-fructosyltransferase* (*1-FFT*) and *sucrose:fructan 6-fructosyltransferase* (*6-SFT*) in wheat and barley (Huynh et al. 2012; Veenstra et al. 2017) residing on the short arm of chromosome 1R and the long arm of chromosome 4R, respectively (Fig. 9.1). For breeding rye as livestock feed a major antinutritional factor residing on rye chromosome 6R, as deduced from a unique phenotyping strategy of aneuploid wheat/rye stocks (Thiele et al. 1989), is a promising target for further research.

9.4 Understanding the Rye Genome —The Way Forward

With the 'Weining' (Li et al. 2021) as well as the 'Lo7' (Rabanus-Wallace et al. 2021) high-quality genome assemblies, rye has finally reached the genomics era, enabling the integration and advancement of both, fundamental and applied breeding research to understand how the genome builds, maintains, and operates rye. For this purpose further progress in rye phenomics and functional genomics research is necessary to associate genome sequence information with phenotypes related to rye growth and development. A roadmap to determining the biological function of almost every gene in a cereal genome includes (1) the development of tools and genetic resources for an international community of scientists to conduct functional genomics research, (2) assignment of biological functions to every annotated gene, (3) the description of systems-wide epigenomes, gene expression profiles and regulatory networks, (4) global analyses of the proteome and protein–protein interactions, (5) the assessment of natural variation, and (6) investment in publicly accessible bioinformatics tools and databases, to the level that has been achieved for rice (Zhang et al. 2008). Indeed, due to its limited genome size and diploidy, rice is an excellent choice among

Table 9.6 GT43, GT47 and GT61 genes encoding glycosyltransferases from rye and wheat

Chr	Gene model[a]	Position (bp)[a]	N° of SNPs[b]	Wheat Gene ID[c]	Alias[c]	Gene ID Wheat A[d]	Gene ID Wheat B[d]	Gene ID Wheat D[d]
1R	SECCE1Rv1G0009720	56,625,323		TaGT43_6	TaIRX9a	TraesCS1A02G064000	TraesCS1B02G081900	TraesCS1D02G064500
1R	SECCE1Rv1G0031430	442,320,289	5			TraesCS1A02G206400	TraesCS1B02G220000	TraesCS1D02G209700
1R	SECCE1Rv1G0056380	686,738,085	6	TaGT43_4		TraesCS1A02G391000	TraesCS1B02G419100	TraesCS1D02G399000
2R	SECCE2Rv1G0069010	26,021,610	3	TaGT43_7		TraesCS2A02G082500	TraesCS2B02G096700	TraesCS2D02G080400
2R	SECCE2Rv1G0069030	26,119,584	4	TaGT43_8		TraesCS2A02G082600	TraesCS2B02G096800	TraesCS2D02G080500
2R	SECCE2Rv1G0101510	566,045,873	1			TraesCS2A02G288300	TraesCS2B02G305100	TraesCS2D02G286600
2R	SECCE2Rv1G0129990	872,186,303		TaGT43_10		TraesCS2A02G671400	TraesCS2B02G539600	TraesCS2D02G512800
3R	SECCE3Rv1G0165060	169,082,663	3	TaGT43_9		TraesCS3A02G309600	TraesCS3B02G152600	TraesCS3D02G135300
3R	SECCE3Rv1G0186860	691,836,452	3	TaGT43_3		TraesCS3A02G270100	TraesCS3B02G304000	TraesCS3D02G269800
3R	SECCE3Rv1G0209180	936,956,898	4			TraesCS3A02G439900	TraesCS3B02G474000	TraesCS3D02G432700
3R	SECCE3Rv1G0209200	937,163,674	8	TaGT47_13		TraesCS3A02G440100	TraesCS3B02G474200	TraesCS3D02G432900
3R	SECCE3Rv1G0209270	937,414,633	8	TaGT47_12		TraesCS3A02G440800	TraesCS3B02G474900	TraesCS3D02G433400
5R	SECCE5Rv1G0338670	602,232,701	4			TraesCS5A02G299100	TraesCS5B02G298600	TraesCS5D02G306000
5R	SECCEUnv1G0540090	78,347,884	4			TraesCS5A02G546600	TraesCS4B02G380300	TraesCSU02G130500
6R	SECCE6Rv1G0408940	578,546,159	5	TaGT61		TraesCS6A02G309400	TraesCS6B02G339100	TraesCS6D02G288800
7R	SECCE7Rv1G0479050	246,539,612		TaGT43_2	TaIRX9b	TraesCS4A02G107400	TraesCS4B02G197000	TraesCS4D02G197300
7R	SECCE7Rv1G0509920	785,056,752	1	TaGT43_1		TraesCS7A02G441400	TraesCS7B02G340100	TraesCS7D02G430700

[a]According to Rabanus-Wallace et al. (2021)
[b]According to Bauer et al. (2017)
[c]Mitchell et al. (2007), Pellny et al. (2019)
[d]IWGSC (2018)

cereals for genomic studies and serves as a model species for crop biology and agricultural research (Song et al. 2018). Correspondingly, the common evolutionary origin of the grasses (Pont et al. 2019) enables the identification of yield-related genes in wheat and barley using the rice genome as a blueprint (see Nadolska-Orczyk et al. 2017 for review). As exemplified in this chapter, thus, a straightforward initial approach to classifying part of the deciphered rye genes in both genome assemblies is to integrate sequence information from cloned genes and QTL from rice for subsequent validation by SMART breeding in rye. Naturally, such a strategy will include the steadily increasing number of identified plant genes, particularly from barley (Hansson et al. 2018) and wheat (Jia et al. 2018). However and most importantly, high-quality genome assemblies (Li et al. 2021; Rabanus-Wallace et al. 2021) facilitate the application of next-generation sequencing-based approaches like mapping-by-sequencing (James et al. 2013)) to accelerate the discovery of the genetic basis for species-specific traits and unique selling points of rye like its reproduction biology and tolerance of biotic and abiotic stress. A highly coordinated effort that connects scientists and resources globally is necessary to efficiently reach this goal. We thus propose an International Functional Genomics Project on Rye (IFGPR) that should, in a first instance, develop highly replicable genetic stocks in the public domain. For example, both sequenced inbred lines are predestined to induce random mutations caused by chemicals or physical agents, that can subsequently be detected by the well-established high-throughput TILLING (Targeting Induced Local Lesion IN Genomes) technique (Tadele 2016). Important experiences on the use of ethyl methanesulfonate (EMS) as a chemical mutagen in rye have been reported by Müntzing and Bose (1969). Likewise, sodium azide successfully served as a chemical mutagen in rye as well (Adolf and Riemann 1989). TILLING populations in rye will allow for reverse genetic approaches on genes with unknown function that have been inactivated by induced mutations, and which can be studied to characterize the phenotypic effects of particular genic mutations. Furthermore, exome sequencing (Mascher et al. 2013) of allelic mutants from TILLING populations is a validated approach to verifying that the correct gene has been identified in gene isolation projects (Komatsuda et al. 2007; Saintenac et al. 2018). Likewise, genome editing techniques will need to be adapted to rye as they provide a precise and rapid approach to functionally validate genes and facilitate gene discovery (Chao et al. 2019; Zhan et al. 2020; Liu et al. 2021).

Alternatively, a genetic approach exploiting the DNA mismatch repair system (MMR) has proven to have comparable efficacy in generating mutants (Bronner et al. 1994; Hoffman et al. 2004; Xu et al. 2012; Cui et al. 2017; Rakosy-Tican et al. 2019; Jiang et al. 2020). MMR is a highly conserved pathway responsible for maintaining the genomic integrity of different organisms by recognizing and correcting single nucleotide mismatches and unpaired nucleotides that arise through replication errors, deamination of 5-methylcytosine, and recombination between divergent sequences (Manova and Gruszka 2015; Spampinato 2017). Among the eight Mutator S homologs (MSH) subunits that recognize DNA lesions, *MSH2*, *MSH3* and *MSH6* have been mapped to the short arm of chromosome 1R, the long arm of chromosome 2R and to the long arm of chromosome 5R in rye, respectively (Korzun et al. 1999; Fig. 9.1). In total 49 genes are predicted to encode DNA mismatch repair proteins mapping to each of the seven rye chromosomes in the ‘Lo7’ genome assembly (Rabanus-Wallace et al. 2021), thus emphasizing the importance of MMR in the outbreeding rye. Encouraged by the substantial natural genetic diversity of these gene models identified in the 600 k SNP array (Bauer et al. 2017), allele mining in random mating populations appears to be a promising approach to capturing alleles that can be used to genetically induce mutations for the purposes of research and breeding in rye.

Both sequenced rye inbred lines (Li et al. 2021; Rabanus-Wallace et al. 2021) could serve as a stimulus for research activities to establish a library of rye mutants to be stored in genetic seed stock centers—a resource, that is currently not available for rye. For this purpose, a sophisticated strategy based on genomic signatures of long-term random mating populations, that has recently been developed in maize (Mayer et al. 2017, 2020) may provide a valuable means to capture native genetic diversity for genome-based studies and breeding in rye. This may include the creation of double haploid (DH) libraries from landraces (Hölker et al. 2019), as a recently published protocol reported promising progress (Zieliński et al. 2020) in overcoming genotype dependency with respect to tissue culture responses of rye (Thomas and Wenzel 1975; Wenzel et al. 1977; Hoffmann and Wenzel 1981; Flehinghaus et al. 1991; Daniel 1993; Fleinghaus-Roux et al. 1995; Rakoczy-Trojanowska et al. 1997; Guo and Pulli 2000; Tenhola-Roininen et al. 2006; Gruszczyńska and Rakoczy-Trojanowska 2011), that currently limits routine application of DH technology in rye. A straightforward approach to capture the full complement of sequence diversity are multiple high-quality sequence assemblies. The technically feasible approach of a rye pan-genome infrastructure ultimately would enable to detect selection signals along the genome differentiating the two rye elite breeding pools, as has been demonstrated in an initial scan based on a draft rye genome sequence (Bauer et al. 2017).

9.5 Conclusions

Hybrid breeding has been identified as a key technology for increasing and securing cereal production on finite arable land without increasing water and fertilizer use (Whitford et al. 2013). The natural genetic diversity in rye was the fundamental basis to achieve a series of technological advances over a century of breeding and research, that ultimately facilitated the establishment of hybrid breeding. These advances include self-fertility mutations, which enable the development of inbred lines to capture and manage valuable genetic diversity, nuclear-cytoplasmic gynodioecy to establish a natural, reliable, environmentally friendly and cost-effective production of hybrid seed, effective *Rf* genes to enable grain production in CMS-based hybrids and minimize ergot contamination in the harvest, as well as two genetically divergent gene pools necessary to exploit heterosis. Despite these formidable achievements, major challenges in rye production remain, in particular concerning tailor-made grain qualities that might further advance rye from its current role as an all-rounder, to an authentic high-performance crop with diverse certified end-uses. For this purpose, improved breeding efforts are strategically important for enhancing the competitiveness of rye in modern agricultural production systems. In view of current international efforts to combat global climate change and increasing demand for high quality and healthy food, rye offers sustainable options to help alleviate hunger in the face of a steadily growing population. The release of two high-quality rye genome assemblies (Li et al. 2021; Rabanus-Wallace et al. 2021) will accelerate the transition from merely phenotypic to a haplotype-based breeding (Bevan et al. 2017; Brinton et al. 2020) and substantially increase the efficiency, precision and flexibility of rye breeding. Driven by the targeted improvement of complex inherited traits with challenging phenotypes like end-use quality, the rich genetic diversity of the outbreeding rye will render forward genetic approaches a promising field in cereal research. The examples outlined in the present chapter demonstrate how both published physical rye maps will foster gene isolation projects in rye. As bi-parental populations help to control the rate of false positive candidates in gene and QTL mapping experiments, this classical approach will henceforth take on central importance in ongoing efforts to isolate and characterize specific loci to bridge the genotype–phenotype gap for precision breeding in rye.

Acknowledgements We would like to express our deep gratitudes to the family von Lochow for their consent to recognize Ferdinand III. von Lochow by making an image from their private archive available for this publication. Sincere thanks are given to both anonymous reviewers for helpful comments on the manuscript.

References

Abbai R, Singh VK, Nachimuthu VV, Sinha P, Selvaraj R, Vipparla AK, Singh AK, Singh UM, Varshney RK, Kumar A (2019) Haplotype analysis of key genes governing grain yield and quality traits across 3K RG panel reveals scope for the development of tailor-made rice with enhanced genetic gains. Plant Biotechnol J 17:1612–1622. https://doi.org/10.1111/pbi.13087

Abbai R, Singh VK, Snowdon RJ, Kumar A, Schnurbusch T (2020) Seeking crops with balanced parts for the ideal whole. Trends Plant Sci 12:1189–1193. https://doi.org/10.1016/j.tplants.2020.08.011

Abdullah S, Sehgal SK, Glover KD, Ali S (2017) Reaction of global collection of rye (*Secale cereale* L.) to tan spot and *Pyrenophora tritici-repentis* races in South Dakota. Plant Pathol J 33:229–237. https://doi.org/10.5423/PPJ.OA.12.2016.0265

Adhikari KB, Tanwir F, Gregersen PL, Steffensen SK, Jensen BM, Poulsen LK, Nielsen CH, Høyer S, Borre M, Fomsgaard IS (2015) Benzoxazinoids: cereal phytochemicals with putative therapeutic and health-protecting properties. Mol Nutr Food Res 59:1324–1438. https://doi.org/10.1002/mnfr.201400717

Adolf K, Riemann KH (1989) Züchtung und Produktion von Winterroggen. In: Akad. Landw.-Wiss. DDR (ed) Fortschrittsbericht für die landwirtschaft und nahrungsgüterwirtschaft, vol. 27, pp 1–35

Anders N, Wilkinson MD, Lovegrove A, Freeman J, Tryfona T, Pellny TK, Weimar T, Mortimer JC, Stott K, Baker JM, Defoin-Platel M, Shewry PR, Dupree P, Mitchell RA (2012) Glycosyl transferases in family 61 mediate arabinofuranosyl transfer onto xylan in grasses. Proc Natl Acad Sci U S A 109:989–993. https://doi.org/10.1073/pnas.1115858109

Antoniou T, Marquardt RR, Cansfield PE (1981) Isolation, partial characterization, and antinutritional activity of a factor (pentosans) in rye grain. J Agric Food Chem 29:1240–1247. https://doi.org/10.1021/jf00108a035

Auinger HJ, Schönleben M, Lehermeier C, Schmidt M, Korzun V, Geiger HH, Piepho HP, Gordillo A, Wilde P, Bauer E, Schön CC (2016) Model training across multiple breeding cycles significantly improves genomic prediction accuracy in rye (*Secale cereale* L.). Theor Appl Genet 129:2043–2053. https://doi.org/10.1007/s00122-016-2756-5

Bahrani H, Thoms K, Båga M, Larsen J, Graf R, Laroche A, Sammynaiken R, Chibbar RN (2019) Preferential accumulation of glycosylated cyanidins in winter-hardy rye (*Secale cereale* L.) genotypes during cold acclimation. Environ Exp Bot 164:203–212. https://doi.org/10.1016/j.envexpbot.2019.05.006

Bakera B, Makowska B, Groszyk J, Niziołek M, Orczyk W, Bolibok-Brągoszewska H, Hromada-Judycka A, Rakoczy-Trojanowska M (2015) Structural characteristics of *ScBx* genes controlling the biosynthesis of hydroxamic acids in rye (*Secale cereale* L.). J Appl Genet 56:287–298. https://doi.org/10.1007/s13353-015-0271-z

Bakera B, Rakoczy-Trojanowska M (2020) Isolation and structural analysis of the *Bx6* and *Bx7* genes controlling the biosynthesis of benzoxazinoids in rye (*Secale cereale* L.). Acta Physiol Plant 42:56. https://doi.org/10.1007/s11738-020-03046-8

Bauer E, Schmutzer T, Barilar I, Mascher M, Gundlach H, Martis MM, Twardziok SO, Hackauf B, Gordillo A, Wilde P, Schmidt M, Korzun V, Mayer KFX, Schmid K, Schon CC, Scholz U (2017) Towards a whole-genome sequence for rye (*Secale cereale* L.). Plant J 89 (5):853–869.https://doi.org/10.1111/tpj.13436

Bederska-Łojewska D, Świątkiewicz S, Arczewska-Włosek A, Schwarz T (2017) Rye non-starch polysaccharides: their impact on poultry intestinal physiology, nutrients digestibility and performance indices - a review. Ann Anim Sci 17:351–369. https://doi.org/10.1515/aoas-2016-0090

Bednarek PT, Masojć P, Lewandowska R, Myśków B (2003) Saturating rye genetic map with amplified fragment length polymorphism (AFLP) and random amplified polymorphic DNA (RAPD) markers. J Appl Genet 44(1):21–33

Benito C, Zaragoza C, Gallego FJ, de la Peña A, Figueiras AM (1991) A map of rye chromosome 2R using isozyme and morphological markers. Theor Appl Genet 82(1):112–116. https://doi.org/10.1007/BF00231284

Bevan MW, Uauy C, Wulff BB, Zhou J, Krasileva K, Clark MD (2017) Genomic innovation for crop improvement. Nature 543:346–354. https://doi.org/10.1038/nature22011

Bienias A, Góralska M, Masojć P, Milczarski P, Myśków B (2020) The GAMYB gene in rye: sequence, polymorphisms, map location, allele-specific markers, and relationship with α-amylase activity. BMC Genomics 21:578. https://doi.org/10.1186/s12864-020-06991-3

Boden SA, Cavanagh C, Cullis BR, Ramm K, Greenwood J, Jean Finnegan E, Trevaskis B, Swain SM (2015) Ppd-1 is a key regulator of inflorescence architecture and paired spikelet development in wheat. Nat Plants 1:14016. https://doi.org/10.1038/nplants.2014.16

Bolibok-Bragoszewska H, Heller-Uszyńska K, Wenzl P, Uszyński G, Kilian A, Rakoczy-Trojanowska M (2009) DArT markers for the rye genome—genetic diversity and mapping. BMC Genomics 10:578. https://doi.org/10.1186/1471-2164-10-578

Bolibok-Brągoszewska H, Targońska M, Bolibok L, Kilian A, Rakoczy-Trojanowska M (2014) Genome-wide characterization of genetic diversity and population structure in *Secale*. BMC Plant Biol 14:184. https://doi.org/10.1186/1471-2229-14-184

Boros D, Lukaszewski AJ, Aniol A, Ochodzki P (2002) Chromosome location of genes controlling the content of dietary fibre and arabinoxylans in rye. Euphytica 128:1–8. https://doi.org/10.1023/A:1020639601959

Börner A, Melz G (1988) Response of rye genotypes differing in plant height to exogenous gibberellic acid application. Arch Züchtungsforsch 18:71–74

Börner A, Plaschke J, Korzun V, Worland AJ (1996) The relationships between the dwarfing genes of wheat and rye. Euphytica 89:69–75. https://doi.org/10.1007/bf00015721

Börner A, Korzun V, Voylokov AV, Worland AJ, Weber E (2000) Genetic mapping of quantitative trait loci in rye (*Secale cereale* L.). Euphytica 116:203–209. https://doi.org/10.1023/A:1004052505692

Braun EM, Tsvetkova N, Rotter B, Siekmann D, Schwefel K, Krezdorn N, Plieske J, Winter P, Melz G, Voylokov AV, Hackauf B (2019) Gene expression profiling and fine mapping identifies a gibberellin 2-oxidase gene co-segregating with the dominant dwarfing gene *Ddw1* in Rye (*Secale cereale* L.). Front Plant Sci 10:857. https://doi.org/10.3389/fpls.2019.00857

Brinton J, Uauy C (2019) A reductionist approach to dissecting grain weight and yield in wheat. J Integr Plant Biol 61:337–358. https://doi.org/10.1111/jipb.12741

Brinton J, Ramirez-Gonzalez RH, Simmonds J, Wingen L, Orford S, Griffiths S; 10 Wheat Genome Project, Haberer G, Spannagl M, Walkowiak S, Pozniak C, Uauy C (2020) A haplotype-led approach to increase the precision of wheat breeding. Commun Biol 3:712. https://doi.org/10.1038/s42003-020-01413-2

Bronner CE, Baker SM, Morrison PT, Warren G, Smith LG, Lescoe MK, Kane M, Earabino C, Lipford J, Lindblom A, Tannergard P, Bollag RJ, Godwin AR, Nordenskjosolld M, Fishel R, Kolodner R, Liskay RM (1994) Mutation in the DNA mismatch repair gene homologue hMLH1 is associated with hereditary nonpolyposis colon cancer. Nature 368:258–261. https://doi.org/10.1038/368258a0

Buksa K, Nowotna A, Praznik W, Gambuś H, Ziobro R, Krawontka J (2010) The role of pentosans and starch in baking of wholemeal rye bread. Food Res Intern 43:2045–2051. https://doi.org/10.1016/j.foodres.2010.06.005

Buksa K, Praznik W, Loeppert R, Nowotna A (2016) Characterization of water and alkali extractable arabinoxylan from wheat and rye under standardized conditions. J Food Sci Technol 53:1389–1398. https://doi.org/10.1007/s13197-015-2135-2

Campoli C, Matus-Cádiz MA, Pozniak CJ, Cattivelli L, Fowler DB (2009) Comparative expression of *Cbf* genes in the Triticeae under different acclimation induction temperatures. Mol Genet Genomics 282:141–152. https://doi.org/10.1007/s00438-009-0451-9

Camut L, Regnault T, Sirlin-Josserand M, Sakvarelidze-Achard L, Carrera E, Zumsteg J, Heintz D, Leonhardt N, Lange MJP, Lange T, Davière JM, Achard P (2019) Root-derived GA12 contributes to temperature-induced shoot growth in *Arabidopsis*. Nat Plants 5:1216–1221. https://doi.org/10.1038/s41477-019-0568-8

Cao X, Chen X, Liu Y, Xu Z, Li L, Zhou Y, Liu J, Zhao Z, Chen M, Ma Y (2016) An iNTT system for the large-scale screening of differentially expressed, nuclear-targeted proteins: cold-treatment-induced nucleoproteins in rye (*Secale cereale* L.). BMC Genomics 17:189. https://doi.org/10.1186/s12864-016-2548-y

Cao Y, Zeng H, Ku L, Ren Z, Han Y, Su H, Dou D, Liu H, Dong Y, Zhu F, Li T, Zhao Q, Chen Y (2020) *ZmIBH1-1* regulates plant architecture in maize. J Exp Bot 71:2943–2955. https://doi.org/10.1093/jxb/eraa052

Carrillo JM, Vázquez JF, Orellana J (1992) Identification and mapping of the *Gli-R3* locus on chromosome 1R of rye (*Secale cereale* L.). Theor Appl Genet 84:237–241. https://doi.org/10.1007/BF00224005

Chai JF, Liu X, Jia JZ (2005) Homoeologous cloning of omega-secalin gene family in a wheat 1BL/1RS translocation. Cell Res 15:658–664. https://doi.org/10.1038/sj.cr.7290335

Chang TD (1975) Mapping of the gene for hairy peduncle (*Hp*) on rye chromosome 5R. Can J Genet Cytol 17:127–128

Chao T, Liu Z, Zhang Y, Zhang L, Huang R, He L, Gu Y, Chen Z, Zheng Q, Shi L, Zheng W, Qi X, Kong E, Zhang Z, Lawrence T, Liang Y, Lu L (2019) Precise and rapid validation of candidate gene by allele specific knockout with CRISPR/Cas9 in wild mice. Front Genet 10:124. https://doi.org/10.3389/fgene.2019.00124

Chebotar S, Röder MS, Korzun V, Saal B, Weber WE, Börner A (2003) Molecular studies on genetic integrity of open-pollinating species rye (*Secale cereale* L.) after long-term genebank maintenance. Theor Appl Genet 107:1469–1476. https://doi.org/10.1007/s00122-003-1366-1

Chen HI, Li PF, Yang CH (2019a) NAC-Like gene GIBBERELLIN SUPPRESSING FACTOR regulates the gibberellin metabolic pathway in response to cold and drought stresses in arabidopsis. Sci Rep 9:19226. https://doi.org/10.1038/s41598-019-55429-8

Chen Q, Yang CJ, York AM, Xue W, Daskalska LL, DeValk CA, Krueger KW, Lawton SB, Spiegelberg BG, Schnell JM, Neumeyer MA, Perry JS, Peterson AC, Kim B, Bergstrom L, Yang L, Barber IC, Tian F, Doebley JF (2019b) TeoNAM: a nested association mapping population for domestication and agronomic trait analysis in maize. Genetics 213:1065–1078. https://doi.org/10.1534/genetics.119.302594

Choi BS, Kim YJ, Markkandan K, Koo YJ, Song JT, Seo HS (2018) GW2 functions as an E3 ubiquitin ligase for rice expansin-like 1. Int J Mol Sci 19:1904. https://doi.org/10.3390/ijms19071904

Clarke BC, Mukai Y, Appels R (1996) The *Sec-1* locus on the short arm of chromosome 1R of rye (*Secale cereale*). Chromosoma 105:269–275

Cockram J, Jones H, Leigh FJ, O'Sullivan D, Powell W, Laurie DA, Greenland AJ (2007) Control of flowering time in temperate cereals: genes, domestication, and sustainable productivity. J Exp Bot 58:1231–1244. https://doi.org/10.1093/jxb/erm042

Colebrook EH, Thomas SG, Phillips AL, Hedden P (2014) The role of gibberellin signalling in plant responses to abiotic stress. J Exp Biol 217:67–75. https://doi.org/10.1242/jeb.089938

Collins NC, Shirley NJ, Saeed M, Pallotta M, Gustafson JP (2008) An ALMT1 gene cluster controlling aluminum tolerance at the Alt4 locus of rye (*Secale cereale* L). Genetics 179:669–682. https://doi.org/10.1534/genetics.107.083451

Crespo-Herrera LA, Garkava-Gustavsson L, Åhman I (2017) A systematic review of rye (*Secale cereale* L.) as a source of resistance to pathogens and pests in wheat (*Triticum aestivum* L.). Hereditas. 154:14. https://doi.org/10.1186/s41065-017-0033-5

Cui H, Wu Q, Zhu B (2017) Specific-locus amplified fragment sequencing reveals spontaneous single-nucleotide mutations in rice *OsMsh6* mutants. Biomed Res Int 2017:4816973

Cyran M, Rakowska M, Miazga D (1996) Chromosomal location of factors affecting content and composition of non-starch polysaccharides in wheat-rye addition lines. Euphytica 89:153–157

Cyran M, Cygankiewicz A (2004) Variability in the content of water-extractable and water-unextractable non-starch polysaccharides in rye flour and their relationship to baking quality parameters. Cer Res Comm 143-150. https://doi.org/10.1007/BF03543292

Cyran MR, Dynkowska WM (2014) Mode of endosperm and wholemeal arabinoxylans solubilisation during rye breadmaking: genotypic diversity in the level, substitution degree and macromolecular characteristics. Food Chem 145:356–364. https://doi.org/10.1016/j.foodchem.2013.07.093

Danyluk J, Kane NA, Breton G, Limin AE, Fowler DB, Sarhan F (2003) TaVRT-1, a putative transcription factor associated with vegetative to reproductive transition in cereals. Plant Physiol 132:1849–1860. https://doi.org/10.1104/pp.103.023523

De Bustos A, Rubio P, Jouve N (2001) Characterisation of two gene subunits on the 1R chromosome of rye as orthologs of each the *Glu-1* genes of hexaploid wheat. Theor Appl Genet 103:733–742. https://doi.org/10.1007/s001220100669

De Bustos A, Jouve N (2003) Characterisation and analysis of new HMW-glutenin alleles encoded by the *Glu-R1* locus of *Secale cereale*. Theor Appl Genet 107:74–83. https://doi.org/10.1007/s00122-003-1234-z

Daniel G (1993) Anther culture in rye: improved plant regeneration using modified MS-media. Plant Breed 110:259–261. https://doi.org/10.1111/j.1439-0523.1993.tb00587.x

Davison FR, Brewbaker HE, Thompson NA (1924) Brittle straw and other abnormalities in rye. J Agric Res 28:169–172

Deleu LJ, Lemmens E, Redant L, Delcour JA (2020) The major constituents of rye (Secale cereale L.) flour and their role in the production of rye bread, a food product to which a multitude of health aspects are ascribed. Cereal Chem 97:739–754. https://doi.org/10.1002/cche.10306

Devos KM, Atkinson MD, Chinoy CN, Francis HA, Harcourt RL, Koebner RM, Liu CJ, Masojć P, Xie DX, Gale MD (1993) Chromosomal rearrangements in the rye genome relative to that of wheat. Theor Appl Genet 85:673–680. https://doi.org/10.1007/BF00225004

De Vries JN, Sybenga J (1984) Chromosomal location of 17 monogenically inherited morphological markers in rye (*Secale cereale* L.) using the translocation tester set. Z Pflanzenzücht 92:117–139

Dias SC, Franco OL, Magalhães CP, de Oliveira-Neto OB, Laumann RA, Figueira EL, Melo FR, Grossi-De-Sá MF (2005) Molecular cloning and expression of an alpha-amylase inhibitor from rye with potential for controlling insect pests. Protein J 24(2):113–123. https://doi.org/10.1007/s10930-004-1518-4

Dill P (1983) Zur züchterischen Verbesserung der Kornmasse bei Winterroggen (*Secale cereale* L.). Arch Züchtungsforsch 13:157–168

Dill P (1989) Zur züchterischen Verbesserung der Kornmasse bei Winterroggen (*Secale cereale* L.)—Ergebnisse von Drillprüfungen. Arch Züchtungsforsch 20:329–337

Dittmer HJ (1937) A quantitative study of the roots and root hairs of a winter rye plant (*Secale cereale*). Am J Bot 24:417–420

Dobrovolskaya O, Martinek P, Voylokov AV, Korzun V, Röder MS, Börner A (2009) Microsatellite mapping of genes that determine supernumerary spikelets in wheat (*T. aestivum*) and rye (*S. cereale*). Theor Appl Genet 119:867–874. https://doi.org/10.1007/s00122-009-1095-1

Doxey AC, Yaish MW, Griffith M, McConkey BJ (2006) Ordered surface carbons distinguish antifreeze proteins and their ice-binding regions. Nat Biotechnol 24:852–855. https://doi.org/10.1038/nbt1224

Dufay M, Billard E (2012) How much better are females? The occurrence of female advantage, its proximal causes and its variation within and among gynodioecious species. Ann Bot 109(3):505–519. https://doi.org/10.1093/aob/mcr062

Ehdaie B, Whitkus RW, Waines JG (2003) Root biomass, water-use efficiency, and performance of wheat-rye translocations of chromosomes 1 and 2 in spring bread wheat 'Pavon.' Crop Sci 43:710–717. https://doi.org/10.2135/cropsci2003.7100

Erath W, Bauer E, Kastirr U, Schmidt M, Korzun V, Schmiedchen B, Wilde P, Schön CC (2016) Oligogenic control of resistance to soil-borne viruses SBCMV and WSSMV in rye (*Secale cereale* L.). Plant Breed 135:552–559. https://doi.org/10.1111/pbr.12411

Erath W, Bauer E, Fowler DB, Gordillo A, Korzun V, Ponomareva M, Schmidt M, Schmiedchen B, Wilde P, Schön CC (2017) Exploring new alleles for frost tolerance in winter rye. Theor Appl Genet 130:2151–2164. https://doi.org/10.1007/s00122-017-2948-7

Fahy B, Siddiqui H, David LC, Powers SJ, Borrill P, Uauy C, Smith AM (2018) Final grain weight is not limited by the activity of key starch-synthesising enzymes during grain filling in wheat. J Exp Bot 69:5461–5475. https://doi.org/10.1093/jxb/ery314

Falke KC, Sušić Z, Hackauf B, Korzun V, Schondelmaier J, Wilde P, Wehling P, Wortmann H, Mank JR, van der Voort JR, Maurer HP, Miedaner T, Geiger HH (2008) Establishment of introgression libraries in hybrid rye (*Secale cereale* L.) from an Iranian primitive accession as a new tool for rye breeding and genomics. Theor Appl Genet 117:641–652. https://doi.org/10.1007/s00122-008-0808-1

Falke KC, Susić Z, Wilde P, Wortmann H, Möhring J, Piepho HP, Geiger HH, Miedaner T (2009) Testcross performance of rye introgression lines developed by marker-assisted backcrossing using an Iranian accession as donor. Theor Appl Genet 118:1225–1238. https://doi.org/10.1007/s00122-009-0976-7

Fengler A, Marquardt RR (1988) Water-soluble pentosans from rye. II. Effects of rate of dialysis and on the retention of nutrients by the chicks. Cereal Chem 65:298–302

Fernandez R, Lucas E, McGinnis J (1973) Fractionation of a chick growth depressing factor from rye. Poult Sci 52:2252–2259

Fischer S, Melchinger AE, Korzun V, Wilde P, Schmiedchen B, Möhring J, Piepho HP, Dhillon BS, Würschum T, Reif JC (2010) Molecular marker assisted broadening of the Central European heterotic groups in rye with Eastern European germplasm. Theor Appl Genet 120:291–299. https://doi.org/10.1007/s00122-009-1124-0

Flehinghaus T, Deimling S, Geiger HH (1991) Methodogical improvements in rye anther culture. Plant Cell Rep 10:397–400. https://doi.org/10.1007/BF00232610

Fleinghaus-Roux T, Deimling S, Geiger HH (1995) Anther-culture ability in *Secale cereale* L. Plant Breed 114:259–261. https://doi.org/10.1111/j.1439-0523.1995.tb00807.x

Fontecha G, Silva-Navas J, Benito C, Mestres MA, Espino FJ, Hernández-Riquer MV, Gallego FJ (2007) Candidate gene identification of an aluminum-activated organic acid transporter gene at the Alt4 locus for aluminum tolerance in rye (*Secale cereale* L.). Theor Appl Genet 114:249–260. https://doi.org/10.1007/s00122-006-0427-7

Freeman J, Lovegrove A, Wilkinson MD, Saulnier L, Shewry PR, Mitchell RA (2016) Effect of suppression of arabinoxylan synthetic genes in wheat endosperm on chain length of arabinoxylan and extract viscosity. Plant Biotechnol J 14:109–116. https://doi.org/10.1111/pbi.12361

Frey M, Chomet P, Glawischnig E, Stettner C, Grün S, Winklmair A, Eisenreich W, Bacher A, Meeley RB, Briggs SP, Simcox K, Gierl A (1997) Analysis of a chemical plant defense mechanism in grasses. Science 277:696–699. https://doi.org/10.1126/science.277.5326.696

Frey M, Schullehner K, Dick R, Fiesselmann A, Gierl A (2009) Benzoxazinoid biosynthesis, a model for evolution of secondary metabolic pathways in plants. Phytochemistry 70:1645–1651. https://doi.org/10.1016/j.phytochem.2009.05.012

Friesen TL, Faris JD (2004) Molecular mapping of resistance to Pyrenophora tritici-repentis race 5 and sensitivity to Ptr ToxB in wheat. Theor Appl Genet 109:464–471. https://doi.org/10.1007/s00122-004-1678-9

Fuong FT, Voylokov AV, Smirnov VG (1993) Genetic studies of self-fertility in rye (*Secale cereale* L.). 2. The search for isozyme marker genes linked to self-incompatibility loci. Theor Appl Genet 87:619–623. https://doi.org/10.1007/BF00221888

Gabay G, Zhang J, Burguener GF, Howell T, Wang H, Fahima T, Lukaszewski A, Moriconi JI, Santa Maria GE, Dubcovsky J (2021) Structural rearrangements in wheat (1BS)-rye (1RS) recombinant chromosomes affect gene dosage and root length. Plant Genome. 18:e20079. https://doi.org/10.1002/tpg2.20079

Gaikpa DS, Koch S, Fromme FJ, Siekmann D, Würschum T, Miedaner T (2020) Genome-wide association mapping and genomic prediction of Fusarium head blight resistance, heading stage and plant height in winter rye (*Secale cereale*). Plant Breed 139:508–520. https://doi.org/10.1111/pbr.12810

Gallego FJ, Benito C (1997) Genetic control of aluminium tolerance in rye (*Secale cereale* L.). Theor Appl Genet 95:393–399. https://doi.org/10.1007/s001220050575

Gallego FJ, López-Solanilla E, Figueiras AM, Benito C (1998) Chromosomal location of PCR fragments as a source of DNA markers linked to aluminium tolerance genes in rye. Theor Appl Genet 96:426–434. https://doi.org/10.1007/s001220050759

Gan YT, McLeod JG, Scoles GJ, Campbell GL (1998) Relationship between extract viscosity and kernel size in winter rye. Can J Plant Sci 78:423–427. https://doi.org/10.4141/P97-085

Gao F, Ayele BT (2014) Functional genomics of seed dormancy in wheat: advances and prospects. Front Plant Sci 5:458. https://doi.org/10.3389/fpls.2014.00458

Geiger HH, Miedaner T (2009) Rye breeding. In Carena MJ (ed) Handbook of plant breeding, cereals, vol. 3. Springer Publishing, New York, pp 157–181

Geiger HH, Schnell FW (1970a) Cytoplasmic male sterility in rye (*Secale cereale* L.). Crop Sci 10:590–593

Geiger HH, Schnell FW (1970b) Die Züchtung von Roggensorten aus Inzuchtlinien: I. Selbstungsanteile in Polycross-Nachkommenschaften. Theor Appl Genet 40:305–311. https://doi.org/10.1007/BF00264572

Geiger HH (1985) Hybrid breeding in rye. In: Proceding eucarpia meeting of the cereal section on rye, Svalöv, Sweden, pp 237–266

Gertz A, Wricke G (1989) Linkage between the incompatibility locus Z and a ß-glucosidase locus in rye. Plant Breed 102:255–259

Goldberg EE, Kohn JR, Lande R, Robertson KA, Smith SA, Igić B (2010) Species selection maintains self-incompatibility. Science 330(6003):493–495. https://doi.org/10.1126/science.1194513

Goncharenko AA, Timoshchenko AS, Berkutova NS, Lazareva EN (2008) Relation between weighted average molecular mass of water extractive pentosans of winter rye and processing and baking properties of grain. Russ Agricult Sci 34:211–214. https://doi.org/10.3103/S1068367408040010

Goncharenko AA, Timoshchenko AS (2010) Analysis of the correlation potential of traits determining the viscosity potential of a water extract of winter rye grain. Russ Agricult Sci 36:151–155. https://doi.org/10.3103/S1068367410030018

Goncharenko AA, Osipova AV, Ermakov SA, Makarov AV, Semenova TV, Tochilin VN, Davidova EI, Yashina NA, Kondrateva OP, Shcherbakova ZN, Krahmaleva OA (2017) Dynamics of technological and baking qualities of winter rye grain depending on flour yield and water extract viscosity. Russ Agricult Sci 43:361–367. https://doi.org/10.3103/S106836741705007X

Goncharenko AA, Makarov AV, Ermakov SA, Semenova TV, Tochilin VN, Tsygankova NV, Krakhmaleva OA (2019) Selection of winter rye (*Secale cereale* L.) inbred lines for general and specific combining ability and its relationship with valuable traits. Agric Biol 38–46. https://doi.org/10.15389/agrobiology.2019.1.38eng

Góralska M, Bińkowski J, Lenarczyk N, Bienias A, Grądzielewska A, Czyczyło-Mysza I, Kapłoniak K, Stojałowski S, Myśków B (2020) How machine learning methods helped find putative rye wax genes among GBS data. Int J Mol Sci 21(20):7501. https://doi.org/10.3390/ijms21207501

Grądzielewska A, Milczarski P, Molik K, Pawłowska E (2020) Identification and mapping of a new recessive dwarfing gene dw9 on the 6RL rye chromosome and its phenotypic effects. PLoS One. 15:e0229564. https://doi.org/10.1371/journal.pone.0229564

Grochowski L, Kaczmarek J, Kadłubiec W, Bujak H (1995) Using xenia in the breeding of rye hybrids. Acta Soc Bot Pol 64:175–179

Gruner P, Schmitt AK, Flath K, Schmiedchen B, Eifler J, Gordillo A, Schmidt M, Korzun V, Fromme FJ, Siekmann D, Tratwal A, Danielewicz J, Korbas M, Marciniak K, Krysztofik R, Niewińska M, Koch S, Piepho HP, Miedaner T (2020) Mapping stem rust (*Puccinia graminis* f. sp. *secalis*) resistance in self-fertile winter rye populations. Front Plant Sci 11:667. https://doi.org/10.3389/fpls.2020.00667

Gruszczyńska A, Rakoczy-Trojanowska M (2011) Expression analysis of somatic embryogenesis-related *SERK*, *LEC1*, *VP1* and *NiR* ortologues in rye (*Secale cereale* L.). J Appl Genet 52:1–8. https://doi.org/10.1007/s13353-010-0015-z

Guo YD, Pulli S (2000) Isolated microspore culture and plant regeneration in rye (*Secale cereale* L.). Plant Cell Rep 19(9):875–880. https://doi.org/10.1007/s002990000194

Guo B, Sleper DA, Beavis WD (2010) Nested association mapping for identification of functional markers. Genetics 186:373–383. https://doi.org/10.1534/genetics.110.115782

Hackauf B, Wehling P (2002) Identification of microsatellite polymorphisms in an expressed portion of the rye genome. Plant Breed. 121:17–25. https://doi.org/10.1046/j.1439-0523.2002.00649.x

Hackauf B, Wehling P (2003) Development of microsatellite markers in rye: map construction. Plant Breed Seed Sci 48:143–151

Hackauf B, Rudd S, van der Voort JR, Miedaner T, Wehling P (2009) Comparative mapping of DNA sequences in rye (*Secale cereale* L.) in relation to the rice genome. Theor Appl Genet 118:371–384. https://doi.org/10.1007/s00122-008-0906-0

Hackauf B, Korzun V, Wortmann H, Wilde P, Wehling P (2012) Development of conserved ortholog set markers linked to the restorer gene *Rfp1* in rye. Mol Breed 30:1507–1518. https://doi.org/10.1007/s11032-012-9736-5

Hackauf B, Haffke S, Fromme FJ, Roux SR, Kusterer B, Musmann D, Kilian A, Miedaner T (2017a) QTL mapping and comparative genome analysis of agronomic traits including grain yield in winter rye. Theor Appl Genet 130:1801–1817. https://doi.org/10.1007/s00122-017-2926-0

Hackauf B, Bauer E, Korzun V, Miedaner T (2017b) Fine mapping of the restorer gene *Rfp3* from an Iranian primitive rye (*Secale cereale* L.). Theor Appl Genet 130:1179–1189. https://doi.org/10.1007/s00122-017-2879-3

Hamel D, Rozman V, Liška A (2020) Storage of cereals in warehouses with or without pesticides. Insects 11:846. https://doi.org/10.3390/insects11120846

Hansen HB, Møller B, Andersen SB, Jørgensen JR, Hansen A (2004) Grain characteristics, chemical composition, and functional properties of rye (*Secale cereale* L.) as influenced by genotype and harvest year. J Agric Food Chem 52:2282–2291. https://doi.org/10.1021/jf0307191

Hansson M, Komatsuda T, Stein N, Muehlbauer GJ (2018) Molecular mapping and cloning of genes and QTLs. In: Stein N, Muehlbauer G J (eds) The barley genome, 1st ed. (Series: Kole, C (Ed.): compendium of plant Genomes). Springer, Cham, pp 139–154. https://doi.org/10.1007/978-3-319-92528-8_10

Hagenblad J, Oliveira HR, Forsberg NE, Leino MW (2016) Geographical distribution of genetic diversity in *Secale* landrace and wild accessions. BMC Plant Biol 16:23. https://doi.org/10.1186/s12870-016-0710-y

Haseneyer G, Schmutzer T, Seidel M, Zhou R, Mascher M, Schön CC, Taudien S, Scholz U, Stein N, Mayer KF, Bauer E (2011) From RNA-seq to large-scale genotyping - genomics resources for rye (*Secale cereale* L.). BMC Plant Biol 11:131. https://doi.org/10.1186/1471-2229-11-131

Hawliczek A, Bolibok L, Tofil K, Borzęcka E, Jankowicz-Cieślak J, Gawroński P, Kral A, Till BJ, Bolibok-Brągoszewska H (2020) Deep sampling and pooled amplicon sequencing reveals hidden genic variation in heterogeneous rye accessions. BMC Genomics 21(1):845. https://doi.org/10.1186/s12864-020-07240-3

Hepting L (1978) Analyse eines 7 x 7-Sortendialles zur Ermittlung geeigneten Ausgangsmaterials für die

Hybridzüchtung bei Roggen. Z Pflanzenzüchtg 80:188–197

Heribert Nilsson N (1916) Populationsanalysen und Erblichkeitsversuche über die Selbststerilitat, Selbstfertilitat und Sterilitat bei dem Roggen. Ztschr f Plflanzenzuchtung IV:1–44

Heribert Nilsson N (1953) Über die entstehung der selbstfertilität beim roggen. Hereditas 39:65–74

Hoffmann F, Wenzel G (1981) Self compatibility in microspore-derived doubled-haploid rye lines and single grain selection for alkylresorcinol content. Theor Appl Genet 60:129–133

Hoffman PD, Leonard JM, Lindberg GE, Bollmann SR, Hays JB (2004) Rapid accumulation of mutations during seed-to-seed propagation of mismatch-repair-defective Arabidopsis. Genes Dev 18:2676–2685. https://doi.org/10.1101/gad.1217204

Hölker AC, Mayer M, Presterl T, Bolduan T, Bauer E, Ordas B, Brauner PC, Ouzunova M, Melchinger AE, Schön CC (2019) European maize landraces made accessible for plant breeding and genome-based studies. Theor Appl Genet 132(12):3333–3345. https://doi.org/10.1007/s00122-019-03428-8

Howell T, Moriconi JI, Zhao X, Hegarty J, Fahima T, Santa-Maria GE, Dubcovsky J (2019) A wheat/rye polymorphism affects seminal root length and yield across different irrigation regimes. J Exp Bot 70:4027–4037. https://doi.org/10.1093/jxb/erz169

Hsam SLK, Mohler V, Hartl L, Wenzel G, Zeller FJ (2000) Mapping of powdery mildew and leaf rust resistance genes on the wheat rye translocated chromosome T1BL·1RS using molecular and biochemical markers. Plant Breed 119:87–89. 0.1046/j.1439-0523.2000.00444.x

Hurni S, Brunner S, Buchmann G, Herren G, Jordan T, Krukowski P, Wicker T, Yahiaoui N, Mago R, Keller B (2013) Plant J 76(6):957–969. https://doi.org/10.1111/tpj.12345

Huth W (2002) The soil-borne viruses of wheat and rye in Europe—an increasing problem which can be minimised by agricultural measures and growing of resistant cultivars. Gesunde Pflanz 54:51–54

Hübner M, Wilde P, Schmiedchen B, Dopierala P, Gowda M, Reif JC, Miedaner T (2013) Hybrid rye performance under natural drought stress in Europe. Theor Appl Genet 126:475–482. https://doi.org/10.1007/s00122-012-1994-4

Huynh BL, Mather DE, Schreiber AW, Toubia J, Baumann U, Shoaei Z, Stein N, Ariyadasa R, Stangoulis JC, Edwards J, Shirley N, Langridge P, Fleury D (2012) Clusters of genes encoding fructan biosynthesizing enzymes in wheat and barley. Plant Mol Biol 80:299–314. https://doi.org/10.1007/s11103-012-9949-3

International Wheat Genome Sequencing Consortium (IWGSC) (2018) Shifting the limits in wheat research and breeding using a fully annotated reference genome. Science 361:eaar7191. https://doi.org/10.1126/science.aar7191

Isidore E, Scherrer B, Bellec A, Budin K, Faivre-Rampant P, Waugh R, Keller B, Caboche M, Feuillet C, Chalhoub B (2005) Direct targeting and rapid isolation of BAC clones spanning a defined chromosome region. Funct Integr Genomics 5:97–103. https://doi.org/10.1007/s10142-004-0127-9

Jaccoud D, Peng K, Feinstein D, Kilian A (2001) Diversity arrays: a solid state technology for sequence information independent genotyping. Nucleic Acids Res 29(4):E25. https://doi.org/10.1093/nar/29.4.e25

James GV, Patel V, Nordström KJ, Klasen JR, Salomé PA, Weigel D, Schneeberger K (2013) User guide for mapping-by-sequencing in *Arabidopsis*. Genome Biol 14:R61. https://doi.org/10.1186/gb-2013-14-6-r61

Ji X, Jetter R (2008) Very long chain alkylresorcinols accumulate in the intracuticular wax of rye (*Secale cereale* L.) leaves near the tissue surface. Phytochemistry 69:1197–1207. https://doi.org/10.1016/j.phytochem.2007.12.008

Jia M, Guan J, Zhai Z, Geng S, Zhang X, Mao L, Li A (2018) Wheat functional genomics in the era of next generation sequencing: an update. Crop J 6:7–14. https://doi.org/10.1016/j.cj.2017.09.003

Jiang M, Wu X, Song Y, Shen H, Cui H (2020) Effects of *OsMSH6* Mutations on microsatellite stability and homeologous recombination in rice. Front Plant Sci 11:220. https://doi.org/10.3389/fpls.2020.00220

Jürgens HU, Jansen G, Wegener C (2012) Characterisation of several rye cultivars with respect to arabinoxylans and extract viscosity. J Agri Sci 4:1–12. https://doi.org/10.5539/jas.v4n5p1

Kantarek Z, Masojć P, Bienias A, Milczarski P (2018) Identification of a novel, dominant dwarfing gene (*Ddw4*) and its effect on morphological traits of rye. PLoS One 13:e0199335. https://doi.org/10.1371/journal.pone.0199335

Khalil HB, Ehdaeivand MR, Xu Y, Laroche A, Gulick PJ (2015) Identification and characterization of rye genes not expressed in allohexaploid triticale. BMC Genomics 16:281. https://doi.org/10.1186/s12864-015-1480-x

Khlestkina EK, Than MH, Pestsova EG, Röder MS, Malyshev SV, Korzun V, Börner A (2004) Mapping of 99 new microsatellite-derived loci in rye (*Secale cereale* L.) including 39 expressed sequence tags. Theor Appl Genet 109:725–732. https://doi.org/10.1007/s00122-004-1659-z

Khlestkina EK, Than MH, Pestsova EG, Röder MS, Malyshev SV, Korzun V, Börner A (2005) Erratum to: Mapping of 99 new microsatellite-derived loci in rye (*Secale cereale* L.) including 39 expressed sequence tags. Theor Appl Genet. 110:990–991. 0.1007/s00122-004-1904-5

Kobylyansky VD (1972) On the genetics of the dominant factor of shortstrawed rye. Genetika 8:12–17

Kobylyansky VD (1988) Initial material for solving the main problems encountered in breeding winter rye in the northwestern zone of the USSR. J Agric Sci Finland 60:215–221. https://doi.org/10.23986/afsci.72294

Kobylyansky VD, Solodukhina OV (1983) Damage of important fungal diseases and methods for resistance breeding of shortstraw rye. In: Voprosy Sal. i Genetiki zernovych Kult (in Russian). Moskva, pp 140–147

Kobylyansky VD, Solodukhina OV (2015) The role of the vavilov institute of plant industry in the initiation and development of new trends in winter rye breeding in Russia. Proc Appl Bot Genet Breed 176(1):5–19. https://doi.org/10.30901/2227-8834-2015-1-5-19

Kobylyansky VD, Kuznetsova LI, Solodukhina OV, Lavrentyeva NS, Timina MA (2019) Prospects of using low-pentosan grain fodder rye for baking purposes Russ. Agricult Sci 45:1–4. https://doi.org/10.3103/S1068367419010063

Komatsuda T, Pourkheirandish M, He C, Azhaguvel P, Kanamori H, Perovic D, Stein N, Graner A, Wicker T, Tagiri A, Lundqvist U, Fujimura T, Matsuoka M, Matsumoto T, Yano M (2007) Six-rowed barley originated from a mutation in a homeodomain-leucine zipper I-class homeobox gene. Proc Natl Acad Sci U S A 104:1424–1429. https://doi.org/10.1073/pnas.0608580104

Konovalov AA, Moiseeva EA, Goncharov NP, Kondratenko EY (2010) The order of the *bs*, *Skdh*, and *Aadh1* genes in chromosome 5R of rye *Secale cereale* L. Russ J Genet 46:666–671. https://doi.org/10.1134/S1022795410060050

Korzun V, Börner A, Melz G (1996) RFLP mapping of the dwarfing (*Ddw1*) and hairy peduncle (*Hp*) genes on chromosome 5 of rye (*Secale cereale* L.). Theor Appl Genet 92:1073–1077. https://doi.org/10.1007/BF00224051

Korzun V, Malyshev S, Voylokov A, Börner A (1997) RFLP-based mapping of three mutant loci in rye (*Secale cereale* L.) and their relation to homoeologous loci within the Gramineae. Theor Appl Genet 95:468–473

Korzun V, Malyshev S, Kartel N, Westermann T, Weber WE, Börner A (1998) A genetic linkage map of rye (*Secale cereale* L.). Theor Appl Genet 96:203–208

Korzun V, Börner A, Siebert R, Malyshev S, Hilpert M, Kunze R, Puchta H (1999) Chromosomal location and genetic mapping of the mismatch repair gene homologs MSH2, MSH3, and MSH6 in rye and wheat. Genome 42:1255–1257. https://doi.org/10.1139/g99-081

Korzun V, Malyshev S, Voylokov AV, Borner A (2001) A genetic map of rye (Secale cereale L.) combining RFLP, isozyme, protein, microsatellite and gene loci. Theor Appl Genet 102:709–717

Kottmann L, Wilde P, Schittenhelm S (2016) How do timing, duration, and intensity of drought stress affect the agronomic performance of winter rye? Europ J Agron 75:25–32. https://doi.org/10.1016/j.eja.2015.12.010

Koumoto T, Shimada H, Kusano H, She K-C, Iwamoto M, Takano M (2013) Rice monoculm mutation moc2, which inhibits outgrowth of the second tillers, is ascribed to lack of a fructose-1,6-bisphosphatase. Plant Biotech 30:47–56. https://doi.org/10.5511/plantbiotechnology.12.1210a

Kukk V, Tupits I (1996) Short-strawed breeding line *Dm* Vortr Pflanzenzüchtg 35:76–77

Kunkulberga D, Linina A, Kronberga A, Kokare A, Lenenkova I (2017) Grain quality of winter rye cultivars grown in Latvia. In: Proceeding of the 10th baltic conference on food science and technology "FOODBALT", pp 121–125. https://doi.org/10.22616/foodbalt.2017.015

Laidig F, Piepho HP, Rentel D, Drobek T, Meyer U, Huesken A (2017) Breeding progress, variation, and correlation of grain and quality traits in winter rye hybrid and population varieties and national on-farm progress in Germany over 26 years. Theor Appl Genet 130(5):981–998. https://doi.org/10.1007/s00122-017-2865-9

Lang Z, Wills DM, Lemmon ZH, Shannon LM, Bukowski R, Wu Y, Messing J, Doebley JF (2014) Defining the role of prolamin-box binding factor1 gene during maize domestication. J Hered 105:576–582. https://doi.org/10.1093/jhered/esu019

Langner T, Göhre V (2016) Fungal chitinases: function, regulation, and potential roles in plant/pathogen interactions. Curr Genet 62:243–254. https://doi.org/10.1007/s00294-015-0530-x

Lawrence GJ, Shepherd KW (1981) Inheritance of glutenin protein subunits of wheat. Mapping the nucleolus organiser region, seed protein loci and isozyme loci on chromosome 1R in rye. Theor Appl Genet 60:333–337

Lázaro R, García M, Medel P, Mateos GG (2003) Influence of enzymes on performance and digestive parameters of broilers fed rye-based diets. Poult Sci 82:132–140. https://doi.org/10.1093/ps/82.1.132

Lázaro R, Latorre MA, Medel P, Gracia M, Mateos GG (2004) Feeding regimen and enzyme supplementation to rye-based diets for broilers. Poult Sci 83:152–160. https://doi.org/10.1093/ps/83.2.152

Lazo GR, Chao S, Hummel DD, Edwards H, Crossman CC, Lui N, Matthews DE, Carollo VL, Hane DL, You FM, Butler GE, Miller RE, Close TJ, Peng JH, Lapitan NL, Gustafson JP, Qi LL, Echalier B, Gill BS, Dilbirligi M, Randhawa HS, Gill KS, Greene RA, Sorrells ME, Akhunov ED, Dvorák J, Linkiewicz AM, Dubcovsky J, Hossain KG, Kalavacharla V, Kianian SF, Mahmoud AA, Miftahudin MXF, Conley EJ, Anderson JA, Pathan MS, Nguyen HT, McGuire PE, Qualset CO, Anderson OD (2004) Development of an expressed sequence tag (EST) resource for wheat (*Triticum aestivum* L.): EST generation, unigene analysis, probe selection and bioinformatics for a 16,000-locus bin-delineated map. Genetics 168:585–593. https://doi.org/10.1534/genetics.104.034777

Lee SK, Jeon JS, Börnke F, Voll L, Cho JI, Goh CH, Jeong SW, Park YI, Kim SJ, Choi SB, Miyao A, Hirochika H, An G, Cho MH, Bhoo SH, Sonnewald U, Hahn TR (2008) Loss of cytosolic fructose-1,6-bisphosphatase limits photosynthetic sucrose synthesis and causes severe growth retardations in rice (Oryza sativa). Plant Cell Environ 31:1851–1863. https://doi.org/10.1111/j.1365-3040.2008.01890.x

Lee TG, Hong MJ, Johnson JW, Bland DE, Kim DY, Seo YW (2009) Development and functional assessment of EST-derived 2RL-specific markers for 2BS.2RL translocations. Theor Appl Genet 119:663–673. https://doi.org/10.1007/s00122-009-1077-3

Li G, Wang L, Yang J, He H, Jin H, Li X, Ren T, Ren Z, Li F, Han X, Zhao X, Dong L, Li Y, Song Z, Yan Z, Zheng N, Shi C, Wang Z, Yang S, Xiong Z, Zhang M, Sun G, Zheng X, Gou M, Ji C, Du J, Zheng H, Dolezel J, Wang Deng X, Stein N, Yang Q, Zhang K, Wang D (2021) Genomic characteristics and agronomically important genes of rye elucidated using a high-quality genome assembly. Nat Genet, provisionally accepted

Li Y, Haseneyer G, Schön CC, Ankerst D, Korzun V, Wilde P, Bauer E (2011a) High levels of nucleotide diversity and fast decline of linkage disequilibrium in rye (*Secale cereale* L.) genes involved in frost response. BMC Plant Biol. 11:6. https://doi.org/10.1186/1471-2229-11-6.0

Li Y, Böck A, Haseneyer G, Korzun V, Wilde P, Schön CC, Ankerst DP, Bauer E (2011b) Association analysis of frost tolerance in rye using candidate genes and phenotypic data from controlled, semi-controlled, and field phenotyping platforms. BMC Plant Biol 11:146. https://doi.org/10.1186/1471-2229-11-146

Li MJ, Li JQ, Zhang N, Shi ZL (2016a) Cloning of the ω-secalin gene family in a wheat 1BL/1RS translocation line using BAC clone sequencing. Electron J Biotechn 21:1–8. https://doi.org/10.1016/j.ejbt.2015.12.001

Li Z, Ren T, Yan B, Tan F, Yang M, Ren Z (2016b) A Mutant with Expression Deletion of Gene *Sec-1* in a 1RS.1BL Line and Its Effect on Production Quality of Wheat. PLoS One. 11:e0146943.https://doi.org/10.1371/journal.pone.0146943

Li D, Huang Z, Song S, Xin Y, Mao D, Lv Q, Zhou M, Tian D, Tang M, Wu Q, Liu X, Chen T, Song X, Fu X, Zhao B, Liang C, Li A, Liu G, Li S, Hu S, Cao X, Yu J, Yuan L, Chen C, Zhu L (2016c) Integrated analysis of phenome, genome, and transcriptome of hybrid rice uncovered multiple heterosis-related loci for yield increase. Proc Natl Acad Sci U S A 113:E6026–E6035. https://doi.org/10.1073/pnas.1610115113

Li Z, Wang P, You C, Yu J, Zhang X, Yan F, Ye Z, Shen C, Li B, Guo K, Liu N, Thyssen GN, Fang DD, Lindsey K, Zhang X, Wang M, Tu L (2020) Combined GWAS and eQTL analysis uncovers a genetic regulatory network orchestrating the initiation of secondary cell wall development in cotton. New Phytol 226:1738–1752. https://doi.org/10.1111/nph.16468

Limin AE, Fowler DB (1991) Breeding for cold hardiness in winter wheat: problems, progress and alien gene expression. Field Crops Res 27:201–218

Liu Q, Harberd NP, Fu X (2016) SQUAMOSA Promoter Binding Protein-like transcription factors: targets for improving cereal grain yield. Mol Plant 9:765–767. https://doi.org/10.1016/j.molp.2016.04.008

Liu S, Huang Y, Xu H, Zhao M, Xu Q, Li F (2018) Genetic enhancement of lodging resistance in rice due to the key cell wall polymer lignin, which affects stem characteristics. Breed Sci 68:508–515. https://doi.org/10.1270/jsbbs.18050

Liu J, Fernie AR, Yan J (2021) Crop breeding—From experience-based selection to precision design. J Plant Physiol 256:153313. https://doi.org/10.1016/j.jplph.2020.153313.

Lo SF, Ho TD, Liu YL, Jiang MJ, Hsieh KT, Chen KT, Yu LC, Lee MH, Chen CY, Huang TP, Kojima M, Sakakibara H, Chen LJ, Yu SM (2017) Ectopic expression of specific GA2 oxidase mutants promotes yield and stress tolerance in rice. Plant Biotechnol J 15:850–864. https://doi.org/10.1111/pbi.12681

Loarce Y, Hueros G, Ferrer E (1996) A molecular linkage map of rye. Theor Appl Genet 93:1112–1118. https://doi.org/10.1007/BF00230133

Lovegrove A, Wilkinson MD, Freeman J, Pellny TK, Tosi P, Saulnier L, Shewry PR, Mitchell RA (2013) RNA interference suppression of genes in glycosyl transferase families 43 and 47 in wheat starchy endosperm causes large decreases in arabinoxylan content. Plant Physiol 163:95–107. https://doi.org/10.1104/pp.113.222653

Lundqvist A (1956) Self-Incompatibility in Rye .1. Genetic Control in the Diploid. Hereditas 42 (3–4):293–348. https://doi.org/10.1111/j.1601-5223.1956.tb03021.x

Lundqvist A (1960) The origin of self-compatibility in rye. Hereditas 46(1–2):1–19

Lykholay AN, Vladimirov IA, Andreeva EA, Smirnov VG, Voylokov AV (2014) Genetics of anthocyaninless rye. Russ J Genet 501102–1106

Ma XF, Wanous MK, Houchins K, Rodriguez Milla MA, Goicoechea PG, Wang Z, Xie M, Gustafson JP (2001) Molecular linkage mapping in rye (*Secale cereale* L.). Theor Appl Genet 102:517–523

MacLeod AM, Preece IA (1954) Studies of the free sugars of the barley grain. V. Comparison of sugars and fructosans with those of other cereals. J Inst Brew 60:46–55

Madej L, Raczynska-Bojanowska K, Rybka K (1990) Variability of the content of soluble non-digestible polysaccharides in rye inbred lines. Plant Breed 104:334–339

Madsen CK, Dionisio G, Holme IB, Holm PB, Brinch-Pedersen H (2013) High mature grain phytase activity in the Triticeae has evolved by duplication followed by neofunctionalization of the purple acid phosphatase phytase (PAPhy) gene. J Exp Bot 64:3111–3123. https://doi.org/10.1093/jxb/ert116

Mago R, Zhang P, Vautrin S, Šimková H, Bansal U, Luo MC, Rouse M, Karaoglu H, Periyannan S, Kolmer J, Jin Y, Ayliffe MA, Bariana H, Park RF, McIntosh R, Doležel J, Bergès H, Spielmeyer W, Lagudah ES, Ellis JG, Dodds PN (2015) The wheat *Sr50* gene reveals rich diversity at a cereal disease resistance locus. Nat Plants 1:15186. https://doi.org/10.1038/nplants.2015

Mahone GS, Frisch M, Miedaner T, Wilde P, Wortmann H, Falke KC (2013) Identification of quantitative trait loci in rye introgression lines carrying multiple donor chromosome segments. Theor Appl Genet 126:49–58. https://doi.org/10.1007/s00122-012-1958-8

Mahone GS, Frisch M, Bauer E, Haseneyer G, Miedaner T, Falke KC (2015) Detection of donor effects in a rye introgression population with genome-wide prediction. Plant Breed 134:406–415. https://doi.org/10.1111/pbr.12283

Malyshev SV, Khmyl TO, Zabenkova KI, Voylokov V, Korzun V, Kartel NV (1998) RFLP-based mapping of the *Sec-2* and *Sec-5* loci encoding 75K γ-secalins of rye. Plant Breeding 117:329–333. https://doi.org/10.1111/j.1439-0523.1998.tb01949.x

Malyshev S, Korzun V, Voylokov V, Smirnov V, Börner A (2001) Linkage mapping of mutant loci in rye (*Secale cereale* L.). Theor Appl Genet 103:70–74

Manova V, Gruszka D (2015) DNA damage and repair in plants—from models to crops. Front Plant Sci 6:885. https://doi.org/10.3389/fpls.2015.00885

Markert CL, Møller F (1959) Multiple forms of enzymes: tissue, ontogenetic, and specific patterns. Proc Natl Acad Sci U S A 45(5):753–763. https://doi.org/10.1073/pnas.45.5.753

Martis MM, Zhou R, Haseneyer G, Schmutzer T, Vrána J, Kubaláková M, König S, Kugler KG, Scholz U, Hackauf B, Korzun V, Schön CC, Dolezel J, Bauer E, Mayer KF, Stein N (2013) Reticulate evolution of the rye genome. Plant Cell 25:3685–3698. https://doi.org/10.1105/tpc.113.114553

Mascher M, Richmond TA, Gerhardt DJ, Himmelbach A, Clissold L, Sampath D, Ayling S, Steuernagel B, Pfeifer M, D'Ascenzo M, Akhunov ED, Hedley PE, Gonzales AM, Morrell PL, Kilian B, Blattner FR, Scholz U, Mayer KF, Flavell AJ, Muehlbauer GJ, Waugh R, Jeddeloh JA, Stein N (2013) Barley whole exome capture: a tool for genomic research in the genus Hordeum and beyond. Plant J 76:494–505. https://doi.org/10.1111/tpj.12294

Mascher M, Jost M, Kuon JE, Himmelbach A, Aßfalg A, Beier S, Scholz U, Graner A, Stein N (2014) Mapping-by-sequencing accelerates forward genetics in barley. Genome Biol 15:R78. https://doi.org/10.1186/gb-2014-15-6-r78

Masojć P, Gale MD (1991) α-Amylase structural genes in rye. Theor Appl Genet 82:771–776

Masojć P, Milczarski P (2005) Mapping QTLs for alpha-amylase activity in rye grain. J Appl Genet 46:115–123

Masojć P, Banek-Tabor A, Milczarski P, Twardowska M (2007) QTLs for resistance to preharvest sprouting in rye (*Secale cereale* L.). J Appl Genet 48:211–217. https://doi.org/10.1007/BF03195215

Masojć P, Milczarski P (2009) Relationship between QTLs for preharvest sprouting and alpha-amylase activity in rye grain. Mol Breeding 23:75–84. https://doi.org/10.1007/s11032-008-9215-1

Matos M, Camacho MV, Pérez-Flores V, Pernaute B, Pinto-Carnide O, Benito C (2005) A new aluminum tolerance gene located on rye chromosome arm 7RS. Theor Appl Genet 111:360–369. https://doi.org/10.1007/s00122-005-2029-1

Mayer M, Unterseer S, Bauer E, de Leon N, Ordas B, Schön CC (2017) Is there an optimum level of diversity in utilization of genetic resources? Theor Appl Genet 130(11):2283–2295. https://doi.org/10.1007/s00122-017-2959-4

Mayer M, Hölker AC, González-Segovia E, Bauer E, Presterl T, Ouzunova M, Melchinger AE, Schön CC (2020) Discovery of beneficial haplotypes for complex traits in maize landraces. Nat Commun 11(1):4954. https://doi.org/10.1038/s41467-020-18683-3

McGhee ML, Stein HH (2018) Apparent and standardized ileal digestibility of AA and starch in hybrid rye, barley, wheat, and corn fed to growing pigs. J Anim Sci 96:3319–3329. https://doi.org/10.1093/jas/sky206

McGhee ML, Stein HH (2020) The apparent ileal digestibility and the apparent total tract digestibility of carbohydrates and energy in hybrid rye are different from some other cereal grains when fed to growing pigs. J Anim Sci 98:skaa218. https://doi.org/10.1093/jas/skaa218.

Melchinger AE, Utz HF, Schön CC (1998) Quantitative trait locus (QTL) mapping using different testers and independent population samples in maize reveals low power of QTL detection and large bias in estimates of QTL effects. Genetics 149:383–403

Melz G (1989) Beiträge zur Genetik des Roggens (*Secale cereale* L.). DSc thesis, Berlin

Melz G, Thiele V (1990) Chomosome locations of genes controlling "purple leaf base" in rye and wheat. Euphytica 49:155–159

Melz G, Kaczmarek J, Szigat G (1990) Genetical analysis of rye (*Secale cereale* L.). Location of self-fertility genes in different inbred lines. Genet Polon 31:1–7

Melz G, Schlegel R, Thiele V (1992) Genetic linkage map of rye (*Secale cereale* L.). Theor Appl Genet 8533–8545. https://doi.org/10.1007/BF00223842

Melz G, Melz G, Hartman F (2003) Genetics of a male-sterile rye of 'G-type' with results of the first F1-hybrids. Plant Breeding and Seed Science 47:47–55

Meng M, Gao X, Han LJ, Li XY, Wu D, Li HZ, Chen QJ (2014) Correlation analysis between starch properties and single nucleotide polymorphisms of waxy genes in common rye (*Secale cereale* L.). Genet Mol Res 13 (2):2574–2589, 2014 Jan 14. https://doi.org/10.4238/2014.January.14.7

Mengersen FG (1951) Die Wirkung der Inzucht auf verschiedene Merkmale beim Roggen. Z Pflanzenzuchtung 30:218–248

Michelmore RW, Paran I, Kesseli RV (1991) Identification of markers linked to disease-resistance genes by bulked segregant analysis: a rapid method to detect markers in specific genomic regions by using segregating populations. Proc Natl Acad Sci U S A 88:9828–9832. https://doi.org/10.1073/pnas.88.21.9828

Middleton CP, Senerchia N, Stein N, Akhunov ED, Keller B, Wicker T, Kilian B (2014) Sequencing of chloroplast genomes from wheat, barley, rye and their relatives provides a detailed insight into the evolution of the Triticeae tribe. PLoS One. 9(3):e85761. https://doi.org/10.1371/journal.pone.0085761

Miedaner T, Glass C, Dreyer F, Wilde P, Wortmann H, Geiger HH (2000) Mapping of genes for male-fertility restoration in 'Pampa' CMS winter rye (*Secale cereale* L.). Theor Appl Genet 101:1226–1233. https://doi.org/10.1007/s001220051601

Miedaner T, Dänicke S, Schmiedchen B, Wilde P, Wortmann H, Dhillon BS, Geiger HH, Mirdita V (2010) Genetic variation for ergot (*Claviceps purpurea*) resistance and alkaloid concentrations in cytoplasmic-male sterile winter rye under pollen isolation. Euphytica 173:299–306. https://doi.org/10.1007/s10681-009-0083-5

Miedaner T, Schwegler DD, Wilde P, Reif JC (2014) Association between line per se and testcross performance for eight agronomic and quality traits in winter rye. Theor Appl Genet 127(1):33–41. https://doi.org/10.1007/s00122-013-2198-2

Miedaner T, Hübner M, Korzun V, Schmiedchen B, Bauer E, Haseneyer G, Wilde P, Reif JC (2012) Genetic architecture of complex agronomic traits examined in two testcross populations of rye (*Secale cereale* L.). BMC Genomics 13:706. https://doi.org/10.1186/1471-2164-13-706

Miedaner T, Geiger HH (2015) Biology, genetics, and management of ergot (Claviceps spp.) in rye, sorghum, and pearl millet. Toxins (Basel). 7(3):659–678. https://doi.org/10.3390/toxins7030659

Miedaner T, Herter CP, Goßlau H, Wilde P, Hackauf B (2017) Correlated effects of exotic pollen-fertility restorer genes on agronomic and quality traits of hybrid rye. Plant Breed 136:224–229. https://doi.org/10.1111/pbr.12465

Miedaner T, Haffke S, Siekmann D, Fromme FJ, Roux SR, Hackauf B (2018) Dynamic quantitative trait loci (QTL) for plant height predict biomass yield in hybrid rye (*Secale cereale* L.). Biomass Bioenerg 115:10–18. https://doi.org/10.1016/j.biombioe.2018.04.001

Milczarski P, Masojć P (2003) Interval mapping of genes controlling growth of rye plants. Plant Breed Seed Sci 48:135–142

Milczarski P, Bolibok-Brągoszewska H, Myśków B, Stojałowski S, Heller-Uszyńska K, Góralska M, Brągoszewski P, Uszyński G, Kilian A, Rakoczy-Trojanowska M (2011) A high density consensus map of rye (*Secale cereale* L.) based on DArT markers. PLoS One 6:e28495. https://doi.org/10.1371/journal.pone.0028495

Milczarski P, Masojć P, Krajewski P, Stochmal A, Kowalczyk M, Angelov M, Ivanova V, Schollenberger M, Wakuliński W, Banaszak Z, Banaszak K, Rakoczy-Trojanowska M (2017) QTL mapping for benzoxazinoid content, preharvest sprouting, α-amylase activity, and leaf rust resistance in rye (*Secale cereale* L.). PLoS One. 12:e0189912. https://doi.org/10.1371/journal.pone.0189912

Milla MA, Butler E, Huete AR, Wilson CF, Anderson O, Gustafson JP (2002) Expressed sequence tag-based gene expression analysis under aluminum stress in rye. Plant Physiol 130:1706–1716. https://doi.org/10.1104/pp.009969

Mitchell RA, Dupree P, Shewry PR (2007) A novel bioinformatics approach identifies candidate genes for the synthesis and feruloylation of arabinoxylan. Plant Physiol 144:43–53. https://doi.org/10.1104/pp.106.094995

Monteiro F, Vidigal P, Barros AB, Monteiro A, Oliveira HR, Viegas W (2016) Genetic distinctiveness of rye in situ accessions from Portugal unveils a new hotspot of unexplored genetic resources. Front Plant Sci 7:1334. https://doi.org/10.3389/fpls.2016.01334

Moreno MA, Harper LC, Krueger RW, Dellaporta SL, Freeling M (1997) *liguleless1* encodes a nuclear-localized protein required for induction of ligules and auricles during maize leaf organogenesis. Genes Dev 11:616–628. https://doi.org/10.1101/gad.11.5.616

Musa GLC, Dyck PL, Samborski DJ (1984) The inheritance of resistance in rye to *Puccinia recondita* f. sp. *secalis* and f.sp. *tritici*. Can J Plant Sci 64:511–519

Muszyński S, Arczewska M, Świątkiewicz S, Arczewska-Włosek A, Dobrowolski P, Świetlicka I, Hułas-Stasiak M, Blicharski T, Donaldson J, Schwarz T, Tomaszewska E (2020) The effect of dietary rye inclusion and xylanase supplementation on structural organization of bone constitutive phases in laying hens fed a wheat-corn diet. Animals (Basel).10:2010. https://doi.org/10.3390/ani10112010

Müntzing A (1963) A case of preserved heterozygosity in rye in spite of long-continued inbreeding. Hereditas 50:377–413

Müntzing A, Bose S (1969) Induced mutations in inbred lines of rye I. EMS Treatments. Hereditas 62:382–408

Myśków B, Masojć P, Banek-Tabor A, Szołkowski A (2001) Genetic diversity of inbred rye lines evaluated by RAPD analysis. J Appl Genet 42(1):1–14

Myśków B, Stojałowski S, Milczarski P, Masojc P (2010) Mapping of sequence-specific markers and loci controlling preharvest sprouting and alpha-amylase activity in rye (*Secale cereale* L.) on the genetic map of an F2 (S120×S76) population. J Appl Genet 51:283–287. https://doi.org/10.1007/BF03208857

Myśków B, Stojałowski S, Lań A, Bolibok-Brągoszewska H, Rakoczy-Trojanowska M, Kilian A (2012) Detection of the quantitative trait loci for α-amylase activity on a high-density genetic map of rye and comparison of their localization to loci controlling preharvest sprouting and earliness. Mol Breed 30:367–376. https://doi.org/10.1007/s11032-011-9627-1

Myśków B, Hanek M, Banek-Tabor A, Maciorowski R, Stojałowski S (2014) The application of high-density genetic maps of rye for the detection of QTLs controlling morphological traits. J Appl Genet 55:15–26. https://doi.org/10.1007/s13353-013-0186-5

Nadolska-Orczyk A, Rajchel IK, Orczyk W, Gasparis S (2017) Major genes determining yield-related traits in wheat and barley. Theor Appl Genet 130:1081–1098. https://doi.org/10.1007/s00122-017-2880-x

Naing AH, Kim CK (2019) A brief review of applications of antifreeze proteins in cryopreservation and metabolic genetic engineering. 3 Biotech. 9:329. https://doi.org/10.1007/s13205-019-1861-y.

Nakamura S (2018) Grain dormancy genes responsible for preventing pre-harvest sprouting in barley and wheat. Breed Sci 68:295–304. https://doi.org/10.1270/jsbbs.17138

NDong C, Anzellotti D, Ibrahim RK, Huner NP, Sarhan F (2003) Daphnetin methylation by a novel O-methyltransferase is associated with cold acclimation and photosystem II excitation pressure in rye. J Biol Chem 278:6854-6861.https://doi.org/10.1074/jbc.M209439200

Niculaes C, Abramov A, Hannemann L, Frey M (2018) Plant protection by Benzoxazinoids—Recent Insights into biosynthesis and function. Agronomy 8:143. https://doi.org/10.3390/agronomy8080143

Niemeyer HM (2009) Hydroxamic acids derived from 2-hydroxy-2H-1,4-benzoxazin-3(4H)-one: key defense chemicals of cereals. J Agric Food Chem 57:1677–1696. https://doi.org/10.1021/jf8034034

Nielsen MM, Damstrup ML, Thomsen AD, Rasmussen SK, Hansen Å (2007) Phytase activity and degradation of phytic acid during rye bread making. Eur Food Res Technol 225:173–181. https://doi.org/10.1007/s00217-006-0397-7

Nilsson M, Åman P, Härkönen H, Hallmans G, Bach Knudsen KE, Mazur W, Adlercreutz H (1997) Nutrient and lignan content, dough properties and baking performance of rye samples used in Scandinavia. Acta Agric Scand Section B—Soil & Plant Sci 47:26–34. https://doi.org/10.1080/09064719709362435

Nomura T, Ishihara A, Imaishi H, Ohkawa H, Endo TR, Iwamura H (2003) Rearrangement of the genes for the biosynthesis of benzoxazinones in the evolution of Triticeae species. Planta 217:776–782. https://doi.org/10.1007/s00425-003-1040-5

Oeser B, Kind S, Schurack S, Schmutzer T, Tudzynski P, Hinsch J (2017) Cross-talk of the biotrophic pathogen *Claviceps purpurea* and its host *Secale cereale*. BMC Genomics 18:273. https://doi.org/10.1186/s12864-017-3619-4

Oest M, Bindrich U, Voß A, Kaiser H, Rohn S (2020) Rye bread defects: analysis of composition and further influence factors as determinants of dry-baking. Foods 9:1900. https://doi.org/10.3390/foods9121900

Ohnuma T, Yagi M, Yamagami T, Taira T, Aso Y, Ishiguro M (2002) Molecular cloning, functional expression, and mutagenesis of cDNA encoding rye (*Secale cereale*) seed chitinase-c. Biosci Biotechnol Biochem 66:277–284. https://doi.org/10.1271/bbb.66.277

Ohnuma T, Taira T, Yamagami T, Aso Y, Ishiguro M (2004) Molecular cloning, functional expression, and mutagenesis of cDNA encoding class I chitinase from rye (*Secale cereale*) seeds. Biosci Biotechnol Biochem 68:324–332. https://doi.org/10.1271/bbb.68.324

Olsen OA (2020) The modular control of cereal endosperm development. Trends Plant Sci 25:279–290. https://doi.org/10.1016/j.tplants.2019.12.003

Ossent HP (1938) Zehn Jahre Roggenzüchtung. Müncheberg Züchter 10:255–261

Owuoche JO, Briggs KG, Taylor GJ (1996) The efficiency of copper use by 5A/5RL wheat-rye translocation lines and wheat (*Triticum aestivum* L.) cultivars. Plant Soil 180:113–120. https://doi.org/10.1007/BF00015417

Paponov IA, Lebedinskai S, Koshkin EI (1999) Growth analysis of solution culture-grown winter rye, wheat and Triticale at different relative rates of nitrogen supply. Ann Bot 84:467–473

Pejcz E, Spychaj R, Gil Z (2020) Technological methods for reducing the content of fructan in rye bread. Eur Food Res Technol 246:1839–1846. https://doi.org/10.1007/s00217-020-03537-5

Pellny TK, Lovegrove A, Freeman J, Tosi P, Love CG, Knox JP, Shewry PR, Mitchell RA (2012) Cell walls of developing wheat starchy endosperm: comparison of composition and RNA-Seq transcriptome. Plant Physiol 158:612–627. https://doi.org/10.1104/pp.111.189191

Pellny TK, Patil A, Wood AJ, Freeman J, Halsey K, Plummer A, Kosik O, Temple H, Collins JD, Dupree P, Berry S, Shewry PR, Lovegrove A, Phillips AL, Mitchell RAC (2020) Loss of TaIRX9b gene function in wheat decreases chain length and amount of arabinoxylan in grain but increases cross-linking. Plant Biotechnol J 8:2316–2327. https://doi.org/10.1111/pbi.13393

Peterson GN (1934) Improvement of rye through breeding. Sci Agric XIV:651–658

Petroni K, Pilu R, Tonelli C (2014) Anthocyanins in corn: a wealth of genes for human health. Planta 240:901–911. https://doi.org/10.1007/s00425-014-2131-1

Philipp U, Wehling P, Wricke G (1994) A linkage map of rye. Theor Appl Genet 88:243–248

Plaschke J, Börner A, Xie DX, Koebner RMD, Schlegel R, Gale MD (1993) RFLP mapping of genes affecting plant height and growth habit in rye. Theor Appl Genet 85:1049–1054

Plaschke J, Korzun V, Koebner RMD, Börner A (1995) Mapping the GA3-insensitive dwarfing gene *ct1* on chromosome 7 in rye. Plant Breed 114:113–116

Plaza-Wüthrich S, Blösch R, Rindisbacher A, Cannarozzi G, Tadele Z (2016) Gibberellin deficiency confers both lodging and drought tolerance in small cereals. Front Plant Sci 7:643. https://doi.org/10.3389/fpls.2016.00643

Ponomareva ML, Ponomarev SN, Tagirov MS, Gilmullina LF, Mannapova GS (2017) Pentosan content and genotypic variability in wintr rye grain. Agric Biol

52.1041–1048. https://doi.org/10.15389/agrobiology.2017.5.1041eng

Pont C, Wagner S, Kremer A, Orlando L, Plomion C, Salse J (2019) Paleogenomics: reconstruction of plant evolutionary trajectories from modern and ancient DNA. Genome Biol 20:29. https://doi.org/10.1186/s13059-019-1627-1

Portwood JL 2nd, Woodhouse MR, Cannon EK, Gardiner JM, Harper LC, Schaeffer ML, Walsh JR, Sen TZ, Cho KT, Schott DA, Braun BL, Dietze M, Dunfee B, Elsik CG, Manchanda N, Coe E, Sachs M, Stinard P, Tolbert J, Zimmerman S, Andorf CM (2019) MaizeGDB 2018: the maize multi-genome genetics and genomics database. Nucleic Acids Res 47:D1146–D1154. https://doi.org/10.1093/nar/gky1046

Preston JC, Hileman LC (2013) Functional evolution in the plant SQUAMOSA-PROMOTER BINDING PROTEIN-LIKE (SPL) gene family. Front Plant Sci 4:80. https://doi.org/10.3389/fpls.2013.00080

Qu J, Xu S, Zhang Z, Chen G, Zhong Y, Liu L, Zhang R, Xue J, Guo D (2018) Evolutionary, structural and expression analysis of core genes involved in starch synthesis. Sci Rep 8:12736. https://doi.org/10.1038/s41598-018-30411-y

Rabanus-Wallace MT, Hackauf B, Mascher M, Lux T, Wicker T, Gundlach H, Báez M, Houben A, Mayer KFX, Guo L, Poland J, Pozniak CJ, Walkowiak S, Melonek J, Praz C, Schreiber M, Budak H, Heuberger M, Steuernagel B, Wulff B, Börner A, Byrns B, Čížková J, Fowler DB, Fritz A, Himmelbach A, Kaithakottil G, Keilwagen J, Keller B, Konkin D, Larsen J, Li Q, Myśków B, Padmarasu S, Rawat N, Sesiz U, Sezgi B, Sharpe A, Šimková H, Small I, Swarbreck D, Toegelová H, Tsvetkova N, Voylokov AV, Vrána J, Bauer E, Bolibok-Bragoszewska H, Doležel J, Hall A, Jia J, Korzun V, Laroche A, Ma X-F, Ordon F, Özkan H, Rakoczy-Trojanowska M, Scholz U, Schulman AH, Siekmann D, Stojałowski S, Tiwari V, Spannagl M, Stein N (2021) Chromosome-scale genome assembly provides insights into rye biology, evolution, and agronomic potential. Nat Genet manuscript accepted for publication

Rakosy-Tican E, Lörincz-Besenyei E, Molnár I, Thieme R, Hartung F, Sprink T, Antonova O, Famelaer I, Angenon G, Aurori A (2019) New phenotypes of potato co-induced by mismatch repair deficiency and somatic hybridization. Front Plant Sci 10:3. https://doi.org/10.3389/fpls.2019.00003

Radchuk V, Sharma R, Potokina E, Radchuk R, Weier D, Munz E, Schreiber M, Mascher M, Stein N, Wicker T, Kilian B, Borisjuk L (2019) The highly divergent Jekyll genes, required for sexual reproduction, are lineage specific for the related grass tribes Triticeae and Bromeae. Plant J 98:961–974. https://doi.org/10.1111/tpj.14363

Rademacher W (2016) Chemical regulators of gibberellin status and their application in plant production. Annual Plant Reviews 49:359–403. https://doi.org/10.1002/9781119312994.apr0541

Ragaee SM, Campbell GL, Scoles GJ, McLeod JG, Tyler RT (2001a) Studies on rye (*Secale cereale* L.) lines exhibiting a range of extract viscosities. 1. Composition, molecular weight distribution of water extracts, and biochemical characteristics of purified water-extractable arabinoxylan. J Agric Food Chem 49:2437–2345. https://doi.org/10.1021/jf001227g

Ragaee SM, Campbell GL, Scoles GJ, McLeod JG, Tyler RT (2001b) Studies on rye (*Secale cereale* L.) lines exhibiting a range of extract viscosities. 2. Rheological and baking characteristics of rye and rye/wheat blends and feeding value for chicks of wholemeals and breads. J Agric Food Chem 49:2446–2453. https://doi.org/10.1021/jf0012289

Rakoczy-Trojanowska M, Śmiech M, Malepszy S (1997) The influence of genotype and medium on rye (*Secale cereale* L.) anther culture. Plant Cell Tiss Org 48:15–21. https://doi.org/10.1023/A:1005792912316

Rakoczy-Trojanowska M, Krajewski P, Bocianowski J, Schollenberger M, Wakuliński W, Milczarski P, Masojć P, Targońska-Karasek M, Banaszak Z, Banaszak K, Brukwiński W, Orczyk W, Kilian A (2017) Identification of single nucleotide polymorphisms associated with brown rust resistance, α-amylase activity and pre-harvest sprouting in rye (*Secale cereale* L.). Plant Mol Biol Rep 35:366–378. https://doi.org/10.1007/s11105-017-1030-6

Rao YC, Yang YL, Xin DD, Li XJ, Zhai KE, Ma BJ, Pan JW, Qian Q, Zeng DL (2013) Characterization and cloning of a brittle culm mutant (bc88) in rice (*Oryza sativa* L.). Chin Sci Bull 58:3000–3006. https://doi.org/10.1007/s11434-013-5806-2

Ravel C, Nagy IJ, Martre P, Sourdille P, Dardevet M, Balfourier F, Pont C, Giancola S, Praud S, Charmet G (2006) Single nucleotide polymorphism, genetic mapping, and expression of genes coding for the DOF wheat prolamin-box binding factor. Funct Integr Genomics 6:310–321. https://doi.org/10.1007/s10142-006-0022-7

Ren T, Sun Z, Ren Z, Tan F, Luo P, Tang Z, Fu S, Li Z (2020) Molecular and cytogenetic characterization of a wheat-rye 7BS.7RL translocation line with resistance to stripe rust, powdery mildew, and Fusarium head blight. Phytopathology 110:1713–1720. https://doi.org/10.1094/PHYTO-02-20-0061-R

Reynolds M, Chapman S, Crespo-Herrera L, Molero G, Mondal S, Pequeno DNL, Pinto F, Pinera-Chavez FJ, Poland J, Rivera-Amado C, Saint Pierre C, Sukumaran S (2020) Breeder friendly phenotyping. Plant Sci. 295:110396. https://doi.org/10.1016/j.plantsci.2019.110396

Riley R, Ewart JAD (1970) The effect of individual rye chromosomes on the amino acid content of wheat grains. Genet. Res. Camb. 15:209–219

Roux SR, Hackauf B, Linz A, Ruge B, Klocke B, Wehling P (2004) Leaf-rust resistance in rye (*Secale cereale* L.). 2. Genetic analysis and mapping of resistance genes *Pr3*, *Pr4*, and *Pr5*. Theor Appl Genet 110:192–201

Roux SR, Hackauf B, Ruge-Wehling B, Linz A, Wehling P (2007) Exploitation and comprehensive characterization of leaf-rust resistance in rye. Vortr Pflanzenzüchtg 71:144–150

Ryser P (2006) The mysterious root length. Plant Soil 286:1–6. https://doi.org/10.1007/s11104-006-9096-1

Saal B, Wricke G (1999) Development of simple sequence repeat markers in rye (*Secale cereale* L.). Genome 42:964–972

Saintenac C, Lee WS, Cambon F, Rudd JJ, King RC, Marande W, Powers SJ, Bergès H, Phillips AL, Uauy C, Hammond-Kosack KE, Langin T, Kanyuka K (2018) Wheat receptor-kinase-like protein *Stb6* controls gene-for-gene resistance to fungal pathogen Zymoseptoria tritici. Nat Genet 50:368–374. https://doi.org/10.1038/s41588-018-0051-x

Sakuma S, Golan G, Guo Z, Ogawa T, Tagiri A, Sugimoto K, Bernhardt N, Brassac J, Mascher M, Hensel G, Ohnishi S, Jinno H, Yamashita Y, Ayalon I, Peleg Z, Schnurbusch T, Komatsuda T (2019) Unleashing floret fertility in wheat through the mutation of a homeobox gene. Proc Natl Acad Sci U S A 116:5182–5187. https://doi.org/10.1073/pnas.1815465116

Santos E, Matos M, Benito C (2020) Isolation and characterization of a new MATE gene located in the same chromosome arm of the aluminium tolerance (*Alt1*) rye locus. Plant Biol (stuttg) 22:691–700. https://doi.org/10.1111/plb.13107

Sato K (2020) History and future perspectives of barley genomics. DNA Res. 27(4):dsaa023. https://doi.org/10.1093/dnares/dsaa023

Schittenhelm S, Kraft M, Wittich KP (2015) Performance of winter cereals on field-stored soil moisture only. Eur J Agron 52:247–258. https://doi.org/10.1016/j.eja.2013.08.010

Schlegel R, Melz G, Mettin D (1986) Rye cytology, cytogenetics, and genetics. Current status. Theor Appl Genet 72:721–734

Schlegel R, Melz G, Korzun V (1998) Genes, marker and linkage data of rye (*Secale cereale* L.): 5th updated inventory. Euphytica 101:23–67

Schlegel R, Korzun V (2008) Notes on the origin of 4BL-5RL rye translocations in common wheat (*Triticum aestivum* L.). Cereal Res Comm 36:373–385

Schlegel R, Eifler J, Schmidt M, Schmiedchen B, Ordon F, Kastirr U (2020) Screening and genetic studies on resistance to soil-born cereal mosaic virus (SBWMV) in rye. Cereal Res Comm. https://doi.org/10.1007/s42976-020-00105-1

Schneider HM, Lynch JP (2020) Should root plasticity be a crop breeding target? Front Plant Sci 11:546. https://doi.org/10.3389/fpls.2020.00546

Schreiber M, Himmelbach A, Börner A, Mascher M (2018) Genetic diversity and relationship between domesticated rye and its wild relatives as revealed through genotyping-by-sequencing. Evol Appl 12 (1):66–77. https://doi.org/10.1111/eva.12624

Schulz M, Marocco A, Tabaglio V, Macias FA, Molinillo JM (2013) Benzoxazinoids in rye allelopathy—from discovery to application in sustainable weed control and organic farming. J Chem Ecol 39:154–174. https://doi.org/10.1007/s10886-013-0235-x

Scoles GJ, Campbell GL, Mc Leod JG (1993) Variability for grain extract viscosity in bred lines and an F2 population of rye (*Secale cereale* L.). Can J Plant Sci 73:1–6

Scott MF, Ladejobi O, Amer S, Bentley AR, Biernaskie J, Boden SA, Clark M, Dell'Acqua M, Dixon LE, Filippi CV, Fradgley N, Gardner KA, Mackay IJ, O'Sullivan D, Percival-Alwyn L, Roorkiwal M, Singh RK, Thudi M, Varshney RK, Venturini L, Whan A, Cockram J, Mott R (2020) Multi-parent populations in crops: a toolbox integrating genomics and genetic mapping with breeding. Heredity (edinb) 125:396–416. https://doi.org/10.1038/s41437-020-0336-6

Seiler C, Harshavardhan VT, Reddy PS, Hensel G, Kumlehn J, Eschen-Lippold L, Rajesh K, Korzun V, Wobus U, Lee J, Selvaraj G, Sreenivasulu N (2014) Abscisic acid flux alterations result in differential abscisic acid signaling responses and impact assimilation efficiency in barley under terminal drought stress. Plant Physiol 164:1677–1696. https://doi.org/10.1104/pp.113.229062

Senft P, Wricke G (1996) An extended genetic map of rye (*Secale cereale* L.). Plant Breed 115:508–510. https://doi.org/10.1111/j.1439-0523.1996.tb00966.x

Sharma S, Bhat PR, Ehdaie B, Close TJ, Lukaszewski AJ, Waines JG (2009) Integrated genetic map and genetic analysis of a region associated with root traits on the short arm of rye chromosome 1 in bread wheat. Theor Appl Genet 119:783–793. https://doi.org/10.1007/s00122-009-1088-0

Sharma S, Demason DA, Ehdaie B, Lukaszewski AJ, Waines JG (2010) Dosage effect of the short arm of chromosome 1 of rye on root morphology and anatomy in bread wheat. J Exp Bot 61:2623–2633. https://doi.org/10.1093/jxb/erq097

Sharma S, Xu S, Ehdaie B, Hoops A, Close TJ, Lukaszewski AJ, Waines JG (2011) Dissection of QTL effects for root traits using a chromosome arm-specific mapping population in bread wheat. Theor Appl Genet 122:759–769. https://doi.org/10.1007/s00122-010-1484-5

Sheehan H, Bentley A (2020) Changing times: Opportunities for altering winter wheat phenology. Plants, People, Planet 00:1–11. https://doi.org/10.1002/ppp3.10163

Shewry PR, Kreis M, Burgess SR, Parmar S, Miflin BJ (1983) The synthesis and deposition of the prolamin storage proteins (secalins) of rye. Planta 159:439–445. https://doi.org/10.1007/BF00392080

Shewry PR, Bradberry D, Franklin J, White RP (1984) The chromosomal locations and linkage relationships of the structural genes for the prolamin storage proteins (secalins) of rye. Theor Appl Genet 69:63–69. https://doi.org/10.1007/BF00262541

Shi BJ, Gustafson JP, Button J, Miyazaki J, Pallotta M, Gustafson N, Zhou H, Langridge P, Collins NC (2009) Physical analysis of the complex rye (*Secale cereale* L.) *Alt4* aluminium (aluminum) tolerance locus using a whole-genome BAC library of rye cv. Blanco Theor Appl Genet 119:695–704. https://doi.org/10.1007/s00122-009-1080-8

Shi MZ, Xie DY (2014) Biosynthesis and metabolic engineering of anthocyanins in *Arabidopsis thaliana*. Recent Pat Biotechnol 8:47–60. https://doi.org/10.2174/1872208307666131218123538

Shimizu Y, Nasuda S, Endo TR (1997) Detection of the *Sec-1* locus of rye by a PCR-based method. Genes Genet Syst 72:197–203. https://doi.org/10.1266/ggs.72.197

Shull GH (1908) The composition of a field of maize. Am. Breeders Assoc Rep 4:296–301. https://doi.org/10.1093/jhered/os-4.1.296

Sidhu JS, Ramakrishnan SM, Ali S, Bernardo A, Bai G, Abdullah S, Ayana G, Sehgal SK (2019) Assessing the genetic diversity and characterizing genomic regions conferring Tan Spot resistance in cultivated rye. PLoS One. 14:e0214519. https://doi.org/10.1371/journal.pone.0214519

Singh NK, Shepherd KW (1984) Mapping of the genes controlling high-molecular-weight glutenin subunits of rye on the long arm of chromosome 1R. Genet Res 44:117–123. https://doi.org/10.1017/S001667230002632X

Singh NK, Shepherd KW, McIntosh RA (1990) Linkage mapping of genes for resistance to leaf, stem and stripe rusts and ω-secalins on the short arm of rye chromosome 1R. Theor Appl Genet 80:609–616. https://doi.org/10.1007/BF00224219

Skoryk VV, Skoryk VV, Simonenko NV, Skoryk OP (2010) Genetics characteristics of the donor for dominant short stem and large grain winter rye (*Secale cereale* L.). Plant Var Stud Prot J Appl Res 1:5–12. https://doi.org/10.21498/2518-1017.1(11).2010.59362

Smit MN, Zhou X, Landero JL, Young MG, Beltranena E (2019) Increasing hybrid rye level substituting wheat grain with or without enzyme on growth performance and carcass traits of growing-finishing barrows and gilts. Transl Anim Sci 3:1561–1574. https://doi.org/10.1093/tas/txz141

Smirnov WG, Sosnichina SP (1984) Genetika rzi. Leningrad, pp 1–156

Solodukhina OV (2002) Genetic characterization of rye accessions with regard to leaf rust resistance. Russ J Genet 38:399–407. https://doi.org/10.1023/A:1015202303392

Song S, Tian D, Zhang Z, Hu S, Yu J (2018) Rice genomics: over the past two decades and into the future. Genom Proteom Bioinf 16:397–404. https://doi.org/10.1016/j.gpb.2019.01.001

Spampinato CP (2017) Protecting DNA from errors and damage: an overview of DNA repair mechanisms in plants compared to mammals. Cell Mol Life Sci 74:1693–1709. https://doi.org/10.1007/s00018-016-2436-2

Sprague GF, Tatum AL (1942) General *vs.* specific combining ability in single crosses of corn. J Amer Soc Agron 34:923–932

Starzycki S (1976) Diseases, pests and physiology of rye. In: Bushuk W (ed) Rye: production, chemistry and technology, pp 27–61. Am Assoc Cereal Chemists, Inc., St. Paul, MN, USA

Stępniewska S, Hassoon WH, Szafrańska A, Cacak-Pietrzak G, Dziki D (2019) Procedures for breadmaking quality assessment of rye wholemeal flour. Foods 8:331. https://doi.org/10.3390/foods8080331

Stojałowski S, Myśków B, Hanek M (2015) Phenotypic effect and chromosomal localization of *Ddw3*, the dominant dwarfing gene in rye (*Secale cereale* L.). Euphytica 201:43–52. https://doi.org/10.1007/s10681-014-1173-6

Stracke S, Schilling AG, Förster J, Weiss C, Glass C, Miedaner T, Geiger HH (2003) Development of PCR-based markers linked to dominant genes for male-fertility restoration in Pampa CMS of rye (*Secale cereale* L.). Theor Appl Genet 106:1184–1190. https://doi.org/10.1007/s00122-002-1153-4

Subira J, Ammar K, Alvaro F, Garcia del Moral LF, Dreisigacker S, Royo C (2016) Changes in durum wheat root and aerial biomass caused by the introduction of the *Rht-B1b* dwarfing allele and their effects on yield formation. Plant Soil 403:291–304

Sue M, Nakamura C, Nomura T (2011) Dispersed benzoxazinone gene cluster: molecular characterization and chromosomal localization of glucosyltransferase and glucosidase genes in wheat and rye. Plant Physiol 157:985–997. https://doi.org/10.1104/pp.111.182378

Surikov IM (1971) Inheritance of two chlorophyll aberrations in rye. Tr Prikl Bot Genet Se 146:122–130

Su'udi M, Cha JY, Jung MH, Ermawati N, Han CD, Kim MG, Woo YM, Son D (2012) Potential role of the rice OsCCS52A gene in endoreduplication. Planta 235:387–397. https://doi.org/10.1007/s00425-011-1515-8

Święcka S, Berdzik M, Myśków B (2014) Genetic mapping of the ScHd1 gene in rye and an assessment of its relationship with earliness per se and plant morphology. J Appl Genet 55:469–473. https://doi.org/10.1007/s13353-014-0223-z

Sybenga J (1958) Inbreeding effects in rye. Z Vererbungsl 89:338–354. https://doi.org/10.1007/BF00888640

Tadele Z (2016) Mutagenesis and TILLING to dissect gene function in plants. Curr Genomics 17:499–508. https://doi.org/10.2174/1389202917666160520104158

Tai L, Wang HJ, Xu XJ, Sun WH, Ju L, Liu WT, Li WQ, Sun J, Chen KM (2021) Cereal pre-harvest sprouting: a global agricultural disaster regulated by complex genetic and biochemical mechanisms. J Exp Bot. erab024, 20 Jan 2021. https://doi.org/10.1093/jxb/erab024

Taira T, Ohnuma T, Yamagami T, Aso Y, Ishiguro M, Ishihara M (2002) Antifungal activity of rye (*Secale cereale*) seed chitinases: the different binding manner of class I and class II chitinases to the fungal cell walls. Biosci Biotechnol Biochem 66:970–977. https://doi.org/10.1271/bbb.66.970

Tanwir F, Dionisio G, Adhikari KB, Fomsgaard IS, Gregersen PL (2017) Biosynthesis and chemical transformation of benzoxazinoids in rye during seed germination and the identification of a rye Bx6-like gene. Phytochemistry 140:95–107. https://doi.org/10.1016/j.phytochem.2017.04.020

Targonska-Karasek M, Boczkowska M, Podyma W, Pasnik M, Niedzielski M, Rucinska A, Nowak-Zyczynska Z, Rakoczy-Trojanowska M (2020) Investigation of obsolete diversity of rye (Secale cereale L.) using multiplexed SSR fingerprinting and evaluation of agronomic traits. J Appl Genet 61(4):513–529. https://doi.org/10.1007/s13353-020-00579-z

Tenhola-Roininen T, Immonen S, Tanhuanpää P (2006) Rye doubled haploids as a research and breeding tool—a practical point of view. Plant Breed 125:584–590. https://doi.org/10.1111/j.1439-0523.2006.01296.x

Tenhola-Roininen T, Kalendar R, Schulman AH, Tanhuanpää P (2011) A doubled haploid rye linkage map with a QTL affecting α-amylase activity. J Appl Genet 52:299–304. https://doi.org/10.1007/s13353-011-0029-1

Thacker PA, McLeod JG, Campbell GL (2002) Performance of growing-finishing pigs fed diets based on normal or low viscosity rye fed with and without enzyme supplementation. Arch Tierernahr 56:361–370. https://doi.org/10.1080/00039420215631

Thiele V, Melz G, Flamme W (1989) Genetic analysis of rye (*Secale cereale* L.). Location of gene Anu controlling an antinutritive factor. Arch Züchtungsforsch 19:239–241

Thomas E, Wenzel G (1975) Embryogenesis from microspores of rye. Naturwissenschaften 62:40–41

Thomas SG, Phillips AL, Hedden P (1999) Molecular cloning and functional expression of gibberellin 2-oxidases, multifunctional enzymes involved in gibberellin deactivation. Proc Natl Acad Sci U S A 96:4698–4703. https://doi.org/10.1073/pnas.96.8.4698

Thomashow MF (2010) Molecular basis of plant cold acclimation: Insights gained from studying the CBF cold response pathway. Plant Physiol 154:571–577. https://doi.org/10.1104/pp.110.161794

Torop AA, Dedyaev VG, Tschaykin VV, Dokuchaev VV (2003) The results of rye breeding in the central-chernosem region of Russia. Plant Breed Seed Sci 47:69–75

Tracy SR, Nagel KA, Postma JA, Fassbender H, Wasson A, Watt M (2020) Crop Improvement from phenotyping roots: highlights reveal expanding opportunities. Trends Plant Sci 25:105–118. https://doi.org/10.1016/j.tplants.2019.10.015

Tranquilli G, Dubcovsky J (2000) Epistatic interaction between vernalization genes *Vrn-Am* 1 and *Vrn-Am2* in diploid wheat. J Hered 91:304–306. https://doi.org/10.1093/jhered/91.4.304

Veenstra LD, Jannink J-L, Sorrells ME (2017) Wheat fructans: a potential breeding target for nutritionally improved, climate-resilient varieties. Crop Sci 57:1624–1640. https://doi.org/10.2135/cropsci2016.11.0955

Vendelbo NM, Sarup P, Orabi J, Kristensen PS, Jahoor A (2020) Genetic structure of a germplasm for hybrid breeding in rye (*Secale cereale* L.). Plos One 15 (10). https://doi.org/10.1371/journal.pone.0239541

Viegland MD, Spannagl M, Asp T, Paina C, Rudi H, Rognli OA, Fjellheim S, Sandve SR (2013) Evidence for adaptive evolution of low-temperature stress response genes in a Pooideae grass ancestor. New Phytol 199:1060–1068. https://doi.org/10.1111/nph.12337

Villegas E, McDonald CE, Gilles KA (1968) Variability in the lysine content of wheat, rye and Triticale proteins. Research Bulletin, International Maize and Wheat Improvement Center, Mexico, 10

Virtanen AI, Hietala PK (1955a) 2(3)-Benzoxazolinone an anti-fusarium factor in rye seedlings. Acta Chem Scand 9:1543–1544

Virtanen AI, Hietala PK (1955b) The structure of the precursors of benzoxazolinone in rye plants. II. Suomen. Kemistilehti 32:252

Vishwakarma K, Upadhyay N, Kumar N, Yadav G, Singh J, Mishra RK, Kumar V, Verma R, Upadhyay RG, Pandey M, Sharma S (2017) Abscisic acid signaling and abiotic stress tolerance in plants: A review on current knowledge and future prospects. Front Plant Sci 8:161. https://doi.org/10.3389/fpls.2017.00161

von Uexküll HR, Mutert E (1995) Global extent, development and economic impact of acid soils. Plant Soil 171:1–15

Voss-Fels KP, Qian L, Parra-Londono S, Uptmoor R, Frisch M, Keeble-Gagnère G, Appels R, Snowdon RJ (2017) Linkage drag constrains the roots of modern wheat. Plant Cell Environ 40:717–725. https://doi.org/10.1111/pce.12888

Voylokov AV, Priiatkina SN (2004) Linkage of genes controlling morphological traits with isozyme markers of rye chromosomes. Russ J Genet 40:56–61. https://doi.org/10.1023/B:RUGE.0000013449.56274.a9

Voylokov AV, Lykholay AN, Smirnov VG (2015) Genetic control of anthocyanin coloration in rye. Russ J Genet Appl Res 5:262–267

Voylokov AV, Sosnikhina SP, Tikhenko ND, Tsvetkova NV, Mikhailova EI, Smirnov VG (2018) Peterhof collection of rye and its use in genetic studies. Ecol Genet 16:40–49. https://doi.org/10.17816/ecogen16240-49

Waines JG, Ehdaie B (2007) Domestication and crop physiology: Roots of green-revolution wheat. Ann Bot 100:991–998

Wang W, Mauleon R, Hu Z, Chebotarov D, Tai S, Wu Z, Li M, Zheng T, Fuentes RR, Zhang F, Mansueto L, Copetti D, Sanciangco M, Palis KC, Xu J, Sun C, Fu B, Zhang H, Gao Y, Zhao X, Shen F, Cui X, Yu H, Li Z, Chen M, Detras J, Zhou Y, Zhang X, Zhao Y, Kudrna D, Wang C, Li R, Jia B, Lu J, He X, Dong Z, Xu J, Li Y, Wang M, Shi J, Li J, Zhang D, Lee S, Hu W, Poliakov A, Dubchak I, Ulat VJ, Borja FN, Mendoza JR, Ali J, Li J, Gao Q, Niu Y, Yue Z, Naredo MEB, Talag J, Wang X, Li J, Fang X, Yin Y,

Glaszmann JC, Zhang J, Li J, Hamilton RS, Wing RA, Ruan J, Zhang G, Wei C, Alexandrov N, McNally KL, Li Z, Leung H (2018) Genomic variation in 3,010 diverse accessions of Asian cultivated rice. Nature 557:43–49. https://doi.org/10.1038/s41586-018-0063-9

Wang J-R, Pu Z-E, Lan X-J, Baum BR, Yan Z-H, Zheng Y-L, Wie Y-M (2010) Phylogenetic analysis of the dimeric alpha-amylase inhibitor sequences from an orthologous region in 21 different genomes of the tribe Triticeae (Poaceae). Biochem Syst Ecol 38:708–714. https://doi.org/10.1016/j.bse.2010.05.004

Wang Q, Sun G, Ren X, Du B, Cheng Y, Wang Y, Li C, Sun D (2019) Dissecting the genetic basis of grain size and weight in barley (*Hordeum vulgare* L.) by QTL and comparative genetic analyses. Front Plant Sci 10:469. https://doi.org/10.3389/fpls.2019.00469

Ward AT, Marquardt RR (1987) Antinutritional activity of a water-soluble pentosan-rich fraction from rye grain. Poult Sci 66:1665–1674. https://doi.org/10.3382/ps.0661665

Wehling P (1985) Electrophoretic analysis of 10 enzyme systems in rye: linkage relationships and chromosomal location of isozyme loci. Proceding EUCARPIA Meeting of the cereal section on rye. Svalöv, Sweden, pp 101–124

Wehling P (1991) Inheritance, linkage relationship and chromosomal localization of the glutamate oxaloacetate transaminase, acid phosphatase and diaphorase isozyme genes in *Secale cereale* L. Theor Appl Genet 82:569–576

Wehling P, Linz A, Hackauf B, Roux SR, Ruge B, Klocke B (2003) Leaf-rust resistance in rye (*Secale cereale* L.). 1. Genetic analysis and mapping of resistance genes *Pr1* and *Pr2*. Theor Appl Genet 107:432–438

Weipert D (1995) Pentosans in rye. In: Poutanen K, Autio K (eds) International rye symposium: technology and products. Helsinki, Espoo, VTT Symposium, 161, pp 39–48, 7–8 Dec 1995

Weipert D (1996) Pentosans as selection traits in rye breeding. Vortr Pflanzenzüchtg 35:109–119

Weipert D (1997) Processing performance of rye as compared to wheat. Cereal Foods World 8:706–712

Wenzel G, Hoffmann F, Thomas E (1977) Increased induction and chromosome doubling of androgenetic haploid rye. Theor Appl Genet 51:81–86

Whitford R, Fleury D, Reif JC, Garcia M, Okada T, Korzun V, Langridge P (2013) Hybrid breeding in wheat: technologies to improve hybrid wheat seed production. J Exp Bot 64:5411–5428. https://doi.org/10.1093/jxb/ert333

Wilde P, Korzun V, Menzel J, Zhou R, Stein N, Hackauf B (2017) Restorer plants. Patent Application WO2017109012A1

Wojcik-Gront E, Bloch-Michalik M (2016) Assessment of greenhouse gas emission from life cycle of basic cereals production in Poland. Zemdirbyste-Agric 103:259–266. https://doi.org/10.13080/z-a.2016.103.033

Wolski T (1970) Studies on the inbreeding of rye. Genetica Polonica 11:1–26

Wolski T, Brykczynski J, Tymieniecka E (1972) Heritability of some characters of rye under open pollination. Theor Appl Genet 42:168–173. https://doi.org/10.1007/BF00280793

Wricke G (1969) Untersuchungen zur Selbstfertilität beim Roggen (*Secale cereale*). Theor Appl Genet 39:371–378

Wricke G (1973) Inzuchtdepression und Genwirkung beim Roggen (*Secale cereale*). Theoret Appl Genet 43:83–87. https://doi.org/10.1007/BF00274962

Wricke G, Wehling P (1985) Linkage between an incompatibility locus and a peroxidase isozyme locus (Prx7) in rye. Theor Appl Genet 71:289–291

Wricke G, Wilde P, Wehling P, Gieselmann C (1993) An isozyme marker for pollen fertility restoration in the Pampa CMS system of rye (*Secale cereale* L.). Plant Breed 111:290–294

Wricke G (2002) Two major genes for kernel weight in rye. Plant Breed 121:26–28

Xu J, Frick M, Laroche A, Ni ZF, Li BY, Lu ZX (2009) Isolation and characterization of the rye Waxy gene. Genome 52:658–664. https://doi.org/10.1139/G09-036

Xu J, Li M, Chen L, Wu G, Li H (2012) Rapid generation of rice mutants via the dominant negative suppression of the mismatch repair protein *OsPMS1*. Theor Appl Genet 125:975–986. https://doi.org/10.1007/s00122-012-1888-5

Yan L, Loukoianov A, Tranquilli G, Helguera M, Fahima T, Dubcovsky J (2003) Positional cloning of the wheat vernalization gene *VRN1*. Proc Natl Acad Sci U S A 100:6263–6268. https://doi.org/10.1073/pnas.0937399100

Yan L, Loukoianov A, Blechl A, Tranquilli G, Ramakrishna W, SanMiguel P, Bennetzen JL, Echenique V, Dubcovsky J (2004) The wheat *VRN2* gene is a flowering repressor down-regulated by vernalization. Science 303:1640–1644. https://doi.org/10.1126/science.1094305

Yan WH, Wang P, Chen HX, Zhou HJ, Li QP, Wang CR, Ding ZH, Zhang YS, Yu SB, Xing YZ, Zhang QF (2011) A major QTL, *Ghd8*, plays pleiotropic roles in regulating grain productivity, plant height, and heading date in rice. Mol Plant 4:319–330. https://doi.org/10.1093/mp/ssq070

Yeats TH, Rose JK (2013) The formation and function of plant cuticles. Plant Physiol 163:5–20. https://doi.org/10.1104/pp.113.222737

Yeh S, Moffatt BA, Griffith M, Xiong F, Yang DS, Wiseman SB, Sarhan F, Danyluk J, Xue YQ, Hew CL, Doherty-Kirby A, Lajoie G (2000) Chitinase genes responsive to cold encode antifreeze proteins in winter cereals. Plant Physiol 124:1251–1264. https://doi.org/10.1104/pp.124.3.1251

Yeo IY, Lee S, Sadeghi AM, Beeson PC, Hively WD, McCarty GW, Lang MW (2014) Assessing winter cover crop nutrient uptake efficiency using a water quality simulation model. Hydrol Earth Syst Sci

18:5239–5253. https://doi.org/10.5194/hess-18-5239-2014

Yokosho K, Yamaji N, Ma JF (2010) Isolation and characterisation of two MATE genes in rye. Funct Plant Biol 37:296–303. https://doi.org/10.1071/FP09265

Yonemaru JI, Yamamoto T, Fukuoka S, Uga Y, Hori K, Yano M (2010) Q-TARO: QTL annotation rice online database. Rice 3:194–203. https://doi.org/10.1007/s12284-010-9041-z

Zeng W, Jiang N, Nadella R, Killen TL, Nadella V, Faik A (2010) A glucurono(arabino)xylan synthase complex from wheat contains members of the GT43, GT47, and GT75 families and functions cooperatively. Plant Physiol 154:78–97. https://doi.org/10.1104/pp.110.159749

Zhan X, Lu Y, Zhu JK, Botella JR (2020) Genome editing for plant research and crop improvement. J Integr Plant Biol. https://doi.org/10.1111/jipb.13063s

Zhang L, Dunn MA, Pearce RS, Hughes MA (1993) Analysis of organ specificity of a low temperature responsive gene family in rye (*Secale cereale* L.). J Exp Bot 44:1787–1793. https://doi.org/10.1093/jxb/44.12.1787

Zhang D, Choi DW, Wanamaker S, Fenton RD, Chin A, Malatrasi M, Turuspekov Y, Walia H, Akhunov ED, Kianian P, Otto C, Simons K, Deal KR, Echenique V, Stamova B, Ross K, Butler GE, Strader L, Verhey SD, Johnson R, Altenbach S, Kothari K, Tanaka C, Shah MM, Laudencia-Chingcuanco D, Han P, Miller RE, Crossman CC, Chao S, Lazo GR, Klueva N, Gustafson JP, Kianian SF, Dubcovsky J, Walker-Simmons MK, Gill KS, Dvorák J, Anderson OD, Sorrells ME, McGuire PE, Qualset CO, Nguyen HT, Close TJ (2004) Construction and evaluation of cDNA libraries for large-scale expressed sequence tag sequencing in wheat (*Triticum aestivum* L.). Genetics 168:595–608. https://doi.org/10.1534/genetics

Zhang Q, Li J, Xue Y, Han B, Deng XW (2008) Rice 2020: a call for an international coordinated effort in rice functional genomics. Mol Plant 1:715–719. https://doi.org/10.1093/mp/ssn043

Zhang X, Chen S, Sun H, Wang Y, Shao L (2009) Root size, distribution and soil water depletion as affected by cultivars and environmental factors. Field Crop Res 114:75–83

Zhang X, Long Y, Huang J, Xia J (2019) Molecular mechanisms for coping with Al toxicity in plants. Int J Mol Sci 20:1551. https://doi.org/10.3390/ijms20071551

Zheng J, Wu H, Zhu H, Huang C, Liu C, Chang Y, Kong Z, Zhou Z, Wang G, Lin Y, Chen H (2019) Determining factors, regulation system, and domestication of anthocyanin biosynthesis in rice leaves. New Phytol 223:705–721. https://doi.org/10.1111/nph.15807

Zhou S, Richter A, Jander G (2018) Beyond defense: multiple functions of benzoxazinoids in maize metabolism. Plant Cell Physiol 59:1528–1537. https://doi.org/10.1093/pcp/pcy064

Zieliński K, Krzewska M, Żur I, Juzoń K, Kopeć P, Nowicka A, Moravčiková J, Skrzypek E, Dubas E (2020) The effect of glutathione and mannitol on androgenesis in anther and isolated microspore cultures of rye (*Secale cereale* L.). Plant Cell Tiss Organ Cult 140:577–592. https://doi.org/10.1007/s11240-019-01754-9

Zykin PA, Andreeva EA, Lykholay AN, Tsvetkova NV, Voylokov AV (2018) Anthocyanin composition and content in rye plants with different grain color. Molecules 23(4):948. https://doi.org/10.3390/molecules23040948

Genomics of Self-Incompatibility and Male-Fertility Restoration in Rye

10

Joanna Melonek, Viktor Korzun, and Bernd Hackauf

Abstract

Allogamous grasses like rye (*Secale cereale* L.) developed a two-locus gametophytic self-incompatibility (SI) system that forces outcrossing and allows for maintaining a high level of diversity. The strong built-in SI mechanism renders rye the only outbreeding small grain cereal species and an excellent crop to breed hybrids. Hybrid breeding requires self-pollination and hybridization systems in order to perform targeted crosses for a systematic exploitation of heterosis. A managed SI could be applied as a genetic fertilisation control system supplementing systems based on cytoplasmic male sterility and restorer-of-fertility genes. The ability to develop inbred lines to capture and manage the genetic diversity is a crucial precondition for efficient hybrid breeding. Indeed, spontaneous self-fertility detected in some rye mutants overcomes this limitation. Other mechanisms of reproductive isolation affect the production of interspecific wheat (*Triticum aestivum* L.) and rye hybrids and face problems of low crossability, embryo lethality and overall hybrid weakness. This limits the transfer of agronomically important traits from rye to wheat and impairs the development of triticale (X *Triticosecale* Wittmack) lines, which combine the yield potential and grain quality of wheat with the disease and environmental tolerance of rye. This chapter will discuss the most current knowledge of the molecular basis of SI in rye as compared to better understood SI in other plant systems, wheat-rye crossability as well as male sterility in rye and their possible applications to advance rye breeding.

J. Melonek
ARC Centre of Excellence in Plant Energy Biology, School of Molecular Sciences, The University of Western Australia, 6009 Crawley, WA, Australia
e-mail: joanna.melonek@uwa.edu.au

V. Korzun
KWS SAAT SE & Co. KGaA, Grimsehlstr. 31, 37574 Einbeck, Germany
e-mail: viktor.korzun@kws.com

B. Hackauf (✉)
Institute for Breeding Research on Agricultural Crops, Julius Kühn-Institute, Rudolf-Schick-Platz 3a, 18190 Groß-Lüsewitz, Germany
e-mail: bernd.hackauf@julius-kuehn.de

10.1 Self-Incompatibility (SI)—Cell-to-Cell Communication for Fertilisation Control

Self-incompatibility (SI) provides a selective recognition and rejection mechanism for genetically identical (self) pollen and makes the interaction between pollen and pistil a sophisticated fertilisation control mechanism. SI prevents plants from inbreeding and allows maintaining a high level of diversity. Rye (*Secale cereale* L.)

© The Author(s), under exclusive license to Springer Nature Switzerland AG 2021

M. T. Rabanus-Wallace and N. Stein (eds.), *The Rye Genome*, Compendium of Plant Genomes,
https://doi.org/10.1007/978-3-030-83383-1_10

shares a common ancestry with highly inbreeding wheat (*Triticum aestivum* L.) and barley (*Hordeum vulgare* L.) (Martis et al. 2013) but is the only outcrossing small grain cereal in the Triticeae tribe of the grasses. Because of strong self-incompatibility (Fig. 10.1) and wind pollination, traditional rye varieties are panmictic populations, characterised by high levels of heterozygosity and heterogeneity (Geiger and Miedaner 2009). Genetic diversity was found within single rye accessions and its levels were seen to further increase from wild to feral to domesticated rye (Hagenblad et al. 2016). This unexpected difference could be due to ascertainment bias as the SNP panel used in that study was originally developed for elite inbred lines of cultivated rye. Indeed, subsequent research inferred from a genotyping-by-sequencing approach revealed a larger gene pool from wild relatives accessible to rye pre-breeding programs than that reported for its close relative barley, also a diploid Triticeae species (Schreiber et al. 2019).

In Poaceae species, SI is controlled by two unlinked loci, referred to as the *S*- and *Z*-locus (Langridge and Baumann 2008). The *S*-*Z* system is also present in rye where both loci are gametophytically expressed (Lundqvist 1954, 1956). The *S*- and *Z*-loci were shown to be multiallelic with 6–7 alleles estimated at the first locus and 12–13 alleles at the other locus, respectively, in the variety 'Halo' (Trang et al. 1982). As an SI species, outcrossing is notoriously favoured in rye but for efficient hybrid breeding, achieving and maintaining homozygosity in inbred lines to capture genetic diversity is essential. Indeed, the ability to set self-seed was found to be highly heritable in rye and self-offspring with high self-set were recovered from outbreeding rye population by the Swedish botanist and geneticist Nils Heribert Nilsson more than 100 years ago (Heribert Nilsson 1916; Lundqvist 1960). Gene variants have been selected by enforced self-pollination of rye (Peterson 1934; Ossent 1938; Mengersen 1951; Voylokov et al. 1993) and enabled the development of purebred parental genotypes bearing the desired characteristics. Mutations to self-fertility (*sf*) in otherwise highly-outcrossing rye were mapped to chromosomes 1R, 2R, 4R, 5R and 6R using isozymes (Wricke and Wehling 1985; Gertz and Wricke 1989) and DNA markers (Voylokov et al. 1998; Hackauf

Fig. 10.1 Floral architecture and sophisticated genetic mechanisms foster cross-pollination in rye. **a** A single male-fertile rye spike produces ca. 4 million pollen grains that are dispersed throughout a population by wind. **b** Receptive stigma branches, like the one highlighted in an isolated spike, form a large area for pollen interception and demonstrate a functional self-incompatibility system that prevents seed setting after forced self-pollination. **c** The male-sterility-inducing 'Pampa' cytoplasm results in non-dehiscent degenerated anthers

and Wehling 2005) as well as the trisomic set of rye cv. 'Esto' (Melz and Schlegel 1987; Melz and Thiele 1989). Mapping results suggested that the mutations affected both the cell-to-cell recognition response between pollen grain and stigma surface (*S*- and *Z*-locus on chromosomes 1R and 2R, respectively) as well as genes involved in signal transduction preventing self-pollination (Fuong et al. 1993; Voylokov et al. 1998).

Neither the *S*- nor the *Z*-gene has been isolated in rye or any other grass. However, in perennial ryegrass (*Lolium perenne* L.) a gene encoding a Domain of Unknown Function 247 (DUF247) protein as well as, like in rye (Hackauf and Wehling 2005), an ubiquitin-specific protease was found to co-segregate with the *Z*-locus (Shinozuka et al. 2010). Interestingly, in subsequent research another DUF247 protein encoding gene (*LpSDUF247*) was proposed as a candidate for the *S*-gene in the species (Manzanares et al. 2016). In addition, Studer and Asp (2014) reported two glycerol kinase-like linked genes (*LpGK1* and *LpGK2*) as *Z*-candidates in perennial ryegrass. *Z*-allele specific variable regions in these genes were proposed to be used for prediction of *Z*-locus incompatibility phenotype and thus could be applied to control pollination in hybrid breeding systems in *Lolium* (Studer and Asp 2014; Munoz-Sanz et al. 2020).

The integration of the *S*- and *Z*-genomic regions in the recently released rye reference genome of the 'Lo7' inbred line (Rabanus-Wallace et al. 2021) offered the opportunity to revisit and to shed new light on the underlying molecular genetic basis of SI in rye. All information given for the 'Lo7' genome assembly should translate to the 'Weining' genome assembly (Li et al. 2021) as almost perfect collinearity between both reference genome sequences was reported (for more details see Chap. 7). The pattern of segregation distortion observed for 28 out of 30 genetic markers enabled mapping the rye *S*-locus as a gamete selecting factor to an interval on chromosome 1R delimited by *SECCE1Rv1G0014520* and *SECCE1Rv1G0014770* that spans a DNA segment of 3 Mbp with 19 annotated genes (Table 10.1) (Rabanus-Wallace et al. 2021). In contrast to previous analyses (Voylokov et al. 1998), for each of the 28 loci revealing distorted segregation the excess of the marker allele originating from the self-fertile inbred line could be unambiguously attributed to gametic selection caused by the functional SI allele, i.e. only pollen grains carrying the self-fertility mutation were able to grow and penetrate the stigmatic tissue in the selfed F_1 plant. The frequency of recombination between the segregation distortion locus (SDL) and these markers ranged from zero to 31.4% and enabled precise mapping of the SDL in the 'Lo7' genome sequence.

Fine mapping in the perennial grass *Phalaris coerulescens* identified the markers *Bm2* and *BCD762* that co-segregate with *S* (Bian et al. 2004). *Bm2* represents a thioredoxin-encoding gene and its rye ortholog, the gene *SECCE1Rv1G0016380*, maps 22.4 Mbp distal from the gene *SECCE1Rv1G0014550* encoding a 40S ribosomal protein that is represented by *BCD762* as well as the sequence-tagged site (STS) markers TC76051, tcos163 and tcos164, respectively, the latter of which being absolutely linked to the SDL in rye (Rabanus-Wallace et al. 2021). A gene encoding a thioredoxin-h-like protein has been reported to co-segregate with *S* in *Hordeum bulbosum* as well (Kakeda et al. 2008). Notably, Kakeda and co-workers observed substantial suppression of recombination for the chromosomal region encompassing the *S*-locus in this wild self-incompatible species that is the only member of the secondary gene pool of cultivated barley (von Bothmer et al. 1995). Recombination between the SDL and the microsatellite marker *SCM1* (Hackauf and Wehling 2002), which represents the thioredoxin-like gene *SECCE1Rv1G0016380* just like *Bm2*, provides empirical evidence that the recombination frequency at the SDL in rye is higher than in *Phalaris* and *H. bulbosum* (Table 10.1). The reduced recombination in *Phalaris* and *H. bulbosum* identifies the *S*-gene as a 'supergene' (Schwander et al. 2014) that is inherited as one unit, as recombination between different *S*-haplotypes would allow for self-pollination in

Table 10.1 Anchoring the *S*-locus to the rye reference genome of 'Lo7'. SNP markers originate from the rye 600k array (Bauer et al. 2017). References: 1—Rabanus-Wallace et al. (2021), 2—Bian et al. (2004), 3—Kakeda et al. (2008), 4—Manzanares et al. (2016)

Marker	Rye gene model	Position (Mbp)	No of SNPs	Description	Refs.
TC68622	SECCE1Rv1G0002600	9.7		Mei2-like protein	1
TC68078, BCD98	SECCE1Rv1G0006670	35.2		Evolutionarily conserved C-terminal region 2	1, 3
tcos227	SECCE1Rv1G0007060	38.5		E3 UFM1-protein ligase 1-like protein	1
tcos145	SECCE1Rv1G0007720	42.2		Arginine/serine-rich splicing factor, putative	1
HAS175	SECCE1Rv1G0007940	43.0		F-box protein	3
tcos214, tcos220	SECCE1Rv1G0010880	66.9		Glutamate–cysteine ligase, chloroplastic	1
tcos131	SECCE1Rv1G0012230	90.4		Aspartic proteinase	1
TC76893	SECCE1Rv1G0012660	95.4		Aminotransferase	1
TC230399	SECCE1Rv1G0012960	98.8		Alpha-mannosidase	1
	SECCE1Rv1G0014190	111.0	18	NBS-LRR disease resistance protein-like protein	
	SECCE1Rv1G0014220	111.1	18	NBS-LRR disease resistance protein-like protein	
	SECCE1Rv1G0014230	111.2	1	Proline synthase co-transcribed bacterial	
	SECCE1Rv1G0014240	111.2	3	Transmembrane protein, putative (DUF247)	4
	SECCE1Rv1G0014300	112.0		Transmembrane protein, putative (DUF247)	4
	SECCE1Rv1G0014310	112.1	3	RING/U-box superfamily protein, putative	
	SECCE1Rv1G0014320	112.4	8	Chromatin remodelling factor, putative	
	SECCE1Rv1G0014330	112.4	3	Double stranded RNA-binding protein 3	
	SECCE1Rv1G0014340	113.5	1	Transport inhibitor response 1	
	SECCE1Rv1G0014360	114.0	17	ATP-dependent RNA helicase DeaD	
	SECCE1Rv1G0014370	114.2	8	Starch synthase family protein	
	SECCE1Rv1G0014400	114.3	8	Detoxification superfamily protein	
	SECCE1Rv1G0014410	114.4	11	Kinase family	
	SECCE1Rv1G0014430	114.5	10	Cytochrome P450	
	SECCE1Rv1G0014450	114.6	5	Pathogenesis-related thaumatin family protein	
	SECCE1Rv1G0014460	114.6	13	Kinase-like	
	SECCE1Rv1G0014470	114.6	12	Kinase-like	

(continued)

Table 10.1 (continued)

Marker	Rye gene model	Position (Mbp)	No of SNPs	Description	Refs.
tcos3511, tcos137	SECCE1Rv1G0014520	115.1	5	Bifunctional protein FolD	1
	SECCE1Rv1G0014530	115.1	14	Histidine–tRNA ligase	
	SECCE1Rv1G0014540	115.2	1	Desiccation-related protein PCC13-62	
BCD762, TC76051, tcos163, tcos164	SECCE1Rv1G0014550	115.2	10	40S ribosomal protein S4	1, 2
	SECCE1Rv1G0014560	115.3		Translocator protein-like protein	
	SECCE1Rv1G0014570	115.4	15	BTB/POZ and MATH domain-containing protein 2	
	SECCE1Rv1G0014580	115.4	4	Pre-rRNA-processing protein TSR2	
	SECCE1Rv1G0014590	115.5		Acetylglutamate kinase	
	SECCE1Rv1G0014600	115.6	14	BTB/POZ and MATH domain-containing protein 2	
	SECCE1Rv1G0014610	115.6	14	BTB/POZ and MATH domain-containing protein 2	
	SECCE1Rv1G0014620	115.6	10	BTB/POZ and MATH domain-containing protein 2	
tcos162	SECCE1Rv1G0014630	115.6	5	Rubber elongation factor protein, putative	1
	SECCE1Rv1G0014640	115.6	5	Translocase of chloroplast 159, chloroplastic	
	SECCE1Rv1G0014650	115.7	5	Xylosyltransferase 1	
	SECCE1Rv1G0014660	115.7	5	E3 ubiquitin-protein ligase	
	SECCE1Rv1G0014690	116.8		Ubiquitin-conjugating enzyme 23	
	SECCE1Rv1G0014710	117.1	7	cDNA clone:J013058P10, full insert sequence	
	SECCE1Rv1G0014750	117.9	11	Carboxyl-terminal peptidase, putative (DUF239)	
tcos3507, tcos5040	SECCE1Rv1G0014770	118.3		Villin	1
tcos158	SECCE1Rv1G0015090	120.8		Serine/threonine-protein phosphatase	1
LC34	SECCE1Rv1G0015620	126.9		Universal stress protein	3
tcos159	SECCE1Rv1G0015760	129.1		Dihydrolipoyl dehydrogenase	1
TC149368	SECCE1Rv1G0015900	132.3		Receptor-like kinase, putative	1
Bm2, HTL, Xscm1	SECCE1Rv1G0016380	137.6		Thioredoxin-like protein	1, 2, 3
PSR168	SECCE1Rv1G0017560	157.0		FACT complex subunit SSRP1	2
TOP1094	SECCE1Rv1G0030000	422.6		(RAP Annotation release2) NERD domain-containing protein	1

(continued)

Table 10.1 (continued)

Marker	Rye gene model	Position (Mbp)	No of SNPs	Description	Refs.
tcos114	SECCE1Rv1G0031490	443.3		Cytochrome P450 family protein, expressed	1
TOP1088	SECCE1Rv1G0032500	455.9		Calcineurin B-like protein	1
c74941	SECCE1Rv1G0034300	480.2		Mitochondrial transcription termination factor-like	1
PSR653	SECCE1Rv1G0037040	516.6		Transducin/WD-like repeat-protein	2
HPS54	SECCE1Rv1G0037750	530.9		Guanosine nucleotide diphosphate dissociation inhibitor	3
HAS122	SECCE1Rv1G0040080	556.7		S-acyltransferase	3
HAS163	SECCE1Rv1G0042530	581.3		Aminotransferase like protein	3
HAS42	SECCE1Rv1G0051180	651.7		tRNA(Ile)-lysidine synthase	3
tcos16	SECCE1Rv1G0052040	658.4		Farnesyl diphosphate synthase	1
KUR1001	SECCE1Rv1G0054220	673.5		ATP synthase subunit beta	1

individuals with the recombined haplotype and therefore cause the breakdown of the self-incompatibility system. The close linkage between the SDL and the isozyme marker *Prx7* is in accordance with data of Wricke and Wehling (1985) and identifies the SDL tagged in the 'Lo7' genome sequence as the *S*-locus in rye. Thus, the rye reference genome sequences (Li et al. 2021; Rabanus-Wallace et al. 2021) enable integration of data from independent mapping experiments, confirming a high degree of collinearity between homeologous regions of the *S*-locus in rye and *H. bulbosum* (Triticeae), ryegrass (Poeae) and *Phalaris* (Phalaridinae). This result further supports the assumption that self-incompatibility arose early in the diversification of plants and a single self-incompatibility system exists within the Poaceae family (Bian et al. 2004).

Notably, among the 19 genes residing within the *S*-locus genomic region a cluster of four genes, predicted to encode BTB/POZ and MATH domain-containing proteins, maps 195 kb distally of the gene *SECCE1Rv1G0014550* (Table 10.1). Proteins encompassing a BTB (Bric-À-Brac, Tramtrack, Broad-Complex), also known as the POZ (Pox virus and Zinc finger) and a MATH [Meprin and Tumour necrosis factor (TNF)-receptor-associated factor (TRAF) Homology] domain are commonly found in plant and animals, but not in fungi and serve in a wide range of cellular processes regulating cell development and homeostasis (Juranić and Dresselhaus 2014). Both the BTB as well as MATH domain are evolutionarily conserved and involved in protein-protein interactions (Bardwell and Treisman 1994; Park et al. 1999). In addition, the MATH domain has been identified to contribute to the turnover of nucleotide-binding domain and leucine-rich repeat-containing (NLR) immune receptors in *Arabidopsis thaliana* (Huang et al. 2016). A distinctive feature of the MATH-BTB family is its substantial diversification and extensive expansion in the grasses (Gingerich et al. 2007; Juranić et al. 2012; Juranić and Dresselhaus 2014). Plant BTB/POZ are known to act as substrate-specific adaptors of cullin3 (CUL3)-based ubiquitin E3 ligases (Juranić and Dresselhaus 2014) that control the specificity of the ubiquitin proteasome system (Dubiel et al. 2018), which is the major pathway for regulated degradation of cytosolic, nuclear and membrane proteins in all eukaryotic organisms (Livneh et al. 2016). An E3 ubiquitin-protein ligase encoding gene, *SECCE1Rv1G0014660*, maps near the

S-candidate genes in rye as well, just 171.4 kb distal of the BTB/POZ gene cluster (Table 10.1). The ubiquitination/26S proteasome-mediated selective protein degradation pathway has been identified as a precise regulatory mechanism for pollen guidance in gametophytic SI (GSI) systems. In GSI, the *S* phenotype of the pollen (male gametophyte) is determined by its own haploid *S* genotype. According to a widely accepted model for *S*-RNase based GSI, *S*-locus F-box proteins (SLFs) specifically interact as male pollen *S*-determinants with its female partner, the *S*-RNase, in a genotype-dependent manner that leads to the detoxification of *S*-RNase in self-compatible pollen/stigma combinations via ubiquitination/26S proteasome pathway (Munoz-Sanz et al. 2020). However, the GSI mechanism in grasses is assumed to act differently from the *S*-RNase-based GSI in transducing the self-signal into the germinating pollen grain. Grasses reveal a unique pattern of pollen and stigma traits which are otherwise typical for sporophytic SI (SSI) systems, mainly trinucleate pollen grains with a high respiratory rate, "dry type" stigmas and, most importantly, a very fast SI pollen tube inhibition that occurs at the stigma surface or shortly after penetration of the stigmatic tissue (Wehling et al. 1994). A BTB/POZ and MATH-based protein-protein interaction could match these features. Noteworthy in this context, two of the rye BTB/POZ and MATH domain encoding genes, *SECCE1Rv1G0014570* and *SECCE1Rv1G0014610*, can be integrated in the gene expression atlas of Triticeae reproductive development (Tran et al. 2013), illustrating that both genes are expressed in the stigma at a late stage of pollen development (Fig. 10.2). This expression pattern, the mapping position as well as the predicted gene function provide a promising basis for further research to clarify the function of the BTB/POZ and MATH domain-containing protein encoding genes as candidates for the putative female *S*-determinant in the pollen-pistil interaction of self-incompatible rye.

The genomic region carrying the *S*-locus delimited by the marker positions was found to be located 3 Mbp from two genes encoding DUF247 proteins (*SECCE1Rv1G0014240* and *SECCE1Rv1G0014300*), of which the first one shows the highest sequence similarity to the *L.*

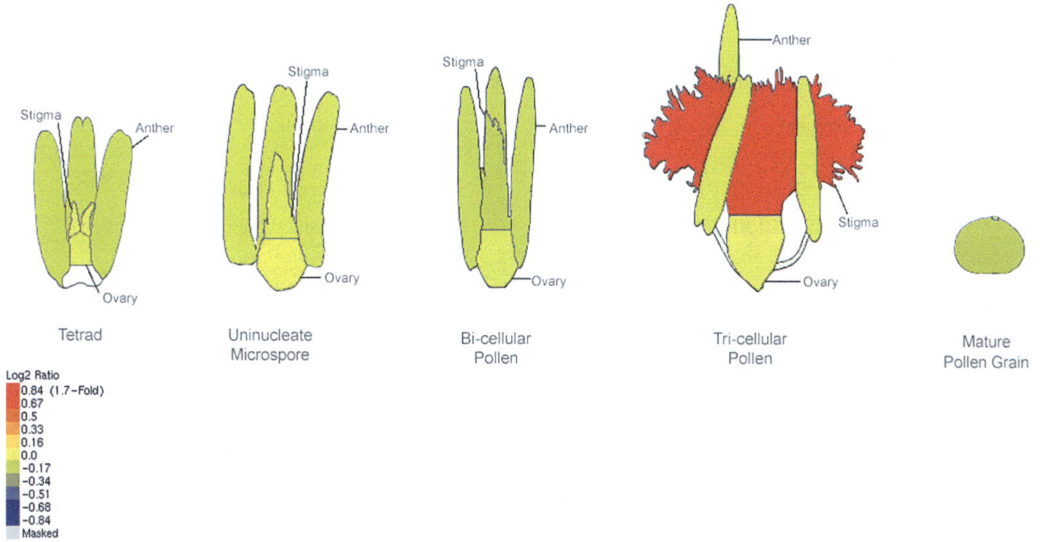

Fig. 10.2 Expression profile of *Ta.19786.1.A1_at* representing the wheat ortholog of rye BTB/POZ and MATH domain-containing protein encoding genes *SECCE1Rv1G0014570* and *SECCE1Rv1G0014610*. The electronic fluorescent pictograph (eFP) browser (Winter et al. 2007) was used to depict the expression pattern in four reproductive tissues at four stages of pollen development (Tran et al. 2013). Red colouring indicates the highest absolute expression of the gene

perennes's DUF247 ortholog (Table 10.1) (Rabanus-Wallace et al. 2021).

Rye is well known as a genetic resource in wheat breeding (Hao et al. 2020) (for more details see Chap. 1) and the activation of the grass SI system in wheat by the introgression of functional *S*- and *Z*-alleles from rye have been discussed as a potential basis for a hybrid system (Whitford et al. 2013). In the wheat reference genome sequence (The International Wheat Genome Sequencing Consortium 2018), *TraesCS1A02G090700* could be identified as an ortholog of the *SECCE1Rv1G0014240* gene, while no gene was predicted and annotated in the *S*-locus genomic regions of the B (pos. 139,075,987–139,076,423) and D (pos. 86,223,644–86,224,039) sub-genomes in the wheat IWGSC RefSeq v1.0 genome (based on BLASTN sequence similarity searches). The expression pattern of *TraesCS1A02G090700* in wheat meets the expectation for a gene encoding the male determinant of *S* to be highly expressed in spikelets and anthers. This is in agreement with the expression pattern of *LOC_Os05g10900* and *BRADI_2g35750*, the orthologs of the *SECCE1Rv1G0014240* in rice (*Oryza sativa*) (Ouyang et al. 2007) and *Brachypodium distachyon* (Papatheodorou et al. 2020) as well as *LpSDUF247* in *L. perenne* (Manzanares et al. 2016). Notably, *BRADI_2g35750* is regulated by *BRADI_1g10047* (Tian et al. 2020), a transcription factor highly expressed in spikelet and floral meristems and encoding a protein most closely related to KNOTTED1 (KN1) of maize (*Zea mays*) and rice HOMEOBOX1 (Derbyshire and Byrne 2013).

Similarly to the *S*-locus, 12 previously developed STS markers spanning a genetic distance of 32.3 cM, with the closest flanking markers mapping at a distance of 0.5 cM and 1.0 cM from *Z*, respectively and one marker co-segregating with *Z*, in a testcross population of 204 progeny (Hackauf and Wehling 2005), served to precisely map the *Z*-locus in the 'Lo7' genome sequence. The anchoring of the flanking markers TC89057 and TC101821 to the genomic region delimited the *Z* interval to a region of 1.3 Mbp with 20 annotated genes in the rye 'Lo7' inbred line (Table 10.2) (Rabanus-Wallace et al. 2021). The *Z*-locus-linked marker TC116908 represents *SECCE2Rv1G0130790* and maps within a 0.2 Mbp distance distal from two

Table 10.2 Anchoring the Z-locus to the rye reference genome of 'Lo7'. SNP markers originate from the rye 600k array (Bauer et al. 2017). Based on Rabanus-Wallace et al. (2021)

Marker	Gene model	Position (Mbp)	No of SNPs	Description
TC89057	SECCE2Rv1G0130710.1	878.2	6	Glycerol kinase
	SECCE2Rv1G0130770.1	878.4		Transmembrane protein, putative (DUF247)
	SECCE2Rv1G0130780.1	878.5		Transmembrane protein, putative (DUF247)
TC116908	SECCE2Rv1G0130790.1	878.7	4	Ubiquitin carboxyl-terminal hydrolase
	SECCE2Rv1G0130800.1	878.7	5	Short chain dehydrogenase/reductase
	SECCE2Rv1G0130810.1	878.8	1	F-box family protein
	SECCE2Rv1G0130820.1	879.0	7	EamA-like transporter family
	SECCE2Rv1G0130830.1	879.0	20	Tryptophan decarboxylase
	SECCE2Rv1G0130840.1	879.0		Methyltransferase-like protein
	SECCE2Rv1G0130870.1	879.2		VQ motif-containing protein, putative
	SECCE2Rv1G0130890.1	879.4	3	Organic cation transporter protein
TC101821	SECCE2Rv1G0130900.1	879.4	7	Clustered mitochondria protein

DUF247 orthologs on chromosome 2R annotated as *SECCE2Rv1G0130770* and *SECCE2Rv1G0130780* (Rabanus-Wallace et al. 2021). Notably, BLASTN sequence similarity searches revealed significant hits of *SECCE2Rv1G0130770* to the *Z*-locus genomic region only in the D sub-genome (pos. 608,440,921–608,442,483) of the wheat IWGSC RefSeq v1.0 but no gene has been predicted in this segment.

The glycerol kinase residing at the *Z*-locus in rye reveals allelic variability (Table 10.2) like the genes *LpGK1* and *LpGK2* (Manzanares et al. 2016). There are six glycerol kinase genes predicted on chromosome 2R, 5R, 6R and 7R in the 'Lo7' genome assembly (Rabanus-Wallace et al. 2021) but only the *SECCE2Rv1G0130710* gene is located in the *Z*-locus genomic region on 2R with similarity to the reported ortholog of *LpGK1* (Studer and Asp 2014). Integration of *SECCE2Rv1G0130710* in the gene expression atlas of Triticeae (Tran et al. 2013) revealed that this rye glycerol kinase gene is expressed in reproductive tissues at early stages of pollen development (Fig. 10.3) as well as in non-reproductive tissues (Tran et al. 2013). Represented by the marker TC89057, *SECCE2Rv1G0130710* was found to recombine with *Z* (Hackauf and Wehling 2005) but to be located just adjacent to the two DUF247 candidates (Rabanus-Wallace et al. 2021). Thus, the observed recombination and the expression pattern largely exclude the glycerol kinase gene as a candidate for *Z* in rye.

Neither the DUF247 candidates at the *Z*- nor at the *S*-locus are represented within the global transcriptional profiles of the anther/pollen, ovary and stigma concurrent developmental stages (Tran et al. 2013). This example demonstrates that the novel rye reference genome sequences (Li et al. 2021; Rabanus-Wallace et al. 2021) close a gap concerning the available genomic resources for studies of reproductive development in Triticeae species and will lead to a deeper understanding of regulatory factors underlying Triticeae floral development and function. The comparatively high recombination frequency at the *S*- and *Z*-loci in rye appears to be an exception among grasses but requires a traditional map-based gene isolation approach to identify the causal mutations over sequence homology searches.

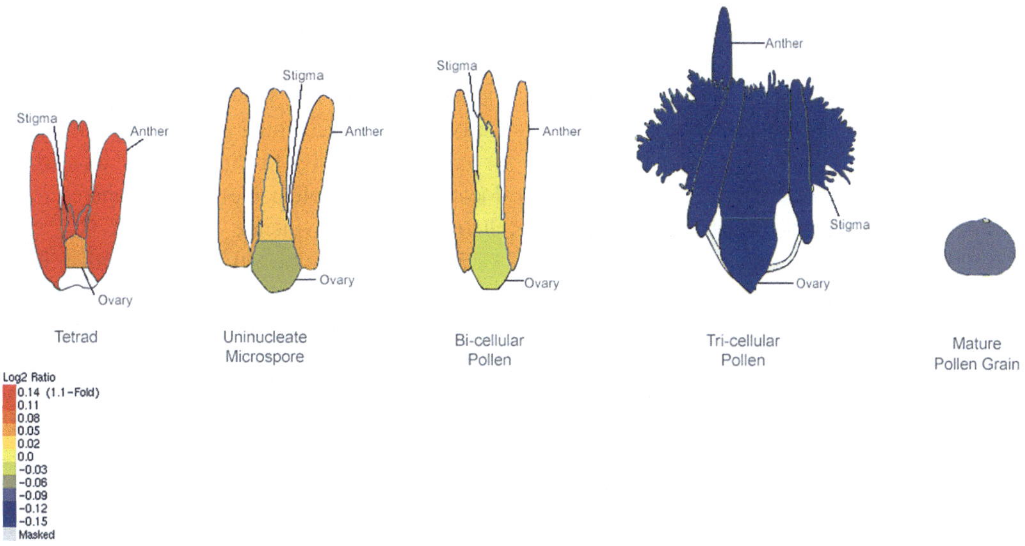

Fig. 10.3 Expression profile of *Ta.9384.2.S1_at* representing the wheat ortholog of rye glycerol kinase gene *SECCE2Rv1G0130710*. The electronic fluorescent pictograph (eFP) browser (Winter et al. 2007) was used to depict the expression pattern in four reproductive tissues at four stages of pollen development (Tran et al. 2013). Red colouring indicates the highest absolute expression of the gene

So far, the best-characterised SI systems are those controlled by a single genetic locus, the *S*-locus only (Langridge and Baumann 2008; Yang et al. 2008). The reference genome sequences enable the search for rye orthologues of *S*-genes that have been already cloned and characterised in other species in sporophytic (Jany et al. 2019) as well as gametophytic (Wang et al. 2019) SI systems. In Brassicaceae, the specific *S*-locus determinants and several modifier factors for the sporophytic self-incompatibility are well studied (Yang et al. 2008). It has been shown that the serine/threonine receptor kinase (*SRK*) (Stein et al. 1991; Takasaki et al. 2000) and cysteine-rich *S*-locus Protein 11 (Sp11) also named SCR (Schopfer et al. 1999) form the female and male *S*-locus determinants, respectively. Both *S*-genes and a third gene encoding a glycoprotein (SLG), which possibly enhances SI expression, are tightly linked and inherited as an *S*-haplotype (Sehgal and Singh 2018). Hundreds of *S*-haplotypes were identified in Brassicaceae (Ruffio-Chable and Gaude 2001), but once the *S*-haplotype in pollen and pistil are identical, the pollen is rejected on the papilla cell surface of the pistil (Murase et al. 2020).

The *S*-alleles were also identified and characterised in the single-locus GSI systems (Rosaceae, Solanaceae, Scrophulariaceae, Rubiaceae) (McCubbin and Kao 1996). They were found to be pistil-expressed and to encode stylar proteins that show RNase activity and therefore are often referred to as *S*-RNases (McCubbin and Kao 1996; Sijacic et al. 2004). The *S*-RNAses act as highly selective cytotoxins and cause rejection of pollen when their single *S*-haplotype matches either of the two *S*-haplotypes in the diploid pistil (Munoz-Sanz et al. 2020). They interact with the F-box-protein encoding gene specifically expressed in pollen, referred to as SLF (*S*-locus F-box) that can be a single gene (Prunus) or a cluster of gene *S*-haplotypes (Solanaceae and in the Rosaceae tribe Maleae). Based on those findings a model of action was proposed where non-self *S*-RNases are degraded in compatible pollinations, but in self-pollinations, self *S*-RNases escape degradation and in turn, are able to degrade the pollen RNA (Munoz-Sanz et al. 2020). In Solanaceae, ubiquitination and degradation of *S*-RNase is attributed to the collective action of the array of 16 to 20 SLF proteins (Kubo et al. 2010, 2015). This is referred to as the collaborative non-self-recognition model (i.e. the array of SLF proteins recognises non-self *S*-RNase) and is currently the most widely accepted model (Munoz-Sanz et al. 2020).

A mechanistically distinct single *S*-locus GSI system has been found in Papaveraceae (Foote et al. 1994). In poppy (*Papaver rhoeas*), the product of the *S*-locus in the pistil is a small protein that interacts with incompatible pollen, triggering a Ca^{2+}-dependent signalling network, resulting in pollen inhibition and programmed cell death (Thomas and Franklin-Tong 2004; Bosch and Franklin-Tong 2008). The *S*-locus comprises two linked genes that encode the male *PrpS* (*Papaver rhoeas* pollen S) *S*-determinant (Foote et al. 1994) and the female *PrsS* (*Papaver rhoeas* stigma) *S*-determinant protein. PrpS is a highly polymorphic receptor expressed in pollen-membrane that interacts via an extracellular loop (Wheeler et al. 2009) with PrsS—a small pistil protein (Foote et al. 1994). The self-interaction triggers a range of responses, including an increase in cytosolic free Ca^{2+}, an influx of Ca^{2+} and K^{+} and the production of reactive oxygen species and nitric oxide. Similarly, the LpSDUF247 protein from perennial grass is predicted to have a C-terminal transmembrane helix and an extracellular domain, indicating that it may act, similarly to PrpS protein, as a ligand located on the pollen exine (Manzanares et al. 2016). In rye, evidence implicated Ca^{2+} and kinase activity in the self-incompatibility interaction and protein phosphorylation to play an important role in signal transduction (Wehling et al. 1994).

Neither of the predicted genes located in the *S*- and *Z*-locus genomic regions of the 'Lo7' reference genome reveal homology to any of the cloned *S*-genes from other species except for a gene belonging to the F-box family residing at the *Z*-locus. This observation suggests that the bi-factorial self-incompatibility mechanism of the grasses represents a further type within GSI systems. The identification of causal mutations in rye was previously hampered by its large genome. In total, 264 and 53 SNP markers from the previously developed rye 600k array (Bauer et al. 2017) map to the *S*- and *Z*-loci, respectively (Tables 10.1 and 10.2) and provide an excellent resource to further increase the resolution of the genetic maps for both loci. Thus, the deciphered blueprint of the rye genome closes a gap in exploring genome sequences of agronomically important cereal crops and offers new prospects to elucidate the molecular basis of SI as a sophisticated cell-to-cell communication system, resulting in rye functioning as the only outbreeding small grain cereal.

10.2 Managing *S* and *Z* Diversity in Rye Breeding Programs

When hybrid breeding in rye was in its infancy, a system allowing for effective selfing without affecting SI was discussed to develop a comparatively simple and straightforward breeding scheme in rye to directly exploit the genotypic variation between random mating populations cultivars. Wricke (1978) developed a method to obtain seeds from self-incompatible rye plants by keeping them at a constant high temperature (30 °C) during anthesis, an observation that was followed by further reports of hypersensitivity of the reproductive development in the Triticeae to high temperature (Sakata et al. 2000; Phillips et al. 2015; Coulton et al. 2020). This pseudo-compatibility of rye plants grown in higher temperatures could be applied to produce inbred progenies of self-incompatible material for hybrid breeding (Gertz and Wricke 1991). However, in contrast to CMS-based hybridization systems, the genotype-phenotype relationships in SI-based breeding schemes are challenging to assess and currently inferred from controlled pollinations or pollen tube growth tests that hamper practical application on a large scale. Functional markers (Andersen and Lubberstedt 2003) that mirror the allelic variation controlling the SI response in rye, are, thus, invaluable tools to manage the genetic diversity of *S* and *Z* in random mating breeding populations. Indeed, superimposing a marker-assisted determination of *S*- and *Z*-genotypes to the suggested recurrent selection scheme for intrapool population improvement in self-incompatible rye (Geiger and Miedaner 2009) will increase the proportion of cross-fertilisation between selected fractions of two genetically distant populations to the maximum of 83% (Wricke 1985). Furthermore, functional *S*- and *Z*-markers can be used to develop genotypes in both heterotic gene pools, that are homozygous either at the *S*- and *Z*-locus. As self-pollination is efficiently prevented in such genotypes (Wricke 1985), they can serve as testers to determine general and specific combining ability (Sprague and Tatum 1942) of the parental genotypes from the opposite pool. The development of suitable tester genotypes takes into consideration that genotypic correlations between line per se and testcross performance decreased with increasing complexity of the trait in rye (Miedaner et al. 2014), just like in maize (Sprague and Tatum 1942). As a consequence, "panmictic-midparent heterosis" (Lamkey and Edwards 1999) can be exploited for enhancing the performance of random mating cultivars, although half of this increase is lost due to a drop in heterozygosity, as Hardy–Weinberg equilibrium is rapidly attained during seed multiplication (Geiger and Miedaner 2009). However, the systematic use of SI for the genetic improvement of random mating populations could increase the competitiveness of open-pollinating cultivars and help to minimise risks of genetic vulnerability of rye caused by the narrow genetic base of plasmotypes in high-performing hybrid varieties. It needs to be mentioned that mutation or migration of novel *S*- or *Z*-alleles will result in novel incompatibility specificities and, thus, substantially reduce the degree of the desired cross-fertilisation (Wricke 1985). Furthermore, rye pollen is efficiently transferred over long distances, Römer (1931) observed almost 3% cross-pollination between plants separated from each other by 662 m. A sophisticated strategy to control undesired pollination of a breeding population is mandatory for the success of an SI-based breeding scheme. This may include the use of recessively inherited morphological markers like the dwarfing gene *ct1* that has been mapped to chromosome 7R (Plaschke et al. 1995) and enables efficient selection of tall plants as illegitimate offspring in a random mating population prior to flowering (Dill 1983).

Hybrid breeding practices have successfully exploited part of the available genetic diversity of rye to increase the genetic gains in agronomic traits including grain yield (Laidig et al. 2017). To safeguard the long-term success of hybrid rye breeding programs, random mating populations provide indispensable genetic resources for the introgression of novel diversity in elite germplasm. Likewise, functional *S*- and *Z*-markers pave the way to reconstitute SI in self-fertile elite inbred lines as a strategy to improve the performance of random mating populations (Voylokov 2007). Indeed, such germplasm resources will tune the rates of recombination between elite alleles and enable dynamic management by cultivation in ecologically contrasting locations, where the populations can evolve over time under environment-specific selection pressure. This concept of evolutionary plant breeding (Döring et al. 2011) may be particularly important for the pollinator genepool, as the probability of finding populations that are genetically diverse from the Petkus pool is reduced due to the dominant role of the 'Petkus' rye as an ancestor of many random mating populations worldwide (Fischer et al. 2010).

10.3 Floral Architecture in Rye Supports Outcrossing and Wind Pollination

In addition to SI, cross-pollination in rye is enhanced by the specific floral architecture (Fig. 10.1). In contrast to wheat with relatively small anthers producing a low number of pollen grains (2.5 thousand pollen grains per anther), rye anthers are large and fully extrude from the floret and each can release up to 19 thousand pollen grains, almost 10 times more than what is estimated for wheat (De Vries 1971; Heslop-Harrison 1979). These traits could be introduced to wheat to enhance cross-pollination for hybrid breeding in wheat (Whitford et al. 2013). Indeed, it has been reported that the addition of the long arm of chromosome 4R increased anther length by 16%, whereas the addition of the short arm of chromosome 4R improved the pollen grain number by 33% in the wheat 'Chinese Spring' cultivar (Nguyen et al. 2015). The increase in pollen grain number conferred by chromosome 4R is mainly associated with the short arm of that chromosome (Nguyen et al. 2015) and largely excludes the known *restorer-of-fertility* (*Rf*) genes residing at the distal end of the long arm of chromosome 4R (see below) as genetic factors determining the described effect in wheat.

Moreover, the newly assembled rye reference genome sequences (Li et al. 2021; Rabanus-Wallace et al. 2021) can be mined to approach the orthologs of genes governing flower habit like *SECCE2Rv1G0130480* for the barley *Cleistogamy 1* gene (*Cly1*) (Wang et al. 2015). In many grass species, the swelling (Miedaner et al. 2019) of the lodicule, a structure found in the floret that is functionally related to the petal, opens up the floret, allowing for pollen dispersal and cross-pollination (Wang et al. 2015). In barley, closed (cleistogamous) and open (chasmogamous) flower variants are known (Honda et al. 2005; Wang et al. 2015). In cleistogamous variants, the flowers shed their pollen before flower opening and this results in autogamy (self-fertilisation) (Honda et al. 2005; Nair et al. 2010). The expression of the *Cly1* gene is controlled by *miRNA172* that keeps its RNA levels low (Nair et al. 2010; Ning et al. 2013; Wang et al. 2015). Once understood, knowledge of this mechanism in rye could be transferred to wheat to make wheat flowers more open thus more accessible for wind pollination.

10.4 Post-zygotic Incompatibility (Hybrid Incompatibility)

Compared to other cultivated small grain cereals, the high intraspecific diversity makes cultivated rye well adapted to harsh growing conditions and more resistant to biotic and abiotic stresses (e.g.

extreme winter-hardiness and growth on poor soils) (Schittenhelm et al. 2014; Myśków et al. 2018; Miedaner and Laidig 2019). In wheat breeding, rye is an excellent source for allele mining of pathogen and pest resistance (Baum and Appels 1991; Crespo-Herrera et al. 2017) and root architecture genes (Howell et al. 2014, 2019). The first attempt to systematically cross bread wheat with rye is documented by Wilson (1876) and to date, genes-conferring resistance to almost 20 pathogens and pests were transferred from rye to wheat (Crespo-Herrera et al. 2017). In particular, rye chromosome 1R has been a rich source of resistance genes for wheat improvement for resistance to stem and leaf rust, yellow rust and powdery mildew (Crespo-Herrera et al. 2017). Through comparison of the chloroplast genome sequences, it has been estimated that rye is more closely related to wheat than barley. Barley diverged from *T. aestivum* 8–9 million years ago whereas rye only 3–4 million years ago (Middleton et al. 2014). Rye is a secondary crop and arose as a weed in *Triticum* spp. representing a case of Vavilovian mimicry (McElroy 2014) (see Chap. 6). However, despite the close kinship and spontaneous introgressions of rye chromatin to chromosome 2D observed in a Portuguese wheat landrace (Ribeiro-Carvalho et al. 2001), post-zygotic reproductive barriers exist between rye and wheat that restrict the pool of germplasms that can be used systematically for gene introgressions in breeding programs and limit the transfer of desirable genes across the two closely related species (Tikhenko et al. 2011). This hybrid incompatibility makes the reintroduction of diversity present in the different gene pools into elite varieties through intra- and interspecific crosses difficult.

10.5 The Post-zygotic Reproductive Isolation (RI) and Embryo Lethality

The post-zygotic reproductive isolation (RI) or hybrid incompatibility often causes tissue necrosis and disturbances in the development of interspecific hybrids. Necrosis can be temperature dependent (Bomblies and Weigel 2007) and cause slow growth, wilting, discolouration and lethality (Tonosaki et al. 2016). The RI-effects have been found to be similar to autoimmune-like reactions caused by epistatic interactions between resistance (*R*) genes (Bomblies et al. 2007; Jeuken et al. 2009) that lower the overall fitness and viability of interspecific hybrids. Those negative effects are known as hybrid weakness. Interestingly, phenotypes associated with hybrid necrosis resemble those observed in response to abiotic and biotic stresses (Bomblies et al. 2007; Bomblies and Weigel 2007) but unlike the processes underlying the disease resistance being reasonably well understood, little is known about the mechanisms underlying hybrid incompatibility and their links to plant development. One of the phenotypes observed in wheat-rye hybrids caused by RI is embryo lethality. It has been reported that RI is caused by an interaction between a pair of *embryo lethality* (*Eml*) genes *Eml-A1* and *Eml-R1* encoded in the wheat and rye genomes, respectively (Tikhenko et al. 2008, 2011, 2017). The interaction happens shortly after embryo axis formation and while the endosperm develops normally, the maintenance of the shoot apical meristem (SAM) is disturbed, eventually leading to abortion of embryo development (Tikhenko et al. 2008). These findings suggest that the *Eml* loci are not only involved in reproductive barriers but also play a role in SAM maintenance (Tikhenko et al. 2008). Initially, based on linkage with microsatellite markers *Xgwm1103/Xgwm732*, the *Eml-A1* locus was mapped to the end of chromosome 6AL in wheat (Tikhenko et al. 2017). A recent study of Tsvetkova et al. (2018) reported the hybrid dwarfness (*Hdw-R1*) locus, which controls the post-zygotic reproductive isolation between wheat and rye, to be linked with *Eml-Rl*. Both genes responsible for abnormal development of wheat-rye hybrids are linked near an evolutionary translocation on chromosome 6R (Tsvetkova et al. 2018). The *Eml-R1* gene was mapped to the long arm of chromosome 6R in rye (Tikhenko et al. 2011). The microsatellite markers GRM0173 and GRM0130 define the target interval carrying *Hdw-R1* and

Eml-R1 (Tsvetkova et al. 2018) and a 67 Mb segment in the 'Lo7' reference genome sequence, as specified by their corresponding gene models *SECCE6Rv1G0421000* and *SECCE6Rv1G0431830*, respectively. The high-quality 'Lo7' reference genome sequence will accelerate the discovery of the genetic basis responsible for phenotypes like *Hdw-R1* and *Eml-R1* using next generation sequencing (NGS) technologies by serving as a blueprint to guide the alignment of reads from whole-genome sequencing of bulked recombinants and subsequent analysis for local skews in the parental allele frequencies, as has been demonstrated in model (James et al. 2013) and crop plants (Mascher et al. 2014; Pankin et al. 2018; Hoseinzadeh et al. 2019).

10.6 Wheat-Rye Crossability

After RI, another challenge that limits the transfer of traits from rye to wheat is the low crossability of most of the adapted wheat germplasms with the species (Alfares et al. 2009). This restricts substantially the pool of germplasm that can be used for rye introgressions in breeding programs. Early crossing studies with wheat as a female parent and rye as a pollen donor pointed towards two genes, named *Kr1* and *Kr2*, as being responsible for the poor crossability between wheat and rye (Lein 1943). The genes were mapped to the long arms of chromosome 5B and 5A, respectively (Riley and Chapman 1967; Sitch et al. 1985; Börner et al. 1996). In later studies, an additional two genes designated as *Kr3* and *Kr4* and located on chromosome 5D and 1A were identified (Krolow 1970; Zheng et al. 1992). Dominant alleles of wheat *Kr1 and Kr2* genes inhibit rye pollen germination and pollen tube growth between the style base and embryo sac (Lange and Wojciechowska 1976; Jalani and Moss 1980, 1981). In the late 1990s, an additional locus, named *SKr*, was reported in studies using mapping populations of double haploid (DH) lines produced by anther culture from F_1 hybrids of a cross between non crossable (NC) French wheat cv. 'Courtot' (Ct) and crossable 'Chinese Spring' (Tixier et al. 1998). *SKr* was identified as a major QTL and was found to be located on the distal end of the short arm of chromosome 5B (Lamoureux et al. 2002). Only recessive alleles of the *kr* genes allow crossability with a wide range of effects, however, out of the several *Kr* genes identified, *Kr1* and *Skr1* have been shown to have a major impact on the inhibition to crossability between wheat and rye (Alfares et al. 2009). The most crossable wheat varieties originate from Asia, i.e. 'Chinese Spring', but those lines are not well adapted to European growth conditions therefore an effort to introduce those alleles into European lines was undertaken (Molnar-Lang et al. 1996). In addition, *Skr*-specific markers were developed to identify new crossable cultivars in European wheat (Bouguennec et al. 2018). Their application shortened the time needed to transfer the crossability trait (*skr*) from 8 to 5 years (Bouguennec et al. 2018). The introduction of *skr* improved the crossability of 'Barok' from 1.4 to 30%. In addition to wheat genes, several studies indicated an influence of rye genes on wheat-rye crossability (Taira et al. 1978; Oettler 1982) that were later confirmed (Bouguennec et al. 2018). Despite the successful mapping of *Kr1* (Bertin et al. 2009) and the *Skr* locus (Alfares et al. 2009) over a decade ago, their genes have not been cloned and their molecular function remains unknown (Tonosaki et al. 2016).

Reproductive barriers and incompatibilities pose challenges to plant breeding that must be overcome to advance the development of elite lines with advantageous interspecific introgressions. Traditional techniques to overcome the reproductive barriers must be developed specifically for a given species and depend strongly on the interspecific crossability of parental lines. Self-incompatibility, embryo lethality and low crossability are all caused by genetic determinants. Their isolation and molecular cloning will help to fully understand the molecular mechanisms underlying the interspecific incompatibilities between wheat and rye and thus will accelerate the transfer of traits from wild or close relatives into elite lines in breeding programs. For this purpose, the 'Lo7' reference genome sequence closes an essential gap in the genomic tools available for rye and enables state-of-the-art approaches for fine mapping and gene cloning (Jaganathan et al. 2020) in this small grain cereal.

10.7 The Mitochondrial Basis of Cytoplasmic Male Sterility in Rye is Unknown

Another type of reproduction barrier originates from epistatic interactions between the nuclear and mitochondrial genomes and is known as cytoplasmic male sterility (CMS) (Chase 2007). CMS manifests itself by a failure to produce functional pollen and/or male reproductive organs by otherwise healthy-looking plants (Budar and Pelletier 2001; Chase 2007). In natural populations, CMS is the underlying cause of gynodioecy, the co-existence of female and hermaphrodite plants within a species (Charlesworth 2002; Hanson and Bentolila 2004). CMS can promote outcrossing and increase the fitness of male-sterile (female) plants through resource re-allocation, an effect termed female advantage or compensation (Darwin 1877). Early on, it was shown that CMS is determined by mitochondrially-encoded genes the effects of which can be counteracted by nuclear-encoded restorer-of-fertility genes (*Rf*) (Pruitt and Hanson 1991; Budar and Pelletier 2001; Hanson and Bentolila 2004). Plant mitochondrial genomes show high conservation at the gene sequence level, but high diversity at the genome organisation level (Galtier 2011; Chen et al. 2017). Most likely, complex recombination events associated with the presence of large DNA repeats frequently give rise to novel open reading frames (*orfs*) which, when expressed, can cause CMS (Galtier 2011; Tang et al. 2017). Often those *orfs* are chimeras made in part by regular mitochondrial gene sequences and in part by *orf*-unique sequences (Chase 2007). They can originate from gene fusions, partial/orphan *orfs* or disruptions in gene orientation/promoter association (Chen et al. 2017; Tang et al. 2017) and usually remain cryptic, buried in the mitochondrial genome, as their negative effects are neutralised by the action of the nuclear *Rf* genes. Therefore, CMS-inducing *orfs* are often discovered only by genetic crossing or somatic hybridizations (protoplast fusions) during which the CMS gene (cytoplasm) is separated from the *Rf* gene (nucleus) that suppresses its expression (Chen and Liu 2014). In rye, CMS was discovered independently in several geographically separated populations. The first case of CMS was documented by Putt (1954) but it was lost shortly after and never used in hybrid rye breeding programs (Milczarski et al. 2016). Later, a cross between Argentinian 'Pampa' rye and a German inbred line revealed the presence of sterilising cytoplasm that was named Pampa-type cytoplasm or CMS-P (Geiger and Schnell 1970). Several other male-sterility-inducing cytoplasms were identified in rye populations in Russia, CMS-R (Kobyljanskij and Katerova 1973), Poland in 'Smolickie' rye CMS-C (Łapiński 1972) and CMS-S (Madej 1975; Warzecha and Salak-Warzecha 2003), Finland (Ahokas 1980) and Germany CMS-G ('Gülzow') in 'Schlägler Alt' rye (Melz et al. 2003). However, it has been concluded that apart from CMS-P the remaining cytoplasms are genetically identical to 'Vavilov' cytoplasm (CMS-V) (Geiger et al. 1995; Stojałowski et al. 2006; Milczarski et al. 2016). The discovery of CMS-P in rye was a groundbreaking observation that enabled selective matings on a large scale and paved the way to the development of first hybrid rye varieties. Today, the majority of grown hybrid rye cultivars are based on a system of pollen disruption using CMS-P cytoplasm (for more details see Chap. 2), but the released CMS-G hybrid cultivars (Melz et al. 2003) demonstrated the general feasibility of hybrid breeding using alternative hybridization systems that serve to reduce the potential genetic vulnerability of rye hybrids.

In all studied CMS systems characterised so far, male sterility is associated with changes in the mitochondrial genomes (Hanson and Bentolila 2004; Chase 2007). Already early studies in rye concluded that the two major types of CMS, CMS-P and CMS-V, differ by genes for plasmon sensitivity and fertility restoration responsiveness to different restorer genes (Łapiński and Stojałowski 2003). This suggested that at least two genetically different CMS-mitotypes exist in the rye populations: one for CMS-P and other for CMS-V. Indeed, it has been shown that the restriction pattern of mitochondrial DNA (mtDNA) of fertile plants is different in plants carrying the sterile 'Pampa' cytoplasm compared to fertile plants (Tudzynski et al. 1986). Restriction patterns of mtDNAs of 'Halo', an open pollinated cultivar and the fertile maintainer line turned out to be identical whereas 'Pampa' mtDNA showed a unique pattern, indicating the involvement of mtDNA rearrangements in the expression of male sterility in rye (Tudzynski et al. 1986). Dohmen et al. (1994) reported detailed RFLP (Restriction fragment length polymorphism) analyses of mitochondrial DNA from fertile and male-sterile CMS-P plants. Using several heterologous mitochondrial genes as probes, differences in the overall structure of the two mitochondrial genomes (sterile *vs* fertile) were observed (Dohmen et al. 1994). The obtained results indicated the presence of extra copies of *cob*, *atpA* and *atp9* genes existing in the 'Pampa' cytoplasm (Dohmen et al. 1994). In particular, the transcript levels of the additional *cob*-gene variants appeared to be strongly reduced in the presence of restorer genes (Dohmen et al. 1994). Those additional copies could be linked to CMS, but this remains to be confirmed (Dohmen et al. 1994). Furthermore, a comparison of the genomic organisation of the mitochondrial genomes between P- and the G-type cytoplasms, the latter belonging to the CMS-V type, has shown differences between the two cytoplasms with respect to mtDNA restriction fragment analysis and DNA hybridisation, confirming the different genetic origin of the two cytoplasms (Steinborn et al. 1993). Based on this knowledge, a sequence-characterised amplified region (SCAR) marker approach was developed and enabled rapid population-wide screens for presence/absence of the two different cytoplasms (Stojałowski et al. 2006). To discriminate between the normal and two sterilising cytoplasms CMS-P and CMS-C, markers were developed for three mitochondrial genes *cox1*, *nad6* and *nad2* (Stojałowski et al. 2006). Their use in several rye populations revealed a prevalence of CMS-P in random mating landraces and primitive rye originating from South America and the Fertile Crescent, while the CMS-V could be detected in 69% of the analysed plants representing the Central European gene pool (Stojałowski et al. 2008). These results are congruent with a comprehensive genetic evaluation of germplasm resources using a universal non-restorer tester to identify a sterility-inducing cytoplasm as well as a second tester to differentiate between CMS-P and CMS-V (Łapiński and Stojałowski 2003). Consistent with theoretical studies (*cf.* Rieseberg and Blackman (2010) for review), the spread of CMS-V in random mating populations of rye indicates that this CMS mutation obviously provides a fitness advantage in female function. While the CMS-P mutant has been statistically shown to increase the fitness of females relative to hermaphrodites in rye (Marker et al. 1985), further studies are necessary to explain the different patterns in the geographical distribution of both mitochondrial genotypes. In fact, the spread of the CMS-V locus generated strong selection for the evolution of one or more nuclear restorer allele(s) (Łapiński and Stojałowski 2003) and makes the development of male-sterile seed parent lines more difficult in hybrid breeding programs using CMS-V (Vendelbo et al. 2020) as compared to CMS-P and explains the widespread use of the latter as a reliable and efficient hybridization system.

Today, roughly thirty CMS genes from thirteen crop species have been identified (Kim and Zhang 2018). The mitochondrial genome of the Boro II-type (BT) cytoplasm, one of the most widely used CMS systems in rice hybrid breeding, contains two copies of the *atp6* gene encoding subunit of the ATP synthase complex (Kadowaki et al. 1990; Akagi et al. 1994). A unique sequence (*orf79*) located downstream of one of the *atp6* copies was found to cause male sterility (Kadowaki et al. 1990; Akagi et al. 1994). The sequence *Orf79* encodes a protein the N-terminus of which shows similarity to subunit I of the rice mitochondrial cytochrome oxidase (COXI) and the C-terminal region is of unknown origin (Kazama and Toriyama 2003; Wang et al. 2006). Early studies of wheat *Triticum timopheevii* cytoplasm (T-CMS) identified a chimeric *orf* named *orf256* located upstream of the *cox1* gene to be the cause of male sterility of plants carrying *T. aestivum* nucleus (Rathburn and Hedgcoth 1991). Detailed sequence analysis revealed that the 5'-flanking sequence and the first 11 codons of *orf256* are identical to the analogous region from *coxI*, whereas the rest of *orf256* is not related to any known mitochondrial gene (Rathburn and Hedgcoth 1991; Song and Hedgcoth 1994). Recently, a new mitochondrial gene—*orf279*—was identified as the genetic basis of cytoplasmic male sterility of plants carrying T-CMS cytoplasm (Small and Melonek 2020; Melonek et al. 2021). The 5'-region and the first 96 codons of *orf279* are identical to *atp8* gene encoding subunit 8 of the mitochondrial ATP synthase and the remaining sequence is unique to *orf279* (Small and Melonek 2020; Melonek et al. 2021). It has been shown that the suppression of the *orf279* transcript, most likely by a targeted cleavage by to date unidentified mitochondrial endonuclease, correlates with the male-fertility restoration phenotype while the cleavage of *orf256* transcript does not as it was detected also in male-sterile genotypes (Small and Melonek 2020; Melonek et al. 2021). In several dicot plants such as sunflower (*Helianthus annuus*), *Brassica* and carrot (*Daucus carota*) CMS-*orfs* were found to contain fragments of the *atp8* gene (Kim and Zhang 2018). In 1991, Bonhomme and colleagues, identified *orf138*—a chimeric *orf* carrying a partial sequence of *atp8* gene—as the genetic cause of the CMS-Ogu and today this cytoplasm is widely used in hybrid breeding of *Brassica* crops (Bonhomme et al. 1991). As many of the identified CMS-causing genes contain fragments of genes encoding different subunits of the mitochondrial ATP synthase, it has been hypothesised that defects in ATP production during flowering could trigger abnormal programmed cell death and lead to abortion of pollen development and plant sterility (Chen and Liu 2014).

Polymorphisms detected in rye mitochondrial genome sequences have been explored to clarify the evolution of rye (Isik et al. 2007). These initial results on mitochondrial genome diversity suggest that the outbreeding rye offers a model to comprehensively elucidate the network of mitochondrial genes and their interaction with nuclear-encoded genes as driving forces for a proper function of mitochondria in small grain cereals. Corresponding research will benefit from the progress achieved in sequencing mitochondrial genomes of wheat (Liu et al. 2011) and barley (Hisano et al. 2016), demonstrating that mitochondrial genomes of Triticeae species are similar to those of other grass species in terms of gene content. In addition, the novel 'Lo7' reference genome sequence describes restorer loci for both CMS-P and CMS-V with unprecedented precision (Rabanus-Wallace et al. 2021). Both advancements open new prospects to determine the frequency of CMS and restorer alleles in random mating rye populations. This framework enables complementary research on gynodioecy in wild species (Bergero et al. 2019) and testing whether the mitochondrial male-sterility types in rye have been maintained for long evolutionary times under balancing selection. Just like in other gynodioecious species (Bergero et al. 2019) the reproduction biology of rye offers advantages to understand the maintenance of females. The Charlesworth (2003) review pointed out that SI prevents inbreeding and, thus, ensures a balanced sequence diversity in hermaphroditic as compared to gynodioecious populations. The characterisation of the two major male-sterility systems in rye (CMS-P and CMS-V) revealed nucleo-cytoplasmic genetic control paving the way for an approach similar to the study conducted by Couvet et al. (1998) to analyse evolutionary advantages of the different types of CMS and factors contributing to their maintenance within populations over a long-time scale. Finally, the relative reproductive fitness of females and hermaphrodites should not depend on pollinator service, as rye is wind-pollinated like many wild species (Bergero et al. 2019). Taken together, the reference genome sequences trigger fundamental research for a complete understanding of gynodioecious systems in rye that will have a strong impact on the application in practical rye improvement programs. Future research should include in depth studies aimed at sequencing, assembly and comparison of mitochondrial genomes from fertile and sterile cytoplasms to identify potential candidates for CMS-associated genes in the CMS-P and CMS-V cytoplasms.

10.8 Rye as a Source of Fertility Restorer Genes for CMS-Based Hybrid Breeding Systems in Cereals

Hybrid breeding in rye is an example of agronomic success story (Laidig et al. 2017). The first hybrid rye varieties were released in 1984 and rapidly adopted in Germany and other European countries (Laidig et al. 2017) (for more details see Chap. 2). Hybrid breeding requires a way to block self-pollination of plants and technology that enables crosses between selected lines to exploit heterosis. CMS is often used in hybrid breeding programs as it ensures complete sterility of female (seed parent) plants. During crosses, *Rf* gene(s) are delivered with the male (pollen parent) plants to switch off male-sterility causing CMS genes present in the cytoplasm of the seed parent plants in hybrids. The availability of strong restorer (pollen parent) lines was a crucial factor that determined the success of hybrid breeding programs based on CMS/*Rf* systems in such important cereals like rice or maize (Chen and Liu 2014) but the lack of restorer lines hampers progress in others like wheat (Whitford et al. 2013). To date, the majority of identified *Rf* genes in crop plants belong to a family of RNA-binding factors known as pentatricopeptide repeat (PPR) proteins (Chen and Liu 2014; Kim and Zhang 2018). Recent studies pointed toward members of the mitochondrial transcription termination factor (mTERF) family to be also involved in fertility restoration in cereals (Wilde et al. 2017; Bernhard et al. 2019).

The PPR proteins are characterised by the presence of tandem repeats of variable 31–36 amino acid motifs and based on motif structure are classified into P and PLS classes (Lurin et al. 2004; Cheng et al. 2016). PLS class PPRs act mostly as RNA editing factors (Kotera et al. 2005; Barkan and Small 2014). The P class PPR proteins are involved in intron-splicing, RNA cleavage and stabilisation (Barkan and Small 2014). The mTERF family was named after its founding member in mammals, the MTERF1 protein which was found to mediate the mitochondrial transcription termination (Kruse et al. 1989). Similar to PPRs, mTERF proteins contain tandem arrays of 35–40 amino acids that fold into two or three α-helices (Hammani et al. 2014). To date, the characterised mTERF factors were found to be essential for plant development and stress responses (Kleine and Leister 2015). The maize Zm-mTERF4 protein was found to be required for the accumulation of plastid ribosomes and for the splicing of several group II introns in chloroplasts (Hammani and Barkan 2014). The *Arabidopsis* MDA1 protein was found to associate with components of the plastid-encoded RNA polymerase and transcriptional active chromosome complexes and to play a dual function in transcription and stabilisation of specific chloroplast transcripts within the *psbE* and *ndhH* operons (Meteignier et al. 2020).

Both PPR and mTERF families are present only in eukaryotes and both families have undergone large expansions in the land plant genomes (Lurin et al. 2004; Kleine 2012; Zhao et al. 2014). On average diploid angiosperm genomes encode ~500 PPR proteins with ~1700 PPR genes identified in the genome of hexaploid wheat (Fujii et al. 2011; The International Wheat Genome Sequencing Consortium 2018). In the 'Lo7' reference genome, 591 PPR genes encoding 278 PLS class and 313 P class proteins were found (Rabanus-Wallace et al. 2021). The number of mTERFs encoded in the plant genomes is an order of magnitude lower, e.g. 35 members in *A. thaliana* and 60 reported in *Malus domestica* (Linder et al. 2005; Kleine 2012) but ~400 in the bread wheat genome (Walkowiak et al. 2020). Apart from *Rfm1* in barley that was described to encode a PLS class PPR protein (Rizzolatti et al. 2017), the majority of cloned PPR-type restorer genes encode proteins from the P class that form a clade known as *Restorer-of-fertility like* (RFLs) (Chen and Liu 2014). Roughly 10% of all P class PPR genes belong to the RFL clade (Fujii et al. 2011; Melonek et al. 2019). A recent study in wheat has reported that a group of members within the mTERF family shares similar characteristics with the RFL clade of PPRs (RFL-PPRs) and by analogy were named RFL-mTERFs (Walkowiak et al. 2020). Both RFL-PPR and RFL-mTERF genes share a high level of sequence similarity among members of each family, intron-less gene models and location in clusters at two to three genomic positions (Walkowiak et al. 2020). In rye, within the P class sequences, 41 genes were identified as RFL-type almost double the 26 RFL-PPRs identified in the barley *H. vulgare* cv. 'Morex' genome (Melonek et al. 2019; Rabanus-Wallace et al. 2021). These RFL-PPR genes are organised into clusters located on rye chromosomes 1R, 2R and 4R (Fig. 10.4). The family of the mTERF factors in the rye genome was predicted at 131 genes of which 109 were found to be an RFL-type (Rabanus-Wallace et al. 2021) (Fig. 10.4). The largest cluster of 50 RFL-mTERF genes is located on the distal end of the long arm of chromosome 4R and encompasses a region of 78 Mbp (Fig. 10.4). The density of RFL-PPR and RFL-mTERF members is higher than that of the remaining PPR or mTERF genes (Fig. 10.4) similarly as was observed for the wheat RFL-PPRs (The International Wheat Genome Sequencing Consortium 2018). The RFL-PPR and RFL-mTERF clusters are located at the ends of chromosomes (Fig. 10.4) where the rates of recombination are the highest (Lukaszewski and Curtis 1992). Most likely such locations drive the creation of new gene variants that will be able to suppresses new *orfs* created during recombination events in the mitochondrial genomes.

The RFL-PPR and RFL-mTERF clusters on chromosomes 4R overlap with each other (Fig. 10.4) and with the genomic regions mapped as carrying *Rfp1*, *Rfp2* and *Rfp3* (Stracke et al. 2003; Hackauf et al. 2012, 2017) as well as *Rfc1* (Stojałowski et al. 2011) restorer loci in rye. Likewise, a conserved linkage disequilibrium (LD) block in a breeding population using CMS-G as hybridization system refined the localization of *Rfg1* on the long arm of chromosome 4RL (Börner et al. 1998) to this sub-genomic region (Vendelbo et al. 2020). So far only the sequence of *Rfp1*, a major fertility restorer for the 'Pampa' cytoplasm (CMS-P), the pre-dominating cytoplasm type used in rye hybrid breeding, has been cloned and identified as an mTERF gene (Wilde et al. 2017) located within the cluster of RFL-mTERFs on rye chromosome 4R (Fig. 10.4). Integration of markers into the 'Lo7' reference genome sequence revealed that *Rfp1* and *Rfc1* (Milczarski et al. 2016) are closely linked but separate genes (Rabanus-Wallace et al. 2021). Based on the overlap of the mapped genomic interval carrying the *Rfc1* restorer with a cluster of RFL-PPR gene suggests that *Rfc1* most likely encodes a PPR-type rather than mTERF-type *Rf* protein (Rabanus-Wallace et al. 2021). The co-localization of functional *Rf* genes with a cluster of RFL-mTERFs and RFL-PPRs qualifies this sub-genomic segment as a 'hotspot of mito-nuclear interaction' in the rye genome and provides an excellent template for the development of functional markers (Andersen and Lubberstedt 2003) for *Rf* genes.

10.9 Expansion of the RFL-PPR and RFL-MTERF Families in Outcrossing Self-Incompatible Species

The total number of RFL-PPR genes identified in the rye genome (41) is almost double the 26 RFLs identified in barley (Melonek et al. 2019) or other diploid cereal species including rice with only 20 RFLs (Melonek et al. 2016). It may

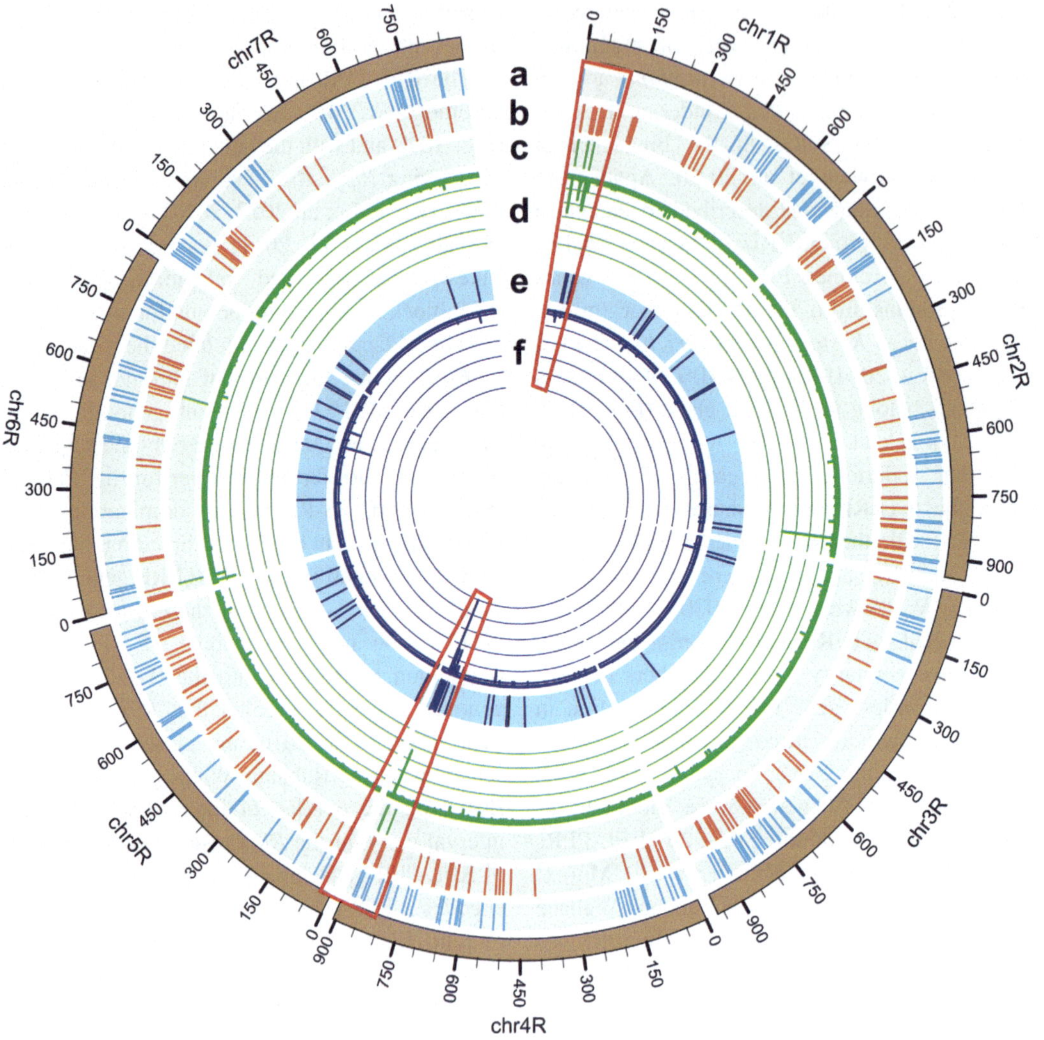

Fig. 10.4 Genome-wide distribution of PPR and mTERF genes in the rye genome. Distribution of genes encoding PLS **a**, P **b** and RFL **c** proteins from the PPR family as well as the mitochondrial transcription termination factor (mTERF) family **d**. Density of P class **e** and mTERF genes **f** along rye chromosomes was calculated for 2 Mbp windows. Red frames indicate the overlapping RFL-PPR and RFL-mTERF gene clusters on chromosomes 1R and 4R

appear that the genomes of self-incompatible outcrossing species could contain more PPR and mTERF gene variants compared to inbreeding species due to being genetically more diverse species compared to highly inbreeding species. Segmental duplications were proposed to be the major driver of RFL-PPR expansions observed in the genome of outcrossing *A. lyrata* compared to highly inbreeding *A. thaliana* (Gaborieau and Brown 2016). Retention of the new functional RFL-PPR and RFL-mTERF gene variants in the genomes of outcrossing species could suggest their biological importance, making the surplus of genes an attractive source of alleles for plant breeding. From an evolutionary perspective, the expanded number of RFL-PPR and RFL-mTERF genes in self-incompatible species may be highly advantageous to counter-balance reproductive barriers arising within a population from genomic conflicts between biparentally inherited nuclear genes and uniparentally inherited mitochondria. As selfish cytoplasmic elements use CMS as a means to enhance their propagation via meiotic drive (Ågren and Clark 2018), an outcrossing species are likely to encounter a more expansive selection of such elements accompanied by proliferation of compensatory nuclear alleles, restorer genes which repress the effects of the CMS genes, in a co-evolutionary "arms-race" (Touzet and Budar 2004).

10.10 Possible Applications of Rye Restorer Genes in Hybrid Breeding in Other Triticeae Species

The lack of strong restorer (pollinator) lines is currently one of the factors limiting the development of hybrid varieties on a commercial scale in wheat (Whitford et al. 2013). It has been reported that rye restorer genes are able to overcome sterility caused by CMS derived from *Aegilops kotschyi* or *T. timopheevii*, the two most commonly used cytoplasms in wheat hybrid breeding programs (Curtis and Lukaszewski 1993; Tsunewaki 2015). The *Rfc3* and *Rfc4* restorer genes located on rye chromosomes 6R and 4R, respectively, can restore *T. timopheevii* cytoplasm in wheat-rye addition lines. The *Rfc3* is stronger than *Rfc4*. These restorer genes, particularly *Rfc3*, have the potential to be applied in wheat hybrid breeding programs using *T. timopheevii* cytoplasm (Curtis and Lukaszewski 1993). In hexaploid triticale (X *Triticosecale* Wittmack), cytoplasms of *T. timopheevii* and *Aegilops sharonensis* were shown to induce different levels of male sterility, e.g. plants with *T. timopheevii* cytoplasm were almost always male-fertile. Thus, the inability of *T. timopheevii* cytoplasm to induce male sterility in triticale was due to restorer-of-fertility genes located in the rye genome. This restoration process remains under the control of several *Rf* genes (Góral et al. 2010; Stojałowski et al. 2013). These studies point to potentially untapped rye genetic diversity and underline the significance of the 'Lo7' reference genome sequence for hybrid breeding not only in rye but also in other Triticeae species like wheat or triticale.

An alternative hybridisation system that could be applied in wheat hybrid breeding is based on cytoplasms from *Ae. kotschyi*, *Aegilops uniaristata* or *Aegilops mutica* (Lukaszewski 2017). Cytogenetic analyses of 1RS.1BL rye translocations revealed the presence of Rf^{multi} (*Restoration of fertility in multiple CMS systems*) locus on the short arm of chromosome 1B in wheat that is able to readily restore all three cytoplasms while its counterpart on the short arm of rye chromosome 1R is not (Tsunewaki 2015; Hohn and Lukaszewski 2016). The interspecific comparison of the genomic region mapped as carrying the Rf^{multi} locus revealed the presence of a PPR-RFL cluster with an almost double number of genes in the bread wheat cv. 'Chinese Spring' reference genome (The International Wheat Genome Sequencing Consortium 2018) compared to the 'Lo7' rye genome (Fig. 10.5) (Rabanus-Wallace et al. 2021). In addition, sequence analysis revealed that only two RFL-PPR genes encoding full-length proteins (*TraesCS1B02G071642* and *TraesCS1B02G072900*) are located in the marker-delimited region in the wheat genome (Fig. 10.5). A highly similar copy of *TraesCS1B02G072900* gene (*SECCE1Rv1G0008410*) but not of *TraesCS1B02G071642* was found to be present in the rye genome, therefore the *TraesCS1B02G071642* gene was proposed as the most likely candidate for the Rf^{multi} gene in wheat (Rabanus-Wallace et al. 2021). Interestingly, a study of the 1RS.1BL introgression in wheat cv. 'Aikang58' (AK58) pointed to another gene in the region, *TraesCS1B02G072300,* as the most likely candidate for Rf^{multi} (Fig. 10.5) (Ru et al. 2020). However, sequence analysis of the *TraesCS1B02G072300* gene revealed that it encodes a C-terminally truncated and thus most likely non-functional RFL-PPR protein (Rabanus-Wallace et al. 2021) (Fig. 10.5).

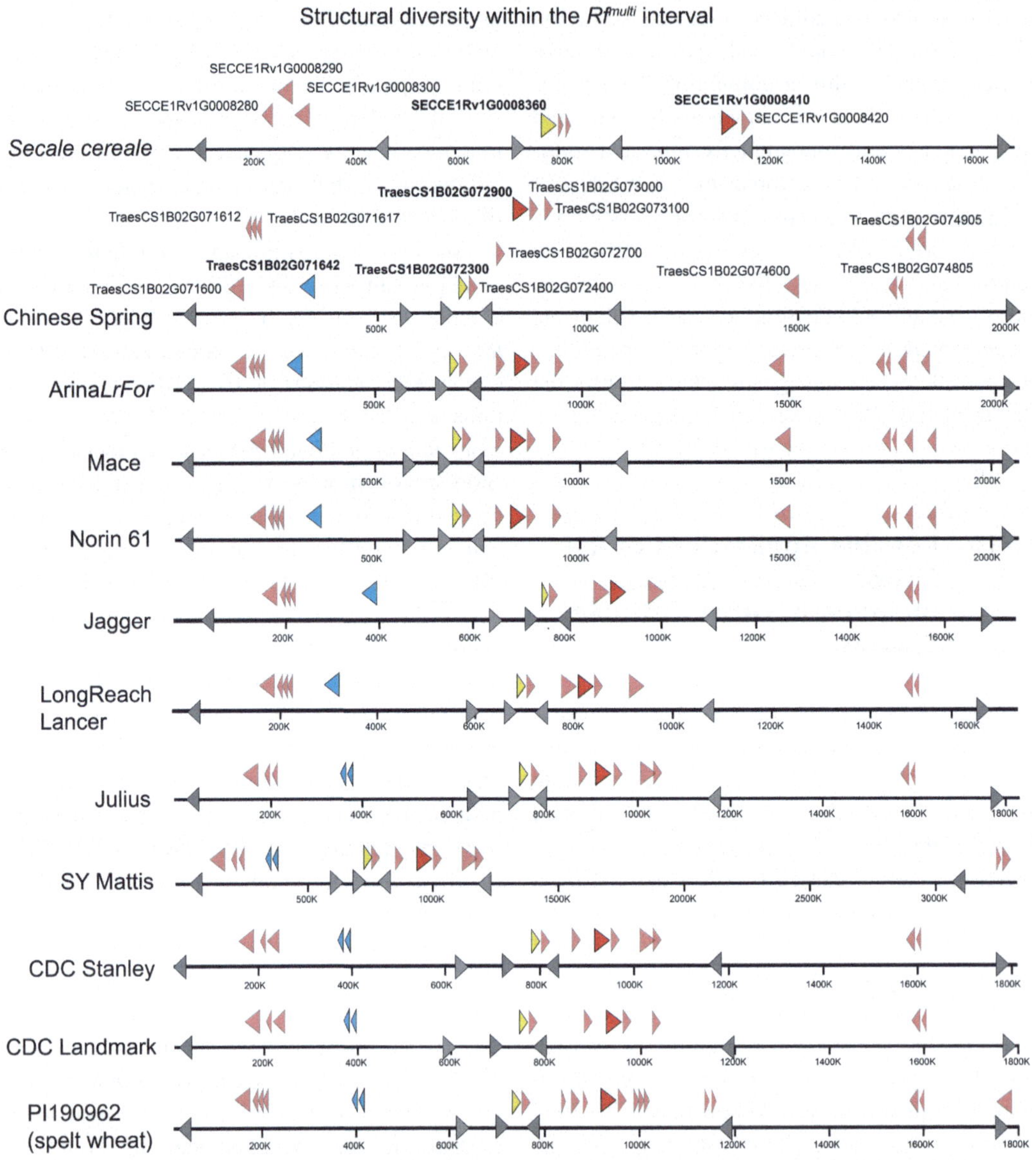

Fig. 10.5 Sequence conservation in the *Rf*^{multi} region between rye and 11 wheat accessions. RFL-PPR genes are shown as light pink triangles above the chromosome scale. In grey, conserved non-PPR genes used as syntenic anchors are shown. The *TraesCS1B02G071642* gene is marked in blue, the *TraesCS1B02G072300* gene in yellow and the *TraesCS1B02G072900* in red

Lukaszewski (2017) concluded that all wheat varieties that carry the 1BS chromosome arm transmit a functional copy of the *Rf*^{multi} gene to their progenies when used as pollinators. To analyse the sequence conservation among the *Rf*^{multi} gene candidates, we analysed the genomic regions carrying the *Rf*^{multi} locus in pseudomolecule level assemblies of 11 wheat varieties (Walkowiak et al. 2020) (Fig. 10.5). Interestingly, the comparison shows that only six wheat cultivars carry a full copy of the *TraesCS1B02G071642* gene while the remaining five varieties carry a truncated version of the gene (Fig. 10.5). For example, none of the

German (Julius) or Canadian (CDC Stanley, CDC Landmark) wheat cultivars carry a functional version of the *TraesCS1B02G071642* gene (Fig. 10.5). In addition, Tsunewaki (2015) noted that the recessive non-restoring allele (*rf* ^{multi}) would be limited to wheat Spelta (*T. spelta var. duhamelianum*). Indeed, in agreement with that, the *TraesCS1B02G071642* gene is disrupted in *T. aestivum spelta* (Spelt) (Fig. 10.5). These results indicate that maybe not all wheat accessions carry a functional *Rf*^{multi} allele or that *TraesCS1B02G071642* is not the restorer gene. On the other hand, even though a highly similar copy of the gene *TraesCS1B02G072900* is also present in the rye genome (*SECCE1Rv1G0008410)*, its expression may be differently regulated in rye compared to wheat and thus it can be functional in wheat but non-functional in rye. Future studies will validate which of the proposed candidates is indeed *Rf*^{multi}, as the 1RS translocation from rye could be a promising alternative for the development of a genetic hybridization system in wheat in addition to the highly complex *T. timopheevii* system. In the future, the widely available wheat germplasm carrying the 1RS translocation (Jung and Seo 2014; Crespo-Herrera et al. 2017; Schlegel 2020) could be applied to the development of maintainer genotypes in the seed parent pool.

The comparison of the rye and wheat genome sequences will provide a perfect basis for precise identification of molecular markers that will allow to break the genetic linkage between *Rf*^{multi} and loci with negative effects on the quality of wheat grain and flour baking quality (von Bothmer et al. 1995; Lukaszewski 2000) also located on the short arm of chromosome 1R. Although further analyses are needed to confirm the ability of *TraesCS1B02G071642* to restore fertility of CMS wheat plants, this is only another example that points to so far mostly unexploited potential of rye genetic material and underlines the significance of the rye reference genomes (Li et al. 2021; Rabanus-Wallace et al. 2021) for hybrid breeding in rye and closely related species in the Triticeae tribe.

10.11 Conclusion

At a time of unprecedented human population expansion and biodiversity loss, research on plant reproduction, with its potential to help increase crop yields and deliver food security, has never been more important. The validity of this statement given a decade ago by Simon Hiscock (Hiscock 2011) has not expired. Decades of scientific plant breeding and research were necessary to gain fundamental knowledge of the genetic basis of the unique reproductive biology of rye in order to keep this healthy minor cereal competitive in modern agricultural production systems. The unpretentious rye enabled already the post-Roman societies not only to survive but to thrive in Europe and shaped them in a way that pioneered technological developments on an unprecedented scale in world history (Mitterauer 2010). In view of the current international efforts at developing solutions to combat global climate change, rye again offers viable options to help alleviate the hunger of a steadily increasing population. As outlined in this chapter, the novel rye reference genome sequences close a gap in the availability of genomic resources for small grain cereals as important staples for human beings and offer promising perspectives to further advance our understanding of the unique reproduction biology of rye.

References

Ågren JA, Clark AG (2018) Selfish genetic elements. PLoS Genet 14:e1007700. https://doi.org/10.1371/journal.pgen.1007700

Ahokas H (1980) Evidence for cytoplasmic male-sterility in two finnish rye cultivars. Hereditas 92:373–375

Akagi H, Sakamoto M, Shinjyo C, Shimada H, Fujimura T (1994) A unique sequence located downstream from the rice mitochondrial *atp6* may cause male-sterility. Curr Genet 25:52–58. https://doi.org/10.1007/Bf00712968

Alfares W, Bouguennec A, Balfourier F, Gay G, Berges H, Vautrin S, Sourdille P, Bernard M, Feuillet C (2009) Fine mapping and marker development for the crossability gene SKr on chromosome 5BS of hexaploid wheat (*Triticum aestivum* L.). Genetics

183:469–481. https://doi.org/10.1534/genetics.109.107706

Andersen JR, Lubberstedt T (2003) Functional markers in plants. Trends Plant Sci 8:554–560. https://doi.org/10.1016/j.tplants.2003.09.010

Bardwell VJ, Treisman R (1994) The POZ domain: a conserved protein-protein interaction motif. Genes Dev 8:1664–1677. https://doi.org/10.1101/gad.8.14.1664

Barkan A, Small I (2014) Pentatricopeptide repeat proteins in plants. Annu Rev Plant Biol 65:415–442. https://doi.org/10.1146/annurev-arplant-050213-040159

Bauer E, Schmutzer T, Barilar I, Mascher M, Gundlach H, Martis MM, Twardziok SO, Hackauf B, Gordillo A, Wilde P, Schmidt M, Korzun V, Mayer KFX, Schmid K, Schon CC, Scholz U (2017) Towards a whole-genome sequence for rye (*Secale cereale* L.). Plant J 89:853–869. https://doi.org/10.1111/tpj.13436

Baum M, Appels R (1991) The cytogenetic and molecular architecture of chromosome 1R - one of the most widely utilized sources of alien chromatin in wheat varieties. Chromosoma 101:1–10. https://doi.org/10.1007/Bf00360680

Bergero R, Levsen N, Wolff K, Charlesworth D (2019) Arms races with mitochondrial genome soft sweeps in a gynodioecious plant, *Plantago lanceolata*. Mol Ecol 28:2772–2785. https://doi.org/10.1111/mec.15121

Bernhard T, Koch M, Snowdon RJ, Friedt W, Wittkop B (2019) Undesired fertility restoration in *msm1* barley associates with two mTERF genes. Theor Appl Genet 132:1335–1350. https://doi.org/10.1007/s00122-019-03281-9

Bertin I, Fish L, Foote TN, Knight E, Snape J, Moore G (2009) Development of consistently crossable wheat genotypes for alien wheat gene transfer through fine-mapping of the *Kr1* locus. Theor Appl Genet 119:1371–1381. https://doi.org/10.1007/s00122-009-1141-z

Bian XY, Friedrich A, Bai JR, Baumann U, Hayman DL, Barker SJ, Langridge P (2004) High-resolution mapping of the *S* and *Z* loci of *Phalaris coerulescens*. Genome 47:918–930. https://doi.org/10.1139/G04-017

Bomblies K, Lempe J, Epple P, Warthmann N, Lanz C, Dangl JL, Weigel D (2007) Autoimmune response as a mechanism for a Dobzhansky-Muller-type incompatibility syndrome in plants. Plos Biol 5:1962–1972. https://doi.org/10.1371/journal.pbio.0050236

Bomblies K, Weigel D (2007) Hybrid necrosis: autoimmunity as a potential gene-flow barrier in plant species. Nat Rev Genet 8:382–393. https://doi.org/10.1038/nrg2082

Bonhomme S, Budar F, Ferault M, Pelletier G (1991) A 2.5 kb *Nco*I fragment of Ogura radish mitochondrial DNA is correlated with cytoplasmic male-sterility in *Brassica* cybrids. Curr Genet 19:121–127. https://doi.org/10.1007/Bf00326293

Börner A, Korzun V, Ben Amer IM, Stracke S, Fürste B (1996) RFLP mapping of genes controlling dwarfness and crossability with rye. Ann Wheat Newsletter 42:80

Börner A, Korzun V, Polley A, Malyshev S, Melz G (1998) Genetics and molecular mapping of a male fertility restoration locus (*Rfg1*) in rye (*Secale cereale* L.). Theor Appl Genet 97:99–102. https://doi.org/10.1007/s001220050871

Bosch M, Franklin-Tong VE (2008) Self-incompatibility in Papaver: signalling to trigger PCD in incompatible pollen. J Exp Bot 59:481–490. https://doi.org/10.1093/jxb/erm195

Bouguennec A, Lesage VS, Gateau I, Sourdille P, Jahier J, Lonnet P (2018) Transfer of recessive *skr* crossability trait into well-adapted French wheat cultivar Barok through marker-assisted backcrossing method. Cereal Res Commun 46:604–615. https://doi.org/10.1556/0806.46.2018.043

Budar F, Pelletier G (2001) Male sterility in plants: occurrence, determinism, significance and use. C R Acad Sci III 324:543–550. https://doi.org/10.1016/S0764-4469(01)01324-5

Charlesworth D (2002) What maintains male-sterility factors in plant populations? Heredity 89:408–409. https://doi.org/10.1038/sj.hdy.6800193

Charlesworth D (2003) Effects of inbreeding on the genetic diversity of populations. Philos Trans R Soc Lond B Biol Sci 358:1051–1070. https://doi.org/10.1098/rstb.2003.1296

Chase CD (2007) Cytoplasmic male sterility: a window to the world of plant mitochondrial-nuclear interactions. Trends Genet 23:81–90. https://doi.org/10.1016/j.tig.2006.12.004

Chen LT, Liu YG (2014) Male sterility and fertility restoration in crops. Annu Rev Plant Biol 65:579–606. https://doi.org/10.1146/annurev-arplant-050213-040119

Chen ZW, Zhao N, Li SS, Grover CE, Nie HS, Wendel JF, Hua JP (2017) Plant mitochondrial genome evolution and cytoplasmic male sterility. Crit Rev Plant Sci 36:55–69. https://doi.org/10.1080/07352689.2017.1327762

Cheng SF, Gutmann B, Zhong X, Ye YT, Fisher MF, Bai FQ, Castleden I, Song Y, Song B, Huang JY, Liu X, Xu X, Lim BL, Bond CS, Yiu SM, Small I (2016) Redefining the structural motifs that determine RNA binding and RNA editing by pentatricopeptide repeat proteins in land plants. Plant J 85:532–547. https://doi.org/10.1111/tpj.13121

Coulton A, Burridge AJ, Edwards KJ (2020) Examining the effects of temperature on recombination in wheat. Front Plant Sci 11:230. https://doi.org/10.3389/fpls.2020.00230

Couvet D, Ronce O, Gliddon C (1998) The maintenance of nucleocytoplasmic polymorphism in a metapopulation: the case of gynodioecy. Am Nat 152:59–70. https://doi.org/10.1086/286149

Crespo-Herrera LA, Garkava-Gustavsson L, Ahman I (2017) A systematic review of rye (*Secale cereale* L.) as a source of resistance to pathogens and pests in wheat (*Triticum aestivum* L.). Hereditas 154:1–9. https://doi.org/10.1186/s41065-017-0033-5

Curtis CA, Lukaszewski AJ (1993) Localization of genes in rye that restore male-fertility to hexaploid wheat with *timopheevi* cytoplasm. Plant Breeding 111:106-112. https://doi.org/10.1111/j.1439-0523.1993.tb00615.x

Darwin CR (1877) The different forms of flowers on plants of the same species. John Murray, London

De Vries AP (1971) Flowering biology of wheat, particularly in view of hybrid seed production—a review. Euphytica 20:152–170

Derbyshire P, Byrne ME (2013) *MORE SPIKELETS1* is required for spikelet fate in the inflorescence of *Brachypodium*. Plant Physiol 161:1291–1302. https://doi.org/10.1104/pp.112.212340

Dill P (1983) Zur züchterischen Verbesserung der Kornmasse bei Winterroggen (*Secale cereale* L.). Arch Züchtungsforsch 13:157–168

Dohmen G, Hessberg H, Geiger HH, Tudzynski P (1994) CMS in rye: comparative RFLP and transcript analyses of mitochondria from fertile and male-sterile plants. Theor Appl Genet 89:1014–1018. https://doi.org/10.1007/BF00224532

Döring TF, Knapp S, Kovacs G, Murphy K, Wolfe MS (2011) Evolutionary plant breeding in cereals-into a new era. Sustainability 3:1944–1971. https://doi.org/10.3390/su3101944

Dubiel W, Dubiel D, Wolf DA, Naumann M (2018) Cullin 3-based ubiquitin ligases as master regulators of mammalian cell differentiation. Trends Biochem Sci 43:95–107. https://doi.org/10.1016/j.tibs.2017.11.010

Fischer S, Melchinger AE, Korzun V, Wilde P, Schmiedchen B, Mohring J, Piepho HP, Dhillon BS, Wurschum T, Reif JC (2010) Molecular marker assisted broadening of the Central European heterotic groups in rye with Eastern European germplasm. Theor Appl Genet 120:291–299. https://doi.org/10.1007/s00122-009-1124-0

Foote HCC, Ride JP, Franklintong VE, Walker EA, Lawrence MJ, Franklin FCH (1994) Cloning and expression of a distinctive class of self-incompatibility (*S*) gene from *Papaver rhoeas* L. Proc Natl Acad Sci USA 91:2265–2269. https://doi.org/10.1073/pnas.91.6.2265

Fujii S, Bond CS, Small ID (2011) Selection patterns on restorer-like genes reveal a conflict between nuclear and mitochondrial genomes throughout angiosperm evolution. P Natl Acad Sci USA 108:1723–1728. https://doi.org/10.1073/pnas.1007667108

Fuong FT, Voylokov AV, Smirnov VG (1993) Genetic studies of self-fertility in rye (Secale cereale L.). 2. The search for isozyme marker genes linked to self-incompatibility loci. Theor Appl Genet 87:619–623. https://doi.org/10.1007/BF00221888

Gaborieau L, Brown GG (2016) Comparative genomic analysis of the compound *Brassica napus Rf* locus. BMC Genomics 17:834. https://doi.org/10.1186/s12864-016-3117-0

Galtier N (2011) The intriguing evolutionary dynamics of plant mitochondrial DNA. BMC Biol 9:61. https://doi.org/10.1186/1741-7007-9-61

Geiger HH, Miedaner T (2009) Rye breeding. In: Carena MJ (ed) Cereals. Handbook of plant breeding, vol 3. Springer, New York, pp 157–181

Geiger HH, Schnell FW (1970) Breeding rye varieties from inbred lines. 1. Selfing-proportions in polycross progenies. Theor Appl Genet 40:305–311. https://doi.org/10.1007/Bf00264572

Geiger HH, Yuan Y, Miedaner T, Wilde P (1995) Environmental sensitivity of cytoplasmic genic male sterility (CMS) in *Secale cereale* L. In: Kück U, Wricke G (eds) Genetic mechanisms for hybrid breeding. Adv Plant Breed, vol 18, pp 7–17

Gertz A, Wricke G (1989) Linkage between the incompatibility locus-*Z* and a beta-glucosidase locus in rye. Plant Breed 102:255–259. https://doi.org/10.1111/j.1439-0523.1989.tb00344.x

Gertz A, Wricke G (1991) Inheritance of temperature-induced pseudocompatibility in rye. Plant Breed 107:89–96. https://doi.org/10.1111/j.1439-0523.1991.tb00535.x

Gingerich DJ, Hanada K, Shiu SH, Vierstra RD (2007) Large-scale, lineage-specific expansion of a bric-a-brac/tramtrack/broad complex ubiquitin-ligase gene family in rice. Plant Cell 19:2329–2348. https://doi.org/10.1105/tpc.107.051300

Góral H, Stojałowski S, Tyrka M, Wędzony M (2010) Inheritance of fertility restoration in winter triticale with cytoplasm of *Triticum timopheevi*. Folia Pomer Univ Technol Stetin 276:11–18

Hackauf B, Bauer E, Korzun V, Miedaner T (2017) Fine mapping of the restorer gene *Rfp3* from an Iranian primitive rye (*Secale cereale* L.). Theor Appl Genet 130:1179–1189. https://doi.org/10.1007/s00122-017-2879-3

Hackauf B, Korzun V, Wortmann H, Wilde P, Wehling P (2012) Development of conserved ortholog set markers linked to the restorer gene *Rfp1* in rye. Mol Breeding 30:1507–1518. https://doi.org/10.1007/s11032-012-9736-5

Hackauf B, Wehling P (2002) Identification of microsatellite polymorphisms in an expressed portion of the rye genome. Plant Breed 121:17–25. https://doi.org/10.1046/j.1439-0523.2002.00649.x

Hackauf B, Wehling P (2005) Approaching the self-incompatibility locus *Z* in rye (*Secale cereale* L.) via comparative genetics. Theor Appl Genet 110:832–845. https://doi.org/10.1007/s00122-004-1869-4

Hagenblad J, Oliveira HR, Forsberg NEG, Leino MW (2016) Geographical distribution of genetic diversity in *Secale* landrace and wild accessions. BMC Plant Biol 16:23. https://doi.org/10.1186/s12870-016-0710-y

Hammani K, Barkan A (2014) An mTERF domain protein functions in group II intron splicing in maize chloroplasts. Nucleic Acids Res 42:5033–5042. https://doi.org/10.1093/nar/gku112

Hammani K, Bonnard G, Bouchoucha A, Gobert A, Pinker F, Salinas T, Giege P (2014) Helical repeats modular proteins are major players for organelle gene expression. Biochimie 100:141–150. https://doi.org/10.1016/j.biochi.2013.08.031

Hanson MR, Bentolila S (2004) Interactions of mitochondrial and nuclear genes that affect male gametophyte development. Plant Cell 16:S154–S169. https://doi.org/10.1105/tpc.015966

Hao M, Zhang LQ, Ning SZ, Huang L, Yuan ZW, Wu BH, Yan ZH, Dai SF, Jiang B, Zheng YL, Liu DC (2020) The resurgence of introgression breeding, as exemplified in wheat improvement. Front Plant Sci 11:252. https://doi.org/10.3389/fpls.2020.00252

Heribert Nilsson N (1916) Populationsanalysen und Erblichkeitsversuche über die Selbststerilität, Selbstfertilität und Sterilität bei dem Roggen. Z Plflanzenzüchtg 4:1–44

Heslop-Harrison J (1979) Interpretation of the hydrodynamics of pollen. Am J Bot 66:737–743. https://doi.org/10.2307/2442418

Hisano H, Tsujimura M, Yoshida H, Terachi T, Sato K (2016) Mitochondrial genome sequences from wild and cultivated barley (*Hordeum vulgare*). BMC Genomics 17:824. https://doi.org/10.1186/s12864-016-3159-3

Hiscock SJ (2011) Preface: Sexual plant reproduction. Ann Bot 108:585–587. https://doi.org/10.1093/aob/mcr217

Hohn CE, Lukaszewski AJ (2016) Engineering the 1BS chromosome arm in wheat to remove the Rf^{multi} locus restoring male fertility in cytoplasms of *Aegilops kotschyi*, *Ae. uniaristata* and *Ae. mutica*. Theor Appl Genet 129:1769–1774. https://doi.org/10.1007/s00122-016-2738-7

Honda I, Turuspekov Y, Komatsuda T, Watanabe Y (2005) Morphological and physiological analysis of cleistogamy in barley (*Hordeum vulgare*). Physiol Plant 124:524–531. https://doi.org/10.1111/j.1399-3054.2005.00541.x

Hoseinzadeh P, Zhou R, Mascher M, Himmelbach A, Niks RE, Schweizer P, Stein N (2019) High resolution genetic and physical mapping of a major powdery mildew resistance locus in barley. Front Plant Sci 10:146. https://doi.org/10.3389/fpls.2019.00146

Howell T, Hale I, Jankuloski L, Bonafede M, Gilbert M, Dubcovsky J (2014) Mapping a region within the 1RS.1BL translocation in common wheat affecting grain yield and canopy water status. Theor Appl Genet 127:2695–2709. https://doi.org/10.1007/s00122-014-2408-6

Howell T, Moriconi JI, Zhao X, Hegarty J, Fahima T, Santa-Maria GE, Dubcovsky J (2019) A wheat/rye polymorphism affects seminal root length and yield across different irrigation regimes. J Exp Bot 70:4027–4037. https://doi.org/10.1093/jxb/erz169

Huang S, Chen X, Zhong X, Li M, Ao K, Huang J, Li X (2016) Plant TRAF proteins regulate NLR immune receptor turnover. Cell Host Microbe 20:271. https://doi.org/10.1016/j.chom.2016.07.005

Isik Z, Parmaksiz I, Coruh C, Geylan-Su YS, Cebeci O, Beecher B, Budak H (2007) Organellar genome analysis of rye (*Secale cereale*) representing diverse geographic regions. Genome 50:724–734. https://doi.org/10.1139/G07-052

The International Wheat Genome Sequencing Consortium (2018) Shifting the limits in wheat research and breeding using a fully annotated reference genome. Science 361:eaar7191. https://doi.org/10.1126/science.aar7191

Jaganathan D, Bohra A, Thudi M, Varshney RK (2020) Fine mapping and gene cloning in the post-NGS era: advances and prospects. Theor Appl Genet 133:1791–1810. https://doi.org/10.1007/s00122-020-03560-w

Jalani BS, Moss JP (1980) The site of action of the crossability genes (*Kr1*, *Kr2*) between Triticum and Secale. 1. Pollen germination, pollen-tube growth, and number of pollen tubes. Euphytica 29:571–579. https://doi.org/10.1007/Bf00023204

Jalani BS, Moss JP (1981) The site of action of crossability genes (*Kr1*, *Kr2*) between Triticum and Secale. 2. Proportion of pistils containing pollen tubes and effects of alternate pollination on seed set. Euphytica 30:105–112. https://doi.org/10.1007/Bf00033665

James GV, Patel V, Nordstrom KJV, Klasen JR, Salome PA, Weigel D, Schneeberger K (2013) User guide for mapping-by-sequencing in *Arabidopsis*. Genome Biol 14:R61. https://doi.org/10.1186/gb-2013-14-6-r61

Jany E, Nelles H, Goring DR (2019) The molecular and cellular regulation of *Brassicaceae* self-incompatibility and self-pollen rejection. Int Rev Cel Mol Bio 343:1–35. https://doi.org/10.1016/bs.ircmb.2018.05.011

Jeuken MJW, Zhang NW, McHale LK, Pelgrom K, den Boer E, Lindhout P, Michelmore RW, Visser RGF, Niks RE (2009) *Rin4* causes hybrid necrosis and race-specificresistance in an interspecific lettuce hybrid. Plant Cell 21:3368–3378. https://doi.org/10.1105/tpc.109.070334

Jung WJ, Seo YW (2014) Employment of wheat-rye translocation in wheat improvement and broadening its genetic basis. J Crop Sci Biotechnol 17:305–313

Juranić M, Dresselhaus T (2014) Phylogenetic analysis of the expansion of the MATH-BTB gene family in the grasses. Plant Signal Behav 9:e28242. https://doi.org/10.4161/psb.28242

Juranić M, Srilunchang KO, Krohn NG, Leljak-Levanic D, Sprunck S, Dresselhaus T (2012) Germline-specific MATH-BTB substrate adaptor MAB1 regulates spindle length and nuclei identity in maize. Plant Cell 24:4974–4991. https://doi.org/10.1105/tpc.112.107169

Kadowaki K, Suzuki T, Kazama S (1990) A chimeric gene containing the 5' portion of *atp6* is associated with cytoplasmic male-sterility of rice. Mol Gen Genet 224:10–16

Kakeda K, Ibuki T, Suzuki J, Tadano H, Kurita Y, Hanai Y, Kowyama Y (2008) Molecular and genetic characterization of the *S* locus in *Hordeum bulbosum* L., a wild self-incompatible species related to cultivated barley. Mol Genet Genom 280:509–519. https://doi.org/10.1007/s00438-008-0383-9

Kazama T, Toriyama K (2003) A pentatricopeptide repeat-containing gene that promotes the processing of aberrant *atp6* RNA of cytoplasmic male-sterile rice. FEBS Lett 544:99–102. https://doi.org/10.1016/S0014-5793(03)00480-0

Kim YJ, Zhang DB (2018) Molecular control of male fertility for crop hybrid breeding. Trends Plant Sci 23:53–65. https://doi.org/10.1016/j.tplants.2017.10.001

Kleine T (2012) *Arabidopsis thaliana* mTERF proteins: evolution and functional classification. Front Plant Sci 3:233. https://doi.org/10.3389/fpls.2012.00233

Kleine T, Leister D (2015) Emerging functions of mammalian and plant mTERFs. Biochim Biophys Acta 1847:786–797. https://doi.org/10.1016/j.bbabio.2014.12.009

Kobyljanskij VD, Katerova AG (1973) Inheritance of cytoplasmatic and nuclear male sterility in Oshima diploid rye. Genetika 6:5–11

Kotera E, Tasaka M, Shikanai T (2005) A pentatricopeptide repeat protein is essential for RNA editing in chloroplasts. Nature 433:326–330. https://doi.org/10.1038/nature03229

Krolow KD (1970) Investigations on compatibility between wheat and rye. Z Pflanzenzuchtung 64:44–72

Kruse B, Narasimhan N, Attardi G (1989) Termination of transcription in human mitochondria—identification and purification of a DNA-binding protein factor that promotes termination. Cell 58:391–397. https://doi.org/10.1016/0092-8674(89)90853-2

Kubo K, Entani T, Takara A, Wang N, Fields AM, Hua ZH, Toyoda M, Kawashima S, Ando T, Isogai A, Kao T, Takayama S (2010) Collaborative non-self recognition system in S-RNase-based self-incompatibility. Science 330:796–799. https://doi.org/10.1126/science.1195243

Kubo K, Paape T, Hatakeyama M, Entani T, Takara A, Kajihara K, Tsukahara M, Shimizu-Inatsugi R, Shimizu KK, Takayama S (2015) Gene duplication and genetic exchange drive the evolution of S-RNase-based self-incompatibility in Petunia. Nat Plants 1:14005. https://doi.org/10.1038/Nplants.2014.5

Laidig F, Piepho HP, Rentel D, Drobek T, Meyer U, Huesken A (2017) Breeding progress, variation, and correlation of grain and quality traits in winter rye hybrid and population varieties and national on-farm progress in Germany over 26 years. Theor Appl Genet 130:981–998. https://doi.org/10.1007/s00122-017-2865-9

Lamkey KR, Edwards JW (1999) Quantitative genetics of heterosis. In: Coors JG, Pandey S (eds) The genetics and exploitation of heterosis in crops. American Society of Agronomy, Crop Science Society of America, and Soil Science Society of America, Madison, WI, pp 31–48

Lamoureux D, Boeuf C, Regad F, Garsmeur O, Charmet G, Sourdille P, Lagoda P, Bernard M (2002) Comparative mapping of the wheat 5B short chromosome arm distal region with rice, relative to a crossability locus. Theor Appl Genet 105:759–765. https://doi.org/10.1007/s00122-002-0916-2

Lange W, Wojciechowska B (1976) Crossing of common wheat (*Triticum aestivum* L.) with cultivated rye (*Secale cereale* L.). 1. Crossability, pollen grain germination and pollen-tube growth. Euphytica 25:609–620. https://doi.org/10.1007/Bf00041598

Langridge P, Baumann U (2008) Self-Incompatibility in the Grasses. In: Self-Incompatibility in Flowering Plants. Springer, Berlin, Heidelberg. https://doi.org/10.1007/978-3-540-68486-2_13

Łapiński M (1972) Cytoplasmic-genic type of male sterility in *Secale montanum* Guss. Wheat Inform Serv 35:25–28

Łapiński M, Stojałowski S (2003) Occurrence and genetic identity of male sterility-inducing cytoplasm in rye (*Secale spp.*). Plant Breed Seed Sci 48:7–23

Lein A (1943) The genetical basis of the crossability between wheat and rye. Z Indukt Abstamm Vererbungsl 81:28–59

Li GW, Wang LJ, Yang JP, He H, Jin HB, Li XM, Ren TH, Ren ZL, Li F, Han X, Zhao XG, Dong LL, Li YW, Song ZP, Yan ZH, Zheng NN, Shi CL, Wang ZH, Yang SL, Xiong ZJ, Zhang ML, Sun GH, Zheng X, Gou MY, Ji CM, Du JK, Zheng HK, Dolezel J, Deng XW, Stein N, Yang QH, Zhang KP, Wang DW (2021) A high-quality genome assembly highlights rye genomic characteristics and agronomically important genes. Nat Genet 53:574–584. https://doi.org/10.1038/s41588-021-00808-z

Linder T, Park CB, Asin-Cayuela J, Pellegrini M, Larsson NG, Falkenberg M, Samuelsson T, Gustafsson CM (2005) A family of putative transcription termination factors shared amongst metazoans and plants. Curr Genet 48:265–269. https://doi.org/10.1007/s00294-005-0022-5

Liu HT, Cui P, Zhan KH, Lin Q, Zhuo GY, Guo XL, Ding F, Yang WL, Liu DC, Hu SN, Yu J, Zhang AM (2011) Comparative analysis of mitochondrial genomes between a wheat K-type cytoplasmic male sterility (CMS) line and its maintainer line. BMC Genom 12:163. https://doi.org/10.1186/1471-2164-12-163

Livneh I, Cohen-Kaplan V, Cohen-Rosenzweig C, Avni N, Ciechanover A (2016) The life cycle of the 26S proteasome: from birth, through regulation and function, and onto its death. Cell Res 26:869–885. https://doi.org/10.1038/cr.2016.86

Lukaszewski AJ (2000) Manipulation of the 1RS.1BL translocation in wheat by induced homoeologous recombination. Crop Sci 40:216–225. https://doi.org/10.2135/cropsci2000.401216x

Lukaszewski AJ (2017) Chromosomes 1BS and 1RS for control of male fertility in wheats and triticales with cytoplasms of *Aegilops kotschyi*, *Ae. mutica* and *Ae.*

uniaristata. Theor Appl Genet 130:2521–2526. https://doi.org/10.1007/s00122-017-2973-6

Lukaszewski AJ, Curtis CA (1992) Transfer of the *Glu-D1* Gene from chromosome 1D of bread wheat to chromosome 1R in hexaploid triticale. Plant Breed 109:203–210. https://doi.org/10.1111/j.1439-0523.1992.tb00174.x

Lundqvist A (1954) Studies on self-sterility in rye, *Secale cereale* L. Hereditas 40:278–294

Lundqvist A (1956) Self-incompatibility in rye. 1 Genetic Control in the Diploid. Hereditas 42:293–348. https://doi.org/10.1111/j.1601-5223.1956.tb03021.x

Lundqvist A (1960) The origin of self-compatibility in rye. Hereditas 46:1–19

Lurin C, Andres C, Aubourg S, Bellaoui M, Bitton F, Bruyere C, Caboche M, Debast C, Gualberto J, Hoffmann B, Lecharny A, Le Ret M, Martin-Magniette ML, Mireau H, Peeters N, Renou JP, Szurek B, Taconnat L, Small I (2004) Genome-wide analysis of Arabidopsis pentatricopeptide repeat proteins reveals their essential role in organelle biogenesis. Plant Cell 16:2089–2103. https://doi.org/10.1105/tpc.104.022236

Madej L (1975) Research on male sterility in rye. Hodowla Rosl Aklim Nasienn 19:421–422

Manzanares C, Barth S, Thorogood D, Byrne SL, Yates S, Czaban A, Asp T, Yang BC, Studer B (2016) A gene encoding a DUF247 domain protein cosegregates with the *S* self-incompatibility locus in perennial ryegrass. Mol Biol Evol 33:870–884. https://doi.org/10.1093/molbev/msv335

Marker RM, Geiger HH, Wilde P (1985) Influence of 'Pampa' cytoplasm on agronomic characters of F_2 crosses among inbred lines of winter rye. In: Proceedings Eucarpia meeting of the cereal section on rye. Svalöv, Sweden, pp 279–292

Martis MM, Zhou RN, Haseneyer G, Schmutzer T, Vrana J, Kubalakova M, Konig S, Kugler KG, Scholz U, Hackauf B, Korzun V, Schon CC, Dolezel J, Bauer E, Mayer KFX, Stein N (2013) Reticulate evolution of the rye genome. Plant Cell 25:3685–3698. https://doi.org/10.1105/tpc.113.114553

Mascher M, Jost M, Kuon JE, Himmelbach A, Assfalg A, Beier S, Scholz U, Graner A, Stein N (2014) Mapping-by-sequencing accelerates forward genetics in barley. Genome Biol 15:R78. https://doi.org/10.1186/gb-2014-15-6-r78

McCubbin AG, Kao TT (1996) Molecular mechanisms of self-incompatibility. Curr Opin Biotech 7:150–154. https://doi.org/10.1016/S0958-1669(96)80005-X

McElroy JS (2014) Vavilovian mimicry: Nikolai Vavilov and his little-known impact on weed science. Weed Sci 62:207–216. https://doi.org/10.1614/Ws-D-13-00122.1

Melonek J, Duarte J, Martin J, Beuf L, Murigneux A, Varenne P, Comadran J, Specel S, Levadoux S, Bernath-Levin K, Torney F, Pichon JP, Perez P, Small I (2021) The genetic basis of cytoplasmic male sterility and fertility restoration in wheat. Nat Commun 12:1036. https://doi.org/10.1038/s41467-021-21225-0

Melonek J, Stone JD, Small I (2016) Evolutionary plasticity of restorer-of-fertility-like proteins in rice. Sci Rep 6:35152. https://doi.org/10.1038/srep35152

Melonek J, Zhou R, Bayer PE, Edwards D, Stein N, Small I (2019) High intraspecific diversity of Restorer-of-fertility-like genes in barley. Plant J 97:281–295. https://doi.org/10.1111/tpj.14115

Melz G, Melz G, Hartman F (2003) Genetics of a male-sterile rye of 'G-type' with results of the first F1-hybrids. Plant Breed Seed Sci 2003:47–55

Melz G, Schlegel R (1987) Production and cytogenetical analysis of a rye-cytoplasmic tetraploid triticale. Plant Breed 98:200–204. https://doi.org/10.1111/j.1439-0523.1987.tb01117.x

Melz G, Thiele V (1989) Results of cytogenetic investigations in rye *(Secale cereale L.)*. Arch Züchtungsforsch 19:421–428

Mengersen FG (1951) Die Wirkung der Inzucht auf verschiedene Merkmale beim Roggen. Z Pflanzenzuchtung 30:218–248

Meteignier LV, Ghandour R, Meierhoff K, Zimmerman A, Chicher J, Baumberger N, Alioua A, Meurer J, Zoschke R, Hammani K (2020) The Arabidopsis mTERF-repeat MDA1 protein plays a dual function in transcription and stabilization of specific chloroplast transcripts within the *psbE* and *ndhH* operons. New Phytol 227:1376–1391. https://doi.org/10.1111/nph.16625

Middleton CP, Senerchia N, Stein N, Akhunov ED, Keller B, Wicker T, Kilian B (2014) Sequencing of chloroplast genomes from wheat, barley, rye and their relatives provides a detailed insight into the evolution of the Triticeae tribe. PLoS One 9:e85761. https://doi.org/10.1371/journal.pone.0085761

Miedaner T, Korzun V, Bauer E (2019) Genomics-based hybrid rye breeding. In: Miedaner T, Korzun V (eds) Applications of genetics and genomic research in cereals. Woodhead Publishing Series in Food Science, Technology and Nutrition. Elsevier, pp 329–344. https://doi.org/10.1016/C2016-0-04192-X

Miedaner T, Laidig F (2019) Hybrid breeding in rye (Secale cereale L.). Adv Plant Breed Strategies Cereals 5:343–372. https://doi.org/10.1007/978-3-030-23108-8_9

Miedaner T, Schwegler DD, Wilde P, Reif JC (2014) Association between line per se and testcross performance for eight agronomic and quality traits in winter rye. Theor Appl Genet 127:33–41. https://doi.org/10.1007/s00122-013-2198-2

Milczarski P, Hanek M, Tyrka M, Stojalowski S (2016) The application of GBS markers for extending the dense genetic map of rye (*Secale cereale* L.) and the localization of the *Rfc1* gene restoring male fertility in plants with the C source of sterility-inducing cytoplasm. J Appl Genet 57:439–451. https://doi.org/10.1007/s13353-016-0347-4

Mitterauer M (2010) Why Europe? The medieval origins of its special path. University of Chicago Press, Chicago

Molnar-Lang M, Linc G, Sutka J (1996) Transfer of the recessive crossability allele kr1 from Chinese Spring into the winter wheat variety Martonvasari 9. Euphytica 90:301–305. https://doi.org/10.1007/Bf00027480

Munoz-Sanz JV, Zuriaga E, Cruz-Garcia F, McClure B, Romero C (2020) Self-(in)compatibility systems: target traits for crop-production, plant breeding, and biotechnology. Front Plant Sci 11:195. https://doi.org/10.3389/fpls.2020.00195

Murase K, Moriwaki Y, Mori T, Liu X, Masaka C, Takada Y, Maesaki R, Mishima M, Fujii S, Hirano Y, Kawabe Z, Nagata K, Terada T, Suzuki G, Watanabe M, Shimizu K, Hakoshima T, Takayama S (2020) Mechanism of self/nonself-discrimination in *Brassica* self-incompatibility. Nat Commun 11:4916. https://doi.org/10.1038/s41467-020-18698-w

Myśków B, Góralska M, Lenarczyk N, Czyczyło-Mysza I, Stojałowski S (2018) Putative candidate genes responsible for leaf rolling in rye (*Secale cereale* L.). BMC Genet 19:57. https://doi.org/10.1186/s12863-018-0665-0

Nair SK, Wang N, Turuspekov Y, Pourkheirandish M, Sinsuwongwat S, Chen GX, Sameri M, Tagiri A, Honda I, Watanabe Y, Kanamori H, Wicker T, Stein N, Nagamura Y, Matsumoto T, Komatsuda T (2010) Cleistogamous flowering in barley arises from the suppression of microRNA-guided *HvAP2* mRNA cleavage. P Natl Acad Sci USA 107:490–495. https://doi.org/10.1073/pnas.0909097107

Nguyen V, Fleury D, Timmins A, Laga H, Hayden M, Mather D, Okada T (2015) Addition of rye chromosome 4R to wheat increases anther length and pollen grain number. Theor Appl Genet 128:953–964. https://doi.org/10.1007/s00122-015-2482-4

Ning SZ, Wang N, Sakuma S, Pourkheirandish M, Wu JZ, Matsumoto T, Koba T, Komatsuda T (2013) Structure, transcription and post-transcriptional regulation of the bread wheat orthologs of the barley cleistogamy gene *Cly1*. Theor Appl Genet 126:1273–1283. https://doi.org/10.1007/s00122-013-2052-6

Oettler G (1982) Effect of parental genotype on crossability and response to colchicine treatment in wheat-rye hybrids. Z Pflanzenzucht 88:322–330

Ossent HP (1938) Zehn Jahre Roggenzüchtung. Müncheberg Züchter 10:255–261

Ouyang S, Zhu W, Hamilton J, Lin H, Campbell M, Childs K, Thibaud-Nissen F, Malek RL, Lee Y, Zheng L, Orvis J, Haas B, Wortman J, Buell CR (2007) The TIGR rice genome annotation resource: improvements and new features. Nucleic Acids Res 35:D883–D887. https://doi.org/10.1093/nar/gkl976

Pankin A, Altmuller J, Becker C, von Korff M (2018) Targeted resequencing reveals genomic signatures of barley domestication. New Phytol 218:1247–1259. https://doi.org/10.1111/nph.15077

Papatheodorou I, Moreno P, Manning J, Fuentes AMP, George N, Fexova S, Fonseca NA, Fullgrabe A, Green M, Huang N, Huerta L, Lqbal H, Jianu M, Mohammed S, Zhao LY, Jarnuczak AF, Jupp S, Marioni J, Meyer K, Petryszak R, Medina CAP, Talavera-Lopez C, Teichmann S, Vizcaino JA, Brazma A (2020) Expression atlas update: from tissues to single cells. Nucleic Acids Res 48:D77–D83. https://doi.org/10.1093/nar/gkz947

Park YC, Burkitt V, Villa AR, Tong L, Wu H (1999) Structural basis for self-association and receptor recognition of human TRAF2. Nature 398:533–538. https://doi.org/10.1038/19110

Peterson RF (1934) Improvement of rye through inbreeding. Sci Agric, vol. 14, pp 651–658

Phillips D, Jenkins G, Macaulay M, Nibau C, Wnetrzak J, Fallding D, Colas I, Oakey H, Waugh R, Ramsay L (2015) The effect of temperature on the male and female recombination landscape of barley. New Phytol 208:421–429. https://doi.org/10.1111/nph.13548

Plaschke J, Korzun V, Koebner RMD, Borner A (1995) Mapping the Ga(3)-insensitive dwarfing gene *Ct1* on chromosome 7 in rye. Plant Breed 114:113–116. https://doi.org/10.1111/j.1439-0523.1995.tb00773.x

Pruitt KD, Hanson MR (1991) Transcription of the petunia mitochondrial cms-associated *Pcf* locus in male sterile and fertility-restored lines. Mol Gen Genet 227:348–355. https://doi.org/10.1007/Bf00273922

Putt ED (1954) Cytogenetic studies of sterility in rye. Can J Agric Sci 81:81–119

Rabanus-Wallace MT, Hackauf B, Mascher M, Lux T, Wicker T, Gundlach H, Baez M, Houben A, Mayer KFX, Guo LL, Poland J, Pozniak CJ, Walkowiak S, Melonek J, Praz CR, Schreiber M, Budak H, Heuberger M, Steuernagel B, Wulff B, Borner A, Byrns B, Cizkova J, Fowler DB, Fritz A, Himmelbach A, Kaithakottil G, Keilwagen J, Keller B, Konkin D, Larsen J, Li Q, Myskow B, Padmarasu S, Rawat N, Sesiz U, Biyiklioglu-Kaya S, Sharpe A, Simkova H, Small I, Swarbreck D, Toegelova H, Tsvetkova N, Voylokov AV, Vrana J, Bauer E, Bolibok-Bragoszewska H, Dolezel J, Hall A, Jia JZ, Korzun V, Laroche A, Ma XF, Ordon F, Ozkan H, Rakoczy-Trojanowska M, Scholz U, Schulman AH, Siekmann D, Stojalowski S, Tiwari VK, Spannagl M, Stein N (2021) Chromosome-scale genome assembly provides insights into rye biology, evolution, and agronomic potential. Nat Genet 53:564–573. https://doi.org/10.1038/s41588-021-00807-0

Rathburn HB, Hedgcoth C (1991) A chimeric open reading frame in the 5' flanking region of coxI mitochondrial-DNA from cytoplasmic male-sterile wheat. Plant Mol Biol 16:909–912. https://doi.org/10.1007/Bf00015083

Ribeiro-Carvalho C, Guedes-Pinto H, Heslop-Harrison JS, Schwarzacher T (2001) Introgression of rye chromatin on chromosome 2D in the Portuguese wheat landrace 'Barbela' . Genome 44:1122–1128. https://doi.org/10.1139/gen-44-6-1122

Rieseberg LH, Blackman BK (2010) Speciation genes in plants. Ann Bot 106:439–455. https://doi.org/10.1093/aob/mcq126

Riley R, Chapman V (1967) Inheritance in wheat of crossability with rye. Genet Res 9:259–267. https://doi.org/10.1017/S0016672300010569

Rizzolatti C, Bury P, Tatara E, Pin PA, Rodde N, Berges H, Budar F, Mireau H, Gielen JJL (2017) Map-based cloning of the fertility restoration locus *Rfm1* in cultivated barley (*Hordeum vulgare*). Euphytica 213:276. https://doi.org/10.1007/s10681-017-2056-4

Römer T (1931) Über die Reichweite des Pollens beim Roggen. Z Züchtg A17:14–35

Ru Z, Juhasz A, Li D, Deng P, Zhao J, Gao L, Wang K, Keeble-Gagnere G, Yang Z, Li G, Wang D, Bose U, Colgrave M, Kong C, Zhao G, Zhang X, Liu X, Cui G, Wang Y, Niu Z, Wu L, Cui D, Jia J, Appels R, Kong X (2020) 1RS.1BL molecular resolution provides novel contributions to wheat improvement. Preprint at https://www.biorxiv.org/content/10.1101/2020.09.14.295733v2

Ruffio-Chable V, Gaude T (2001) S-haplotype polymorphism in *Brassica oleracea*. Acta Hortic 546:257–261. https://doi.org/10.17660/ActaHortic.2001.546.31

Sakata T, Takahashi H, Nishiyama I, Higashitani A (2000) Effects of high temperature on the development of pollen mother cells and microspores in barley *Hordeum vulgare* L. J Plant Res 113:395–402. https://doi.org/10.1007/Pl00013947

Schittenhelm S, Kraft M, Wittich KP (2014) Performance of winter cereals grown on field-stored soil moisture only. Eur J Agron 52:247–258. https://doi.org/10.1016/j.eja.2013.08.010

Schlegel R (2020) Current list of wheats with rye and alien introgression. Translocations-Interchanges-Substitutions. Version 08.19. http://www.rye-gene-map.de/rye-introgression/index.html. Accessed 20 Dec 2020

Schopfer CR, Nasrallah ME, Nasrallah JB (1999) The male determinant of self-incompatibility in *Brassica*. Science 286:1697–1700. https://doi.org/10.1126/science.286.5445.1697

Schreiber M, Himmelbach A, Borner A, Mascher M (2019) Genetic diversity and relationship between domesticated rye and its wild relatives as revealed through genotyping-by-sequencing. Evol Appl 12:66–77. https://doi.org/10.1111/eva.12624

Schwander T, Libbrecht R, Keller L (2014) Supergenes and complex phenotypes. Curr Biol 24:R288–R294. https://doi.org/10.1016/j.cub.2014.01.056

Sehgal N, Singh S (2018) Progress on deciphering the molecular aspects of cell-to-cell communication in *Brassica* self-incompatibility response. 3 Biotech 8:347. https://doi.org/10.1007/s13205-018-1372-2

Shinozuka H, Cogan NO, Smith KF, Spangenberg GC, Forster JW (2010) Fine-scale comparative genetic and physical mapping supports map-based cloning strategies for the self-incompatibility loci of perennial ryegrass (*Lolium perenne* L.). Plant Mol Biol 72:343–355. https://doi.org/10.1007/s11103-009-9574-y

Sijacic P, Wang X, Skirpan AL, Wang Y, Dowd PE, McCubbin AG, Huang S, Kao TH (2004) Identification of the pollen determinant of S-RNase-mediated self-incompatibility. Nature 429:302–305. https://doi.org/10.1038/nature02523

Sitch LA, Snape JW, Firman SJ (1985) Intrachromosomal mapping of crossability genes in wheat (*Triticum aestivum*). Theor Appl Genet 70:309–314. https://doi.org/10.1007/Bf00304917

Small I, Melonek J (2020) New gene responsible for cytoplasmic male sterility. Patent application WO/2020/161261

Song JS, Hedgcoth C (1994) Influence of nuclear background on transcription of a chimeric gene (*orf256*) and *coxI* in fertile and cytoplasmic male-sterile wheats. Genome 37:203–209. https://doi.org/10.1139/g94-028

Sprague GF, Tatum AL (1942) General vs. specific combining ability in single crosses of corn. J Amer Soc Agron 34:923–932

Stein JC, Howlett B, Boyes DC, Nasrallah ME, Nasrallah JB (1991) Molecular-cloning of a putative receptor protein-kinase gene encoded at the self-incompatibility locus of *Brassica oleracea*. P Natl Acad Sci USA 88:8816–8820. https://doi.org/10.1073/pnas.88.19.8816

Steinborn R, Schwabe W, Weihe A, Adolf K, Melz G, Borner T (1993) A new type of cytoplasmic male-sterility in rye (*Secale cereale* L.)—analysis of mitochondrial DNA. Theor Appl Genet 85:822–824. https://doi.org/10.1007/Bf00225024

Stojałowski S, Bobrowska A, Hanek M, Myśków B (2013) The importance of chromosomes from the sixth homeologic group in the restoration of male fertility in winter triticale with *Triticum timopheevii* cytoplasm. J Appl Genet 54:179–184. https://doi.org/10.1007/s13353-013-0144-2

Stojałowski S, Kociuba M, Stochmal B, Kondzioła M, Jaciubek M (2008) Determining the plasmotypic structure of rye populations by SCAR markers. J Appl Genet 49:229–232. https://doi.org/10.1007/Bf03195618

Stojałowski S, Łapiński M, Szklarczyk M (2006) Identification of sterility-inducing cytoplasms in rye using the plasmotype-genotype interaction test and newly developed SCAR markers. Theor Appl Genet 112:627–633. https://doi.org/10.1007/s00122-005-0164-3

Stojałowski S, Milczarski P, Hanek M, Bolibok-Bragoszewska H, Myśków B, Kilian A, Rakoczy-Trojanowska M (2011) DArT markers tightly linked with the *Rfc1* gene controlling restoration of male fertility in the CMS-C system in cultivated rye (*Secale cereale* L.). J Appl Genet 52:313–318. https://doi.org/10.1007/s13353-011-0049-x

Stracke S, Schilling AG, Forster J, Weiss C, Glass C, Miedaner T, Geiger HH (2003) Development of PCR-based markers linked to dominant genes for male-fertility restoration in Pampa CMS of rye (*Secale cereale* L.). Theor Appl Genet 106:1184–1190. https://doi.org/10.1007/s00122-002-1153-4

Studer B, Asp T (2014) *Z* locus self-incompatibility alleles in Poaceae. Patent application WO/2014/029861

Taira T, Lelley T, Larter EN (1978) Influence of parental rye on development of embryos and endosperm of wheat-rye hybrids. Can J Bot 56:386–390. https://doi.org/10.1139/b78-047

Takasaki T, Hatakeyama K, Suzuki G, Watanabe M, Isogai A, Hinata K (2000) The *S* receptor kinase determines self-incompatibility in *Brassica* stigma. Nature 403:913–916. https://doi.org/10.1038/35002628

Tang HW, Zheng XM, Li CL, Xie XR, Chen YL, Chen LT, Zhao XC, Zheng HQ, Zhou JJ, Ye S, Guo JX, Liu YG (2017) Multi-step formation, evolution, and functionalization of new cytoplasmic male sterility genes in the plant mitochondrial genomes. Cell Res 27:130–146. https://doi.org/10.1038/cr.2016.115

Thomas SG, Franklin-Tong VE (2004) Self-incompatibility triggers programmed cell death in Papaver pollen. Nature 429:305–309. https://doi.org/10.1038/nature02540

Tian F, Yang DC, Meng YQ, Jin JP, Gao G (2020) PlantRegMap: charting functional regulatory maps in plants. Nucleic Acids Res 48:D1104–D1113. https://doi.org/10.1093/nar/gkz1020

Tikhenko N, Poursarebani N, Rutten T, Schnurbusch T, Börner A (2017) Embryo lethality in wheat-rye hybrids: dosage effect and deletion bin mapping of the responsible wheat locus. Biol Plant 61:342–348. https://doi.org/10.1007/s10535-016-0691-6

Tikhenko N, Rutten T, Voylokov A, Houben A (2008) Analysis of hybrid lethality in F1 wheat-rye hybrid embryos. Euphytica 159:367–375. https://doi.org/10.1007/s10681-007-9528-x

Tikhenko N, Tsvetkova N, Priyatkina S, Voylokov A, Börner A (2011) Gene mutations in rye causing embryo lethality in hybrids with wheat: allelism and chromosomal localisation. Biol Plant 55:448–452. https://doi.org/10.1007/s10535-011-0109-4

Tixier MH, Sourdille P, Charmet G, Gay G, Jaby C, Cadalen T, Bernard S, Nicolas P, Bernard M (1998) Detection of QTLs for crossability in wheat using a doubled haploid population. Theor Appl Genet 97:1076–1082. https://doi.org/10.1007/s001220050994

Tonosaki K, Osabe K, Kawanabe T, Fujimoto R (2016) The importance of reproductive barriers and the effect of allopolyploidization on crop breeding. Breed Sci 66:333–349. https://doi.org/10.1270/jsbbs.15114

Touzet P, Budar F (2004) Unveiling the molecular arms race between two conflicting genomes in cytoplasmic male sterility? Trends Plant Sci 9:568–570. https://doi.org/10.1016/j.tplants.2004.10.001

Tran F, Penniket C, Patel RV, Provart NJ, Laroche A, Rowland O, Robert LS (2013) Developmental transcriptional profiling reveals key insights into Triticeae reproductive development. Plant J 74:971–988. https://doi.org/10.1111/tpj.12206

Trang QS, Wricke G, Weber WE (1982) Number of alleles at the incompatibility loci in *Secale cereale* L. Theor Appl Genet 63:245–248. https://doi.org/10.1007/Bf00304003

Tsunewaki K (2015) Fine mapping of the first multi-fertility-restoring gene, Rf (multi), of wheat for three Aegilops plasmons, using 1BS-1RS recombinant lines. Theor Appl Genet 128:723–732. https://doi.org/10.1007/s00122-015-2467-3

Tsvetkova NV, Tikhenko ND, Hackauf B, Voylokov AV (2018) Two rye genes responsible for abnormal development of wheat-rye hybrids are linked in the vicinity of an evolutionary translocation on chromosome 6R. Plants (basel) 7:55. https://doi.org/10.3390/plants7030055

Tudzynski P, Rogmann P, Geiger HH (1986) Molecular analysis of mitochondrial DNA from rye (*Secale cereale* L.). Theor Appl Genet 72:695–699. https://doi.org/10.1007/BF00289011

Vendelbo NM, Sarup P, Orabi J, Kristensen PS, Jahoor A (2020) Genetic structure of a germplasm for hybrid breeding in rye (*Secale cereale* L). PLoS One 15:e0239541. https://doi.org/10.1371/journal.pone.0239541

von Bothmer R, Jacobsen N, Baden C, Jorgensen RB, Linde-Laursen I (1995) An ecogeographical study of the genus *Hordeum*. In *Systematic and ecogeographic studies on crop genepools 7*. International Plant Genetic Resources Institute, Rome

Voylokov AV (2007) Prospects of using self-fertility in breeding rye populations varieties. Russ J Genet 43:1173–1180. https://doi.org/10.1134/S1022795407100122

Voylokov AV, Fuong FT, Smirnov VG (1993) Genetic-studies of self-fertility in rye (*Secale cereale* L). 1. The identification of genotypes of self-fertile lines for the sf-alleles of self-incompatibility genes. Theor Appl Genet 87:616–618. https://doi.org/10.1007/Bf00221887

Voylokov AV, Korzun V, Börner A (1998) Mapping of three self-fertility mutations in rye (*Secale cereale* L.) using RFLP, isozyme and morphological markers. Theor Appl Genet 97:147–153. https://doi.org/10.1007/s001220050879

Walkowiak S, Gao LL, Monat C, Haberer G, Kassa MT, Brinton J, Ramirez-Gonzalez RH, Kolodziej MC, Delorean E, Thambugala D, Klymiuk V, Byrns B, Gundlach H, Bandi V, Siri JN, Nilsen K, Aquino C, Himmelbach A, Copetti D, Ban T, Venturini L, Bevan M, Clavijo B, Koo DH, Ens J, Wiebe K, N'Diaye A, Fritz AK, Gutwin C, Fiebig A, Fosker C, Fu BX, Accinelli GG, Gardner KA, Fradgley N, Gutierrez-Gonzalez J, Halstead-Nussloch G, Hatakeyama M, Koh CS, Deek J, Costamagna AC, Fobert P, Heavens D, Kanamori H, Kawaura K, Kobayashi F, Krasileva K, Kuo T, McKenzie N, Murata K, Nabeka Y, Paape T, Padmarasu S, Percival-Alwyn L, Kagale S, Scholz U, Sese J, Juliana P, Singh R, Shimizu-Inatsugi R, Swarbreck D, Cockram J, Budak H, Tameshige T, Tanaka T, Tsuji H,

Wright J, Wu JZ, Steuernagel B, Small I, Cloutier S, Keeble-Gagnere G, Muehlbauer G, Tibbets J, Nasuda S, Melonek J, Hucl PJ, Sharpe AG, Clark M, Legg E, Bharti A, Langridge P, Hall A, Uauy C, Mascher M, Krattinger SG, Handa H, Shimizu KK, Distelfeld A, Chalmers K, Keller B, Mayer KFX, Poland J, Stein N, McCartney CA, Spannagl M, Wicker T, Pozniak CJ (2020) Multiple wheat genomes reveal global variation in modern breeding. Nature 588:277–283. https://doi.org/10.1038/s41586-020-2961-x

Wang N, Ning SZ, Wu JZ, Tagiri A, Komatsuda T (2015) An epiallele at *cly1* affects the expression of floret closing (cleistogamy) in barley. Genetics 199:95-U506. https://doi.org/10.1534/genetics.114.171652

Wang QB, Zheng PJ, Zhang L (2019) Identification and classification of S haplotypes in radish (*Raphanus sativus*). Plant Breeding 138:121–130. https://doi.org/10.1111/pbr.12664

Wang ZH, Zou YJ, Li XY, Zhang QY, Chen L, Wu H, Su DH, Chen YL, Guo JX, Luo D, Long YM, Zhong Y, Liu YG (2006) Cytoplasmic male sterility of rice with boro II cytoplasm is caused by a cytotoxic peptide and is restored by two related PPR motif genes via distinct modes of mRNA silencing. Plant Cell 18:676–687. https://doi.org/10.1105/tpc.105.038240

Warzecha R, Salak-Warzecha K (2003) A new source of male sterility in rye (*Secale cereale* L.). Plant Breed Seed Sci 48:61–65

Wehling P, Hackauf B, Wricke G (1994) Phosphorylation of pollen proteins in relation to self-incompatibility in rye (*Secale cereale* L). Sex Plant Reprod 7:67–75

Wheeler MJ, de Graaf BHJ, Hadjiosif N, Perry RM, Poulter NS, Osman K, Vatovec S, Harper A, Franklin FCH, Franklin-Tong VE (2009) Identification of the pollen self-incompatibility determinant in *Papaver rhoeas*. Nature 459:992-U118. https://doi.org/10.1038/nature08027

Whitford R, Fleury D, Reif JC, Garcia M, Okada T, Korzun V, Langridge P (2013) Hybrid breeding in wheat: technologies to improve hybrid wheat seed production. J Exp Bot 64:5411–5428. https://doi.org/10.1093/jxb/ert333

Wilde P, Korzun V, Menzel J, Zhou R, Stein N, Hackauf B (2017) Restorer plants. US patent application 16/064304

Wilson AS (1876) On wheat and rye hybrids. Trans Proc Bot Soc Edinburgh 12:286–288

Winter D, Vinegar B, Nahal H, Ammar R, Wilson GV, Provart NJ (2007) An "Electronic Fluorescent Pictograph" browser for exploring and analyzing large-scale biological data sets. PLoS One 2:e718. https://doi.org/10.1371/journal.pone.0000718

Wricke G (1978) Pseudo-self-incompatibility in rye and its utilization in plant breeding. Z Pflanzenzucht 81:140–148

Wricke G (1985) Hybrid breeding in rye by means of self-incompatibility. In: Proceedings Eucarpia meeting of the cereal section on rye. Svalöv, Sweden, pp 267–278

Wricke G, Wehling P (1985) Linkage between an incompatibility locus and a peroxidase isozyme locus (*Prx 7*) in rye. Theor Appl Genet 71:289–291. https://doi.org/10.1007/Bf00252069

Yang B, Thorogood D, Armstead I, Barth S (2008) How far are we from unravelling self-incompatibility in grasses? New Phytol 178:740–753. https://doi.org/10.1111/j.1469-8137.2008.02421.x

Zhao Y, Cai M, Zhang X, Li Y, Zhang J, Zhao H, Kong F, Zheng Y, Qiu F (2014) Genome-wide identification, evolution and expression analysis of mTERF gene family in maize. PLoS ONE 9:e94126. https://doi.org/10.1371/journal.pone.0094126

Zheng YL, Luo MC, Yen C, Yang JL (1992) Chromosome location of a new crossability gene in common wheat. Wheat Inf Serv 75:36–40

Genetics and Genomics of Stress Tolerance

11

Monika Rakoczy-Trojanowska, Hanna Bolibok-Brągoszewska, Beata Myśków, Magdalena Dzięgielewska, Stefan Stojałowski, Agnieszka Grądzielewska, Maja Boczkowska, and Kinga Moskal

Abstract

Rye (*Secale cereale* L.) is considered the most tolerant cereal crop, both to biotic and abiotic stresses. Below, we present the current knowledge about the genetic background of the tolerance to such stresses as diseases, pests, deficiency and excess of soil minerals (particularly aluminum), drought, and low temperature. Despite recent significant progress in research on the genetic understanding of stress tolerance in rye, genetic factors responsible for tolerance mechanisms in rye are still poorly characterized and further research is needed. We also discuss the role of active compounds in defense against chosen stresses, including brown rust, nematodes, allelochemicals secreted by clover (*Trifolium alexandrinum*), low temperature during vernalization and soil salinity. Furthermore, we describe the use of rye chromatin for improvement of stress tolerance (mainly pathogen and pest resistance) in wheat.

11.1 Introduction

The development of crop varieties which are more tolerant to biotic and abiotic stresses is urgently needed in view of the predicted climatic and demographic changes. A rise of average temperature and a higher frequency of extreme weather events are expected in the near future. It is also estimated that the human population will increase by 26% until 2050 (https://population.un.org/wpp/; Wheeler and von Braun 2013).

Among cereals, rye is characterized by particularly high resistance to various environmental stresses. Therefore, it can be grown in conditions unacceptable by other cereal crops. Rye can be cultivated on poorer soils; it tolerates acidic soils well and saline soils (except for highly saline soils)—relatively well. Fewer pathogens infect rye than other cereals. Furthermore, despite the fact that rye is attacked by many of the same insects as other small grain cereals, the yield losses are not as significant. Another feature that distinguishes rye from other cereals is its exceptionally high frost tolerance. Rye is also more tolerant to drought than other small grain

M. Rakoczy-Trojanowska (✉) · H. Bolibok-Brągoszewska
Institute of Biology, Warsaw University of Life Sciences, Warsaw, Poland
e-mail: monika_rakoczy_trojanowska@sggw.edu.pl

B. Myśków · M. Dzięgielewska · S. Stojałowski
Department of Plant Genetics, Breeding and Biotechnology, West Pomeranian University of Technology in Szczecin, Szczecin, Poland

A. Grądzielewska
Institute of Plant Genetics, Breeding and Biotechnology, University of Life Sciences in Lublin, Lublin, Poland

M. Boczkowska · K. Moskal
The Plant Breeding and Acclimatization Institute - National Research Institute, Błonie, Poland

© The Author(s), under exclusive license to Springer Nature Switzerland AG 2021

M. T. Rabanus-Wallace and N. Stein (eds.), *The Rye Genome*, Compendium of Plant Genomes,
https://doi.org/10.1007/978-3-030-83383-1_11

cereals which is related to its well-developed root system. Due to these properties, rye has been used for more than 50 years as a source of chromatin in wheat breeding, to improve properties, such as resistance to pathogens and pests or nutrient uptake efficiency, through chromosomal substitutions and translocations.

Research on the identification of genes responsible for rye's outstanding tolerance of biotic and abiotic stresses had started several decades ago. Initially, classical genetic methods and cytogenetic approaches, were used. More recently, methods such as genome-wide association studies using diverse unrelated accessions, QTL mapping in biparental populations, transcriptome sequencing based approaches sequence homology-based identification of putative orthologues from closely related species have been applied in studies on genetic control of stress tolerance in rye, providing initial, but still limited data. Below we summarize the current state of knowledge on the genetic control of various biotic and abiotic stresses in rye.

11.2 Genetic Background of Resistance to Biotic Stresses

11.2.1 Diseases

Although rye is considered a crop tolerant to diseases, as many as 37 diseases can attack this crop. To date two bacterial and 35 fungal diseases have been described (https://www.apsnet.org/edcenter/resources/commonnames/Pages/Rye.aspx). However, only some of them, namely rusts (leaf syn. brown, stem and yellow syn. stripe rusts), powdery mildew, eyespot, leaf blotch, root rot, Fusarium head blight, tan spot, leaf blotch, and glume blotch, are of greater economic importance due to the yield losses in quantitative and qualitative terms, which moreover depend on environmental conditions (Nyvall 1989).

The most damaging diseases of rye are rusts, namely leaf rust (LR), stem rust (SR), and, in recent years—yellow rust (YR) caused by, respectively, *Puccinia econdite* f. sp. *secalis*, *P. graminis* f. sp. *secalis* and *P. striiformis* var. *striiformis*). Yield losses due to rusts can reach up to even 70% (Chaves et al. 2008). Figure 11.1 shows symptoms of LR and YR on rye leaves.

Out of the three rusts, the genetic background of resistance is best understood for LR, caused by an airborne pathogen; the obligate biotrophic basidiomycete *P. recondita* f. sp. *secalis* (*Prs*) (Roberge ex Desmaz) (Roux et al. 2010; Miedaner et al. 2012). Rough estimates of up to 40% yield losses due to LR under natural conditions have been reported (Wehling et al. 2003), but, in an early infection or in epidemic years, it can inflict serious yield losses, even as high as 80% (Solodukhina 2002). Yield losses from this rust are usually the result of a decreased number of kernels per head and lower kernel weight.

In rye, resistance to LR can be controlled quantitatively or qualitatively, but qualitative resistance was found in only a few resistant lines (Sperling et al. 1996; Miedaner et al. 2002). Resistance to *Prs* is usually controlled by a single dominant gene and in a few cases by 2 or 3 genes (Musa et al. 1984; Solodukhina and Kobylianski

Fig. 11.1 Leaf (brown spots) and yellow (yellow spots) rusts on rye leaves (phot. A. Grądzielewska)

2003). The LR resistance genes from rye can be divided into two groups. The first group comprises three R genes (*Lr25*, *Lr26* and *Lr45*), which had been transferred from rye into wheat via translocations and the second group—genes which confer resistance to *Prs* only in rye genotypes.

To distinguish rye LR resistance genes from wheat *Lr* genes, the rye genes were renamed *Pr*. To date, 17 dominant *Pr* genes have been identified, including *Pr1–5, Pr-d–f, Pr-i–l, Pr-n, Pr-p, Pr*-r and *Pr-t* (Klocke 2004; Roux et al. 2007) located on 5 of the 7 rye chromosomes (1R, 2R, 4R, 6R, and 7R), on both their short and long arms. Roux et al. (2007) found that the *Pr1–Pr5, Pr-d–f, Pr-n, Pr-p, Pr-r* prevented rust formation both in the seedling and in the adult-plant stage, indicating that these genes are all-stage resistance (ASR) genes (Wu et al. 2020). On the other hand, Solodukhina and Kobylyanski (2003) observed that homozygous plants with the *Pr-d* gene were resistant to LR from seedling to stem growth phases, but the normal fungal pustules occurred in the grain filling phase, as is the case with non-race-specific APRs (Wu et al. 2020).

The *Pr* genes (*Pr1–5, Pr-d-f, n, r*), conferring resistance to a broad range of single pustule isolates (SPI), and the *Pr1, Pr2, Pr-d* and *Pr-r* were suggested as highly valuable for use in rye breeding (Roux et al. 2004, 2007). Most of the identified rye genes represent blocks of linked genes responsible for resistance to individual clones of the fungus (Solodukhina 2002). To improve resistance of rye cultivars to LR and other diseases, some attempts to stack *Pr* genes were conducted, and, as a result, *Lr4*, *Lr6* and *Lr7* have been introduced to 'Estafeta Tatarstana' and *Lr4*, *Lr5* and *Lr6* to 'Era' cultivars (Solodukhina and Kobylyanski 2003).

The majority of the *Pr* genes (*Pr3*, *Pr4*, *Pr5*, *Pr-i*, *Pr-k*, *Pr-n*) were mapped to the 1R chromosome. *Pr3* (syn. *Lr3*; *Lr-c*; Ruge et al. 1999; Roux et al. 2000) was assigned to 1RS and linked 5 cM from the isozyme marker Prx7 (Roux et al. 2004). Although *Pr3* is occupying a locus on 1RS just like *Lr26*, the two genes were found to be different, as *Lr26* is closely linked to Xiag95 and Xsec1 (the seed-storage protein gene Sec-1) near the end of the chromosome arm and *Pr3* is located at a considerable distance from these loci (Hsam et al. 2000; Roux et al. 2004). *Pr-i* and *Pr-k* are also localized to 1RS. It is not known whether these genes are allelic or not to *Pr3* or *Lr26*. Two genes: *Pr4* and *Pr5* (syn *Lr4, Lr-g* and *Lr5, Lr-h*, respectively; Ruge et al. 1999; Roux et al. 2000, 2004), were mapped to 1RL, distally from Xscm107, an EST-derived SSR marker (Roux et al. 2004). *Pr5* was, just like *Pr4*, mapped distally from Xscm107. Nevertheless, the two genes show different reaction patterns to specific SPIs, which indicates that *Pr4* and *Pr5* are different genes. *Pr-n* (syn. *Lr9*) is another gene of Russian origin from 1R, but its specific location on the 1R is unknown, although it displayed an identical reaction pattern to SPIs and chromosomal location as *Pr4* (Roux et al. 2007).

Two *Pr* genes localized on 2RS: *Pr-d* (syn *Lr6*) and *Pr-f* (syn. *Pr8*) (Solodukhina and Kobylyanski 2003; Roux et al. 2007) were distinguished on the basis of their SPI-reaction patterns (Roux et al. 2007). Genes *Pr-j* from the German line 'H26' and *Pr-l* from '94,107' are two genes mapped to 4R at the distance of 1.4 cM and 0.1 cM from SSR markers Xscm47 and Xscm 47, respectively (Klocke 2004). Genes *Pr1* and *Pr3* were both mapped to the 6R. *Pr1* (syn. *Lr1*, *Lr-a*, Roux et al. 2000; Solodukhina 2002) is located in the proximal region of 6RL and linked to the isozyme *Aco1* (*Aconitase 1*) locus at the distance of 1.2 cM (Wehling et al. 2003). *Pr-e* (syn *Lr7*) (Solodukhina 2002) was not assigned to any of the 6R arms (Roux et al. 2007).

Only one gene—*Pr2* (syn. *Lr2*, *Lr-b*; Roux et al. 2000; Solodukhina 2002) originating from the German inbred line 'L2527' (Wehling et al. 2003), was assigned to the 7RL. It was linked to RAPD markers OPO-07 and OPY-11 (Ruge et al. 2001), and mapped distally to markers Xscm122 and cMWG682, in a region homoeologous to *Triticeae* group 2S (Wehling et al. 2003).

The genes *Pr-p*, *Pr-r* and *Pr-t* were found in populations from Argentina, the USA, and Russia, but none of them could be mapped genetically (Roux et al. 2007).

An analysis of the newly released rye genome (Rabanus-Wallace et al. 2019) allowed to find 5 orthologs of wheat *Lr* genes conferring the resistance to LR. Among them are R genes related to seedling stage resistance, namely *Lr1*, *Lr10*, *Lr21* and *Lr22a*, as well as one APR gene, an ortholog of *Lr67*. Święcicka et al. (2019) showed that the expression of the gene *Lr1* increased in rye inbred line 'L318' when infected with a partially compatible *Prs* isolate, between 17 and 48 h post infection (Fig. 11.2). However, further comprehensive research is required to understand the role in the disease response of both this gene and the other genes mentioned above.

Several SR resistance genes have been discovered since the 1970s with a majority of them acting dominantly (Gruner et al. 2020). One of the earliest discovered genes *Sr31*, was first considered as a shared common locus or to be closely linked to genes *Lr26* and *Yr9*, conferring resistance to LR and YR, respectively. A study performed by Mago et al. (2005) showed that *Sr31, Lr26* and *Yr9* are separate, but closely linked genes. Recently, using three biparental populations of inbred lines developed by crossing SR-susceptible with SR-resistant lines, KASP assay markers were used to map a new single SR resistance gene—*Pgs1*. It has been mapped at the distal end of chromosome 7R and appears to be closely linked to a nucleotide-binding leucine-rich repeat (NB-LRR) resistance gene (Gruner et al. 2020). The authors suggested that *Pgs1* is closely linked to the LR resistance gene *Pr2*, reported by Wehling et al. (2003), or that it is a single gene showing a pleiotropic effect, i.e., resistance to both rusts. Besides the *Pgs1* gene, they identified also three APR (adult-plant resistance) related QTLs (QTL-SR1, QTL-SR2, QTL-SR3 on chromosomes 1R, 2R, and 6R, respectively), of which one—QTL-SR3 was inherited dominantly. According to the authors, both the gene *Pgs1* and QTL-SR3 have high potential to be used for SR resistance breeding in rye.

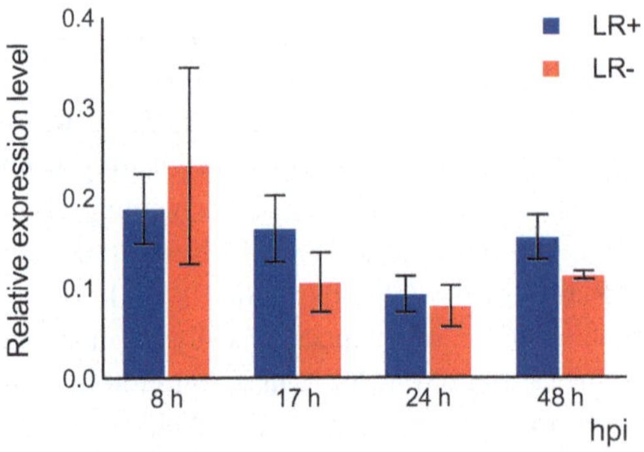

Fig. 11.2 Expression level of the *Lr1* orthologous gene in inbred line 'L318' inoculated with *Prs*. LR+ plants infected with *Prs*; LR− mock-treated plants; I, II, III, IV —8, 16, 24 and 48 h post infection (hpi), (Święcicka et al. 2019, modified)

Although YR is considered to be a disease harmful primarily to wheat, in recent years its frequency and extent of its occurrence in rye has been steadily increasing, mainly due to the warmer climate and the lack of cold and snowy winters. To date, two genes conferring the resistance to YR have been found—*Yr9*, located on chromosome 1RS (Mago et al. 2005), and *Yr83*—on chromosome 6RL (Li et al. 2020a, b). As it was pointed out above, *Yr9* is closely linked to two other YR resistance genes (*Sr31* and *Lr26*).

Ergot caused by *Claviceps purpurea* (Fig. 11.3), is one of the most economically important diseases of rye.

It is harmful due to contamination of grain by sclerotia secreting alkaloids toxic to humans and other mammals, and therefore, is devastating for the use of rye in food and feed production. For

Fig. 11.3 Ergot sclerotia protruding from rye spike (phot. S. Stojałowski)

ergot infection, flowering biology of the host is of fundamental importance. Ergot mimics the pollination of fertilization and for that reason, the specific traits related to flowering have a considerable influence on infection effectiveness and disease symptom development (Miedaner and Geiger 2015). Despite the importance of ergot, the genetic background of the disease response is still poorly understood, mainly due to the complex infection pattern. However, the availability of rye genome sequence data and the development of transcriptome sequencing methods allowed the identification of a relatively large number of differentially expressed genes, as a result of *C. purpurea* infection. Using an RNA-Seq approach Oeser et al. (2017) studied the cross-talk between *C. purpurea* and the host rye plant. They found that infection with a wild-type of the fungus caused a significant change in expression of 55 genes (most of them upregulated) in rye ovaries. Among the upregulated genes, several encoded potential pathogenesis-related proteins, for example, a β-1,3 glucanase and a peroxidase. Applying a similar approach, Mahmood et al. (2020) selected as many as 228 genes linked to ergot infection in two rye hybrids differing in terms of fungus susceptibility. They were associated with different processes, such as metabolic processes, hydrolase and pectinesterase activity, cell wall modification, pollen development, pollen wall assembly, cell wall modification, and pectinesterase activity. The authors concluded that the resistance against ergot in rye resulted from a combination of different pathways, particularly cell wall modification and pectinesterase activity.

Powdery mildew (PM), triggered by the fungus *Blumeria graminis* (DC) E.O. Speer f. sp. *tritici* Em. Marchal (syn. *Erysiphe graminis* DC f. sp. *tritici* Marchal), is a disease affecting living leaf tissues. While it is one of the most important foliar diseases of wheat worldwide, rye is highly tolerant to PM. Some rye cultivars, e.g., the Chinese cultivar 'Qinling', are even considered fully immune to PM (Hao et al. 2018). Thus, rye is a widely used source of resistance genes (single R genes) for wheat improvement. To date, 11 *Pm* genes conferring resistance to this disease have been detected in rye: *Pm1a, Pm1b, Pm7* (1RS), *Pm2* and *Pm8* (2RL), *Pm3* (3RS), *Pm6* (4R), *Pm4* (5RL), *Pm5*, *Pm20* and *Pm(?)* (6RL); (Schlegel et al. 1998; Wang et al. 2010; Crespo-Herrera et al. 2017; Jurkowski and Bujak 2019). A different set of *Pm* genes which differ from previously reported rye genes, were located on chromosome 4R in cvs. 'German White' (An et al. 2013) and 'Kustro' (Fu et al. 2014). Recently, Duan et al. (2017) constructed a new fluorescence in situ hybridization map of chromosome 4R in cv. 'Kustro', where a clearly defined region with the PM resistance gene(s) was identified.

To facilitate the use of R genes in breeding programs and their physical isolation, many of the genes conferring resistance to powdery mildew have been mapped by the use of molecular markers. Jurkowski et al. (2014) found that wheat marker ResPm4 linked to the gene *Pm4* was present in resistant rye genotypes, but it was not present in any of the PM susceptible genotypes. However, in a later work comprising four other markers identified earlier in common wheat, only one marker linked to the *Pm3a* gene was present in three susceptible rye genotypes and the remaining markers, specific for the *Pm3b*, *Pm3c* and *Pm3d* resistance genes, were absent in both resistant and susceptible rye genotypes (Jurkowski and Bujak 2019).

Fusarium head blight (FHB), caused by *Fusarium graminearum*, *F. culmorum* and other *Fusarium* species reduces yield and produces several mycotoxins in all cereals including rye. Rye was found to be the most resistant species to FHB compared to triticale (X *Triticosecale* Wittmack), durum (*T. durum*) and bread wheat (Gaikpa et al. 2020). There are no studies on genetic background of the resistance to FHB in rye. However, studies in triticale identified FHB QTLs on rye chromosomes 3R, 4R, 5R, and 7R (Kalih et al. 2015; Dhariwal et al. 2018; Galiano-Carneiro et al. 2019). Kalih et al. (2014), analyzing a triticale DH mapping population, detected a major QTL on 5R which explained 48, 77, and 71% of genotypic variation for FHB severity, plant height, and heading stage, respectively. The authors suggested that this

QTL resembles the rye gene *Ddw1*, introduced into triticale by Polish breeders, the effect of which was increased resistance to FHB.

Tan spot (TS), resulting from infection by *Pyrenophora tritici-repentis*, does not yet play a similarly important role in rye as it does in wheat. Rye is considered as an alternative host for the fungus and therefore it could be exploited as a source of resistance to TS (Abdullah et al. 2017). *Pyrenophora tritici-repentis* produces three host-selective, disease symptom-related toxins: Ptr ToxA, Ptr ToxB, and Ptr ToxC (Faris et al. 2013). In wheat three host genes *Tsn1, Tsc1, and Tsc2* are responsible for Ptr ToxA, Ptr ToxB, and Ptr ToxC sensitivity (Abdullah et al. 2017). The authors indicated, however, that rye lacks the Ptr ToxA, Ptr ToxB genes and, therefore, Ptr ToxA may not play a significant role in disease development on rye. Recently, Sidhu et al. (2019) identified two QTLs in rye conferring race specific resistance to TS on chromosomes 5R and 2R explaining 13.1% and 11.6% of the phenotypic variation, respectively.

11.2.2 Pests

Rye is attacked by many of the same pests that attack other small grain cereals, but the yield losses are not as severe as for example in wheat. Among them are aphids: bird cherry-oat aphid (*Rhopalosiphum padi*), English grain aphid (*Sitobion avenae*), rose-grass aphid (*Metopolophium dirhodum*), wheat aphid (*Schizaphis graminum*), Russian wheat aphid (*Diuraphis noxia*), cereal leaf beetles (*Oulema melanopus* and *Lema cyanella*), saddle gall midge (*Haplodiplosis equestris*), wheat curl mite (*Aceria tosichella*), Hessian fly (*Mayetiola destructor*), and cereal cyst nematodes (*Heterodera avenae* and *Heterodera filipjevi*). Although rye is well documented and practically confirmed as a rich source of resistance to pests in wheat, the genetic background of the resistance is still poorly understood. To date, genes conferring pest resistance have been reported only in case of five pests, namely Russian wheat aphid (RWA), wheat aphid, Hessian fly, wheat curl mite, and nematodes; all of these genes were successfully introduced to wheat (El Bouhssini et al. 2012; Mondal et al. 2016; Crespo-Herrera et al. 2017; Aguirre-Rojas et al. 2017).

Among those mentioned above, aphids are the most threatening rye pests. They do not only cause damage to the plant by feeding on its tissue, but in many cases, they act passively as vectors of viruses. Moreover, aphids can modify their physiology, behavior, and distribution due to changing climatic conditions, which increases their harmfulness (Mondal et al. 2016). The natural enemy in the fight against aphids are ladybugs (Fig. 11.4). Naturally occurring resistance has only been described in case of two aphids—RWA and wheat aphid. The 15 RWA resistance genes described so far are designated as *Dn1–Dn9*, *Dnx*, *Dny*, and *Dnr1–Dnr4* (Andersson et al. 2015); most of them are single dominant genes. The *Dnr1–Dnr4* and *Dn7* (*Dnr*) genes have been found to be located on chromosomes 1RL, 3RS, 4R, 7R, and 1RS, respectively, and all of them were transferred into wheat through introgressions. The gene *Dn7*, related to all three categories of resistance (antibiosis, antixenosis, and tolerance) is effective against the *RWA-1, 5, 6, 7, 8* aphid biotypes and, as the only one, against *RWA-2* (Haley et al. 2004; Crespo-Herrera 2017).

Two genes—*Gb2* and *Gb6*, both mapped on 1RS, carry the resistance (all three categories—antibiosis; antixenosis, and tolerance) against the

Fig. 11.4 A Ladybug resting on a rye spike where larvae of aphids typically hide. Aphids prefer unusually glaucous rye spikes (phot. B. Myśków)

wheat aphid (greenbug). The *Gb2* and *Gb6* genes confer resistance to biotypes of *S. graminum* B, C, J, and E, G, I, K, respectively (Crespo-Herrera et al. 2017).

Although the Hessian fly prefers wheat as a host, it also may be a serious threat to rye (Crespo-Herrera et al. 2017). During first- and second-instar larval stages, which last for 2–3 weeks, the larvae feed on the stem, which can injure the plant and result in stunted growth, lodging due to weakened stems, and reduced seed production (or even failure to produce seeds at all). Among 35 dominant or partially dominant genes conferring resistance to Hessian fly in wheat, two genes—*H21* and *H25* (determining antibiosis) were translocated from rye; they were mapped on 2RL of *cv.* 'Chaupon' and 6RL of *cv.* 'Balbo', respectively (Hatchett et al. 1993; Li et al. 2015).

The tolerance to wheat curl mite is related to a single dominant gene *Cmc3* located on 1RS (Malik et al. 2003). However, based on data obtained in wheat translocation lines (Harvey et al. 1999) it can be concluded that the effectiveness of *Cmc3* is limited to the weakly virulent biotypes of *A. tosichella*.

Cereal cyst nematode *Heterodera avenae* represents the third group of pests that threaten rye, in addition to insects and the eriophyid curl mite. A rye resistance gene *CreR* was assigned to 6RL (Mondal et al. 2016).

11.3 Genetic Background of Resistance to Abiotic Stresses

11.3.1 Nutrient Stress

Rye is considered to have the highest tolerance of nutrient deficiency among Triticeae, as well as an excellent tolerance of other nutrient stresses such as high aluminum content in soil.

Superior tolerance for the deficiency of nutrients was shown, for example, with respect to elements such as copper (Cu) (Harry and Graham 1981), nitrogen (N) (Paponov et al. 1999) and phosphorus (P) (Pandey et al. 2005). Higher biomass production of rye in low N conditions, when compared to wheat and triticale, was shown to be likely the effect of higher N accumulation in the plant (Paponov et al. 1999). Similarly, in a study involving the same three species, Harry and Graham (1981) observed an outstanding performance of rye in Cu deficient conditions at various pH levels, revealing also the highest concentration of Cu in the rye plants. In case of phosphorus, rye's higher tolerance (in comparison to wheat and triticale) was observed to be the result of an efficient uptake system under deficiency conditions (Pandey 2006) and better efficiency in P utilization (Pandey et al. 2005).

Several studies reported genotypic differences in rye's response to various nutrient stresses. For example, Smolik (2013) observed high variability of response to nitrogen and potassium stress at the seedling stage in a population of 183 RILs derived form a cross between rye inbred line '153/79-1' (tolerant to nutrient stress) and 'Ot1-3' (susceptible to nutrient stress). Various degrees of P-deficiency tolerance were observed by Hawliczek et al. (2019) in a panel of several rye inbred lines.

Genetic factors underlying rye's tolerance of nutrient stresses are mostly unknown. In a few cases, wheat-rye addition and translocation lines were used to identify the chromosomal location of causative genes. For example, genes related to copper use efficiency were found to be located on 5RL. In addition, a 5A/5RL translocation improves the efficiency of Cu use in wheat and is useful for inclusion in breeding programs (Owuoche et al. 1996). With regards to phosphorus, a significantly higher expression of two high affinity phosphate transporters PT6 was observed using RT-PCR in roots of plants grown in low P conditions. Additionally, in the roots of a P-deficiency sensitive line, a significantly higher expression of two other P-utilization related genes, PHO1;2 and TaPHT1.10, was observed (Hawliczek et al. 2019).

Aluminum (Al) tolerance in rye has gained much attention since aluminum toxicity is one of the major constraints to agricultural production on acidic soils, where rye is frequently grown.

While rye is considered to be the most Al-tolerant cereal, there are genotypic differences in the level of tolerance (Shi et al. 2009; De Sousa et al. 2016). Genetic mapping and cytogenetic studies revealed, that there are several loci responsible for Al tolerance in rye. Two genes, *Alt1* and *Alt4* have been mapped on 6RS and 7RS, respectively. Chromosomes 1R, 3R, 5R were identified as the location of additional Al tolerance loci [for details see: (Shi et al. 2009)]. Collins et al. (2008) showed that the *Alt4* locus is a cluster of several genes which are homologues of the wheat *aluminum-activated malate transporter-1* (*ALMT1*). Collins et al. (2008) also showed that the rye *ScMATE1* gene, homologue of the barley *HvMATE1* (*multidrug and toxic compound extrusion, Al-activated citrate transporter*), is located in the vicinity of the *ALMT1* gene cluster on 6RS. Later studies indicated that the *ScMATE1* gene, also known as *ScAACT1* (Santos et al. 2018), and *ScFRDL1* (Yokosho et al. 2010), is also involved in rye Al tolerance (Silva-Navas et al. 2012). Al-induced expression of *ScMATE1*, and the level of Al tolerance varied in the studied rye genotypes, depending on the allelic variant of the *ScMATE1* gene. Results suggestive of copy number variation (with two copies of *ScMATE1* likely present in *S. sylvestre* and *S. cereale* cv 'Riodeva') were also reported (Santos et al. 2016). Recently, Hawliczek et al. (2020), by pooled amplicon sequencing, identified 62 new variants (among them seven putatively deleterious) in the coding sequence of *ScMATE1*, delivering thus further potential targets for functional studies. Additionally, using microarray transcriptome profiling of wheat and rye-wheat addition lines, Salvador-Moreno et al. (2018) identified several additional candidate genes, likely located on 3R, which could contribute to Al tolerance of rye.

11.3.2 Drought

The most important region of rye cultivation is Central and Eastern Europe, which is abundant in sandy soils with poor water retention capacity. Rye is considered the best crop for such soils as its loss on yield is smaller than other cultivated plants (Hlavinka et al. 2009). On the other hand, the reduction of yield under severe drought stress can reach extreme values even for rye. Comparison of naturally-rainfed field plots (with water deficits) to those under controlled irrigation (years 2010–2012, location in Germany and Poland) revealed yield reduction from 2 to 41% (Hübner et al. 2013; Haffke et al. 2015). The losses in yield were moderate in locations with heavy soils and moderate rainfall deficiency, but very large on sandy soils, when rain was scarce. Experiments performed with artificially-induced drought stress (under rain-out shelters) showed that the reduction of yield can exceed 50% (Kottmann et al. 2016), and when the genotype is sensitive to drought and conditions are very unfavorable, the plants may fail to produce grain (Czyczyło-Mysza and Myśków 2017). On the contrary, a resistant genotype shows no reduction in kernel weight even under strong drought stress (Czyczyło-Mysza and Myśków 2017). Farshadfar et al. (2013) distinguished resistance from tolerance. Resistance guarantees yield stability in optimal moisture and under water deficits as well. Tolerance refers to genotypes with acceptable yield performance in unfavorable conditions (when reduction of yield induced by drought is not extreme).

The biological response of rye plants to water deficiency is a very complex process and sometimes leads to avoiding drought through acceleration of maturity (speeding up of the senescence). As a consequence, the vegetative growth period of plants is usually shortened by drought (Kottmann et al. 2016). This strategy is unfavorable for crop production as it results in yield reduction.

Depending on the timepoint/developmental stage, drought may result in different consequences. Early drought (phenological stages: tillering, stem elongation) leads to the reduction in number of spikes per square meter and number of kernels per spike, and late drought (heading, flowering, etc.) affects mainly thousand kernel weight (Kottmann et al. 2016). Plant height and straw yield may be also modified by drought (Haffke et al. 2015; Kottmann et al. 2016). There are likely some common physiological mechanisms for drought response and dwarfness in rye (Braun et al. 2019).

Localization of rye genes responsible for the reaction to drought stress has been investigated mainly in wheat translocation or addition lines. Identification of genes conferring drought tolerance depended strongly on which wheat and rye varieties had been used for development of studied genotypes. The most common translocation used in research involved the short arm of chromosome 1R. The fusion of 1RS with the long arms of wheat chromosomes 1A, 1B, or 1D usually resulted in increased root system development and drought tolerance (Ehdaie et al. 2003; Hoffmann 2008; Sharma 2009; Howell et al. 2014; Karki et al. 2014). Genome-wide experiments with the full set of single chromosome addition lines of wheat 'Chinese Spring'/ rye 'Imperial' were conducted by Farshadfar et al. (2003). The presence of 3R, 5R, and 7R in the wheat genome positively influenced physiological parameters and yield stability under drought conditions. Furthermore, 2R, 4R, and 6R affected the response of plants to water deficit, too. The research mentioned above considered the phenotypic effects of rye whole chromosomes or chromosome arms introduced into the wheat genome. Detailed mapping of genes responsible for drought resistance was not conducted. These studies also provided no insight into potentially existing genetic variation present within rye germplasm.

Interval mapping (localization of QTLs on genetic maps of chromosomes) is a method allowing for relatively precise localization of genes controlling complex traits. Myśków et al. (2018) phenotyped a mapping population consisting of rye genotypes segregating for drought sensitivity under watered and dry conditions and identified 25 quantitative trait loci (QTLs) responsible for the reaction of rye plants to water deficit. These QTLs were distributed among all chromosomes, with thirteen of them co-localized with QTLs for morphology, leaf rolling, preharvest sprouting, α-amylase activity, and heading earliness (Myśków et al. 2018) (Fig. 11.5).

The senescence of triticale plants in drought conditions (Ostrowska et al. 2019). revealed the presence of seven QTLs associated with this process. Among them, three QTLs were located on rye chromosomes: one QTL on chromosome 1R and two on 6R.

The number of experiments on the function of drought-related genes in rye under stress is limited, thus little is known about the physiological mechanisms and proteins associated with drought. Several studies (Yu and Griffith 2001; Yu et al. 2001; Griffith and Yaish 2004) focused on antifreeze proteins (AFP), which have been shown to respond to dehydration conditions. This confirms the well-known phenomenon of overlapping genetic regulation under different stresses. For example, the components of drought and salt stress coincided as both these stresses ultimately result in dehydration of the cell and osmotic imbalance. Dehydrins, also known as group 2 LEA proteins, accumulate in response to both dehydration as well as low temperature (Mahajan and Tuteja 2005). Cryotolerant species, such as wheat and rye, accumulate more heat-stable dehydrins than cryosensitive species, such as maize (Borovskii et al. 2002). The metabolic pathways associated with responses to various stresses are often distinguished into ABA-dependent and ABA-independent signaling pathways. Yu and Griffith (2001) stated that the induction of antifreeze activity and accumulation of AFPs in response to low temperature and drought is not mediated by ABA. Yu et al. (2001) also concluded that ethylene is involved in regulating antifreeze activity in winter rye in response to cold and drought.

A few reports indicate a relationship exists between drought and phenolics content in rye. A slight decrease in the level of total phenolics content was observed in both the rye caryopses of various ripeness (Weidner et al. 2000) and mature plants of inbred lines (Czyczyło-Mysza and Myśków 2017) in conditions of reduced water availability. The phenolic acid fraction (consisting of both bound and free forms) decreased more during dehydration, especially at the initial stage of caryopses development. Cells

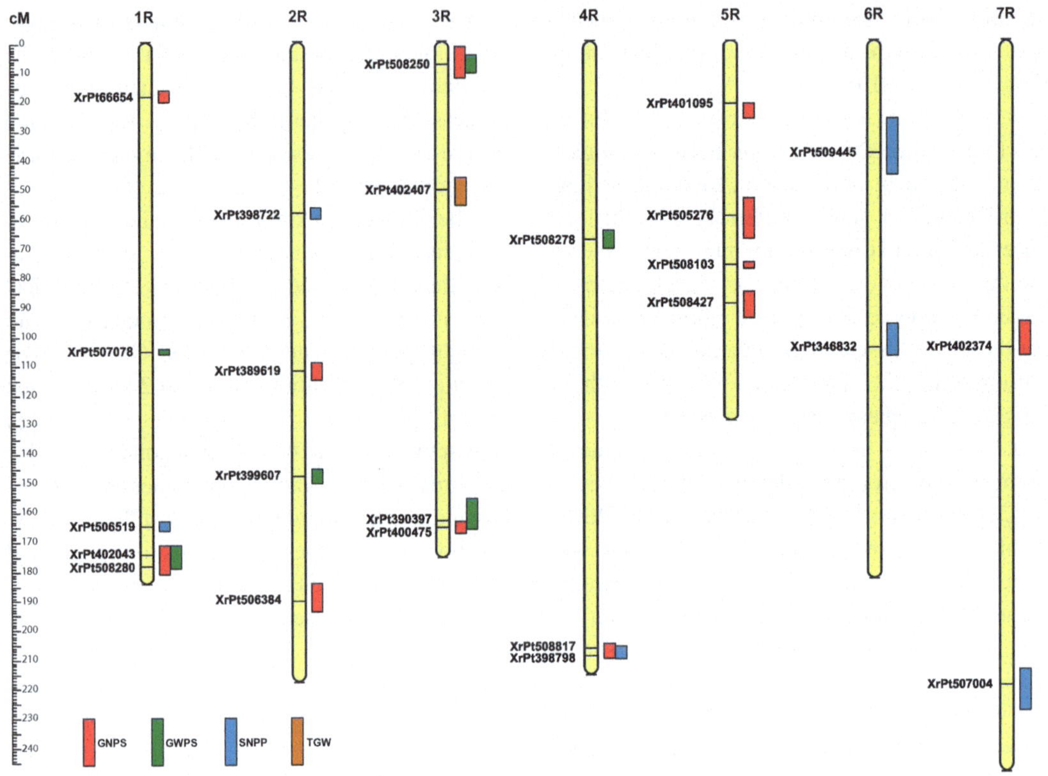

Fig. 11.5 Localization of QTLs for drought index of grain number per spike (GNPS), grain weight per spike (GWPS), spike number per plant (SNPP), and thousand grain weight (TGW) on high-density genetic map of rye population RIL-M (Myśków et al. 2018, modified)

of immature rye grains reacted to water stress by lowering the level of total phenolic compounds, and especially the content of phenolic acids (Weidner et al. 2000). Phenolics are known antioxidants helping to prevent cellular damages caused by oxidative stress and can act as metal chelators, directly scavenging molecular species of active oxygen; their exact role, however, in plant stress responses in nature is still under debate.

The availability of large databases can serve as a resource and tool for researchers to gather information about orthologs of drought stress associated genes in many plant species. A few years ago, DroughtDB (http://pgsb.helmholtz-muenchen.de/droughtdb/drought_db.html), a compilation of genes involved in drought stress response was created to support scientists working with agriculturally important species. This database includes information about the originally identified gene, its function, and mutant phenotypes, and provides detailed information about bioinformatically identified orthologous genes in nine model and crop plant species, including rye. The study provides information on 199 orthologs divided into two main categories: physiological and molecular adaptation genes (Alter et al. 2015). Recently, a rye transcriptome sequencing experiment was performed and analyzed in a comparative context, resulting in the identification of 75 rye transcripts homologous to 43 functionally characterized rice genes, conferring drought tolerance (Braun et al. 2019). Some of these genes have a specific chromosomal position on a high-density map. The increasing collections of genomic sequences from genome-wide sequencing and RNA-seq projects for rye (Haseneyer et al. 2011; Bauer et al. 2017; Braun et al. 2019; Bienias et al. 2020) will allow further identification of genes involved in the response to drought and other stresses based on their orthology.

11.3.3 Low Temperature

Exposure to low temperature can manifest growth, development, and yield disorders (Thakur and Nayyar 2013). At the cellular level, it blocks metabolism, reduces the photosynthetic ability, and changes the permeability of cell membranes, and in its extreme form causes injury and can lead to death (Chinnusamy et al. 2007; Gusta et al. 2009). Depending on the ambient temperature, low temperature stress is divided into chilling stress (between 10 and 0 °C) and natural freezing (below 0 °C) (Ritonga and Chen 2020). Moreover, the tolerance to frost and its mechanisms varies depending on whether it is freezing stress with previous cold acclimatization or a sudden freezing shock (Ball et al. 2012). The complex physiological changes that occur during cold acclimatization and development of frost tolerance are based on comprehensive transcriptomic changes (Hincha et al. 2012).

Rye has the highest resistance to low temperature among small grain cereals and can survive temperatures as low as −35 °C (Fowler and Limin 1987). Frost resistant rye plants can be characterized by a slow growth rate in the autumn, a short mesocotyl and tillering node located deeply in the soil, short and narrow leaves, and high dry matter content (Schlegel 2013). Among winter cereals, rye starts to cold-acclimate at the highest temperature, therefore making the acclimation period the longest and the plants less sensitive to temperature fluctuations during this time. Therefore, rye is able to assimilate more sugar and is better adapted to overcome photoinhibition (Fowler 2008). Sugars are not only a source of energy but also have a cryoprotective and signaling effect that triggers the induction of genes, proteins, and metabolites increasing resistance to frost (Stitt and Hurry 2002). After cold acclimation, winter rye endogenously produces antifreeze protein (AFP) that can improve cold tolerance in plants by inhibiting the growth and recrystallization of ice in intercellular spaces (Griffith et al. 1992). The accumulation of AFP in the rye apoplast is in response to cold, short-day length, dehydration, and ethylene. Antifreeze activity involves glucanases, chitinases, and thaumatin-like proteins (Griffith et al. 1992). The formation of frost tolerance during cold adaptation is partly the result of metabolic changes that are strongly linked to the genotype. Tolerance is associated with an increase in total protein, free amino acids, soluble sugars, and glycine betaine (Janmohammadi et al. 2018). Increased frost tolerance is also induced by brassinosteroids and is related to carbohydrate metabolism (Pociecha et al. 2016). In rye, ethylene also regulates antifreeze activity in response to low temperatures and drought (Yu et al. 2001).

The C-repeat binding factor (*CBF*) gene family regulates cold-responsive genes expression under low temperature stress in *Arabidopsis* (Jia et al. 2016). So far, 27 *ScCBF* genes associated with low temperature stress have been identified in rye (Campoli et al. 2009; Jung and Seo 2019). The majority of these genes were assigned to the long arm of chromosome 5R and another three were mapped on chromosome 2R (*ScCBF7a, ScCBFIa-11.2, ScCbfIa-11*), and two on chromosome 6R (*ScCBF1, ScCBF7b*) (Jung and Seo 2019; Campoli et al. 2009). Regulation of low temperature resistance and vernalization are associated in cereals (Kosová et al. 2008; Galiba et al. 2009). *Vernalization locus 1* (*Vrn1*), an additional important gene of frost tolerance in *Triticeae*, was mapped close to the frost tolerance locus, *Fr1*, on the long arm of homoeologous group 5 near the *Fr2* locus in wheat (Galiba et al. 1995). In European rye populations, a total of 147 single nucleotide polymorphisms (SNP) and 9 in-dels were described for eleven candidate genes (*ScCbf2, ScCbf6, ScCbf9b, ScCbf11, ScCbf12, ScCbf14, ScCbf15, ScVrn1, ScIce2, ScDhn1, and ScDhn3*). A single haplotype was identified for three of them (*ScCbf14, ScVrn1 and ScDhn1*). The remaining genes had a more balanced distribution regarding their haplotype frequency. Two SNP markers XScCbf15 and XScCbf12 were associated with frost tolerance (Li et al. 2011).

In cereals, locus *Fr2* is associated with acclimatization to low temperatures (Fowler 2008; Knox et al. 2008; Campoli et al. 2009). In the mapping population 'Lo157' x 'Puma-SK', low temperature tolerance QTLs were detected on chromosomes 4R, 5R, and 7R. On 5R, a QTL coinciding with locus *Fr-R2* represents variation in the initial rate of cold acclimation (Erath 2017). An *Fr-R2* allele, identified in cv. 'Puma-SK' significantly increased frost tolerance. In the *Triticeae*, *Fr-2* contains a *CBF* gene cluster. In rye, nine putative full-length *CBF* genes and one pseudo *CBF* gene were found at *Fr-R2* and the *XPG-I* and *MatE* genes defined the borders of the locus (Erath 2017). The latest study revealed a cluster of 21 *CBF*-related genes at the *Fr2* locus on chromosome 5R and copy number variation was also identified for four of the *CBF* genes (Rabanus-Wallace et al. 2019). These genes belonged to the same *CBF* subfamily in which copy number variation associated with low temperature tolerance, which was previously detected in wheat (Würschum et al. 2017).

Analysis of the rye transcriptome under cold stress resulted in identifying 29,874 differentially expressed genes. These were related to photosynthesis, plasma membrane stability, glucose, and energy metabolism, as well as cold-response transcription factors (Kong et al. 2020). The Gene Ontology analysis revealed that extracellular components such as waxes, cutin, and suberine are synthesized to protect against low temperatures. *MNS1* and *MNS3*, oligosaccharides and chitin synthesis-related genes were indicated as candidates for playing a key role in cold resistance. MYB and bHLH transcription factors families showed a close linkage with low temperature resistance. The expression of other common low temperature stress genes such as *HSPs*, *NAC*, *bZIPs*, *C2H2* and *CBF* changed after cold stimulation.

11.4 The Genetic Background and Role of Allelochemicals in Defense Against Biotic and Abiotic Stresses

One of the key elements of plant defense strategies against various environmental threats are specialized metabolites (SM), such as phenolics, terpenoids, alkaloids, and nitrogen-containing chemicals (Kong et al. 2019). In rye, the main protective SMs are benzoxazinoids (BXs) and, to a much lesser degree—phenolic acids. The BXs are primarily involved in the defense of plants against fungi, insects, and weeds, high soil salinity, the phenolic acids show mainly antimicrobial activity (Carlsen et al. 2009). BXs are the group of SMs that have been most fully characterized in terms of their biosynthesis, genetic background, and in terms of their role in defense reactions.

11.4.1 Benzoxazinoids (BXs)

The BXs are synthesized mainly by species belonging to the family Poaceae, including rye (Frey et al. 2009; Makowska et al. 2015). In rye, BXs are synthesized at a very high level, thus playing a significant role in processes such as allelopathy (Gavazzi et al. 2010; Schulz et al. 2013; Tabaglio et al. 2013), passive defense against nematodes (Zasada et al. 2005, 2007; Meyer et al. 2009), active defense against LR infection (Rakoczy-Trojanowska et al. 2017; Święcicka et al. 2019), allelochemicals secreted by Berseem clover (*Trifolium alexandrinum* L.) growing in the proximity to rye plants (Rakoczy-Trojanowska et al. 2020) and reaction to soil salinity (Makleit 2005; Bakera and Rakoczy-Trojanowska 2017; Rakoczy-Trojanowska et al. 2019). Their synthesis level is also affected by the low temperature during vernalization (Bakera et al. 2020).

In all cereal species synthetizing BXs, their biosynthesis proceeds in a similar way. The first step is the conversion of indole-3-glycerol phosphate to indole, which is catalyzed by indole-3-glycerol phosphate lyase. Four consecutive cytochrome P450 monooxygenases then convert indole to 2,4-dihydroxy-2*H*-1,4-benzoxazin-3 (4*H*)-one (DIBOA). Other reactions of the BX biosynthesis pathway include the glycosylation of DIBOA to produce 2-O-β-glucoside and O-methylation to generate 2,4,7-trihydroxy-1,4-benzoxazin-3-one (TRIBOA) glucoside as well as the glycosylation of 2,4-dihydroxy-7-methoxy-1,4-benzoxazin-3-one (DIMBOA) to produce 2,4-dihydroxy-7-methoxy-1,4-benzoxazin-3-one (DIMBOA) glucoside and O-methylation to generate 4,7-dimethoxy-2-{[3,4,5-trihydroxy-6-(hydroxymethyl)oxan-2-yl]oxy}-3,4-dihydro-2H-1,4-benzoxazin-3-one (HDMBOA) glucoside. Hydroxylations convert GDIBOA and GDIMBOA to DIBOA and DIMBOA, respectively (Frey et al. 2009; Niculaes et al. 2018). To date, numerous genes that control BX biosynthesis have been isolated and characterized in several Poaceae species, with the highest number 14, found in maize (Frey et al. 2009; Niculaes et al. 2018). The ten rye genes involved in BX biosynthesis identified in the last decade are *ScBx1 (Acc.No* KF636828.1), *ScBx2 (Acc.No* KF620524.1), *ScBx3 (Acc.No* KF636827.1), *ScBx4 (Acc.No* KF636826.1), *ScBx5 (Acc.No* KF636825.1), *ScBx6 (Acc.No* MG516219.1), *ScBx7 (Acc.No* MG519859.1), *Scglu (Acc.No* AY586531.2), *ScGT (Acc.No* AB548283.1), and *ScIgl (Acc.No* MN120476.1) (Sue et al. 2011; Bakera et al. 2015; Tanwir et al. 2017; Bakera and Rakoczy-Trojanowska 2020; Wlazło et al. 2020).

To date, rye BXs, including the genetic background of their biosynthesis, were mainly studied in respect of their role in allelopathy and a toxic impact on nematodes and to a lesser extent—diseases and abiotic stresses.

BXs secreted by rye roots effectively inhibit or reduce the germination, growth, and development not only of many problematic agronomic grasses and broadleaf weeds; e.g., *Chenopodium album* L., *Amaranthus retroflexus* L., and *Portulaca oleracea* L., but also some crops, e.g., cucumber (*Cucumis sativus* L.), melon (*C. melo* L.), tomato (*Solanum lycopersicum* L.), lettuce (*Lactuca sativa* L.), maize (*Zea mays* L.), tobacco (*Nicotiana tabacum* L.), even up to 98% (Gavazzi et al. 2010; Schulz et al. 2013; Tabaglio et al. 2013). In contrast, only a few species, *inter alia* three clovers, namely Berseem clover (*T. alexandrinum L.*), Persian clover (*T. resupinatum* L.), and white clover (*T. repens* L.), secrete phenolic compounds that are mildly allelopathic against rye (Maighany et al. 2007; Carlsen et al. 2012). More recently, Rakoczy-Trojanowska et al. (2020) showed that the co-cultivation of rye with Berseem clover significantly influenced its BX content and expression of related genes. The response was strongly dependent on the individual rye genotype, plant part, time point, gene, and metabolite. The most frequently observed changes were increased levels of *ScBx3* gene expression and GDIMBOA and DIMBOA synthesis, in roots, six weeks after seed germination.

Several previous studies showed a relationship between BXs and resistance to LR. However, the correlation between BX content and composition, and disease resistance was not always consistent, additionally, it depended on the infection place and pathogen character. Based on results of an association study, Rakoczy-Trojanowska et al. (2017) found one environmentally stable SNP (ScBx4_1583) associated with the disease resistance. It was located in the first intron of the gene *ScBx4* encoding a cytochrome P450 monooxygenase. Recently, Święcicka et al. (2019) showed that in rye, both syntheses of BXs and expression of related genes were affected by infection with LR. The key components of the defense response against LR of all analyzed rye inbred lines were as follows: *ScBx1*, *ScBx2*, *ScBx4*, and *Scglu* as well as GDIBOA and GDIMBOA, especially at two of the most critical time-points, 17 and 24 h post-treatment. The changes in gene expression levels and BX content were usually positively associated with disease resistance. The intensity of the reaction depended on the genotype, with the most resistant genotypes mobilizing their defense mechanisms more effectively, in a more coordinated manner, and earlier than the less resistant ones. Interestingly, the mock treatment itself also induced BX synthesis, but in this case, it was MBOA. Therefore, it may be concluded that GDIBOA and GDIMBOA are important components of rye defense responses to LR. Furthermore, along with MBOA, they protect rye against the stress associated with the inoculation procedure and against other stresses of this kind.

Low temperature, is essential for vernalisation of winter crops and is one of the few abiotic stresses influencing BX synthesis (and expression of related genes) in rye. The research conducted by Bakera et al. (2020) showed that after cultivation for seven weeks at 4 °C, the content of all analyzed BXs (HBOA, DIBOA, GDIBOA, DIMBOA, GDIMBOA, and MBOA) and the expression level of six genes *ScBx1–ScBx5* and *ScIgl*) decreased, when compared to those at the initiation of the treatment (21 days after germination) in control and cold-treated plants. At this time point, the decrease in BX concentrations and gene expression was lower in cold-treated plants than in untreated plants. In contrast, on the 77th day after germination, the gene expression levels and BX concentrations in untreated plants had generally increased. These results may be used in breeding aimed at obtaining rye cultivars with increased BX content and thus showing improved stress tolerance.

Makleit (2005) investigated the relationship between DIBOA content in two *Secale* species (*S. cereale* and *S. cereanum*) and the reaction to soil salinity. DIBOA content was shown in all tested varieties to increase between the fourth and eighth hour to between 184.94 and 208.46% after treatment with 100 mM NaCl compared with the control (untreated) plants. It was also reported that the expression of five *ScBx* genes (*ScBx1–ScBx5*) increased significantly under the same salinity stress as applied by Makleit (2005); (Bakera and Rakoczy-Trojanowska 2017; Rakoczy-Trojanowska et al. 2019).

11.5 Rye as a Source of Complex Resistance for Wheat Improvement

For many years, rye has been the primary and the most valuable source of resistance genes used to improve wheat cultivars. The chromosome homoeology enables transferring rye chromatin to wheat. The list of wheat lines and cultivars containing rye introgressions can be found in the database developed by Schlegel (2019).

So far, rye chromosome 1R has been the most commonly used in wheat breeding programs. Over several decades, many lines containing translocations of 1RS.1AL, 1RS.1BL, and 1RS.1DL and substitutions of 1R(1A), 1R(1B), and 1R(1D) have been developed. The 1RS.1AL translocation enriched the wheat gene pool with resistance genes to fungal diseases such as SR (*Sr1RS*^{Amigo}) and PM (*Pm17* allelic to *Pm8*) and pests such as Bird cherry-oat aphid (unknown gene), wheat aphid (*Gb2* and *Gb6*) and wheat curl mite (*Cmc3*) (Friebe et al. 1996; Mohler et al. 2001; Malik et al. 2003; Lu et al. 2010; Crespo-Herrera et al. 2013). The 1RS.1BL translocation has substantially contributed to the global increase in wheat production by the introduction of several resistance genes (Johansson et al. 2020). It has been commonly used in breeding programs around the world since the mid 1980s (Dong et al. 2020). It provides resistance genes to pathogens, i.e., LR (*Lr26*), SR (*Sr31*), YR (*Yr9, YrCn17, YrR212 and yrCH45-1b*), PM (*Pm8, PmCn17*), and pest RWA (*Dn7*) (McIntosh et al. 1995; Anderson et al. 2003; Lapitan et al. 2007; Luo et al. 2008; Yang et al. 2016; Ren et al. 2017). Some studies also indicate an increased resistance to drought in wheat with the 1RS.1BL translocation (Hoffmann 2008; Golkari and Hasaniani 2017; Jang et al. 2017). The 1RS.1DL translocation involves the introduction of another SR resistance gene (*Sr50*) (Mago et al. 2015). In wheat lines containing substitution 1R(1B), an increased allelopathic activity against weeds was observed

(Bertholdsson et al. 2012). The chromosome 1R is also a source of genes that can increase wheat zinc efficiency (Cakmak et al. 1997).

The 2R chromosome contains resistance genes against biotic and abiotic stressors including pathogenic fungi, pests, weeds, and water deficiency (Johansson et al. 2020). The 2BS.2RL translocations improved resistance to Hessian fly (determined by *H21* gene) as well as to LR, SR, and PM. However, the genetic background of resistance to these diseases remains unknown (Friebe et al. 1990; Hysing et al. 2007). Other 2R translocations to wheat made it possible to confer resistance to fungal pathogens resulting from introgression of: *Lr25* and *Pm7* (4BS.4BL-2RL), *Lr45* (2AS-2RS.2RL) and *Sr59* (2DS.2RL) (Friebe et al. 1996; Rahmatov et al. 2016a; Wu et al. 2020). The disomic substitution 1R(1D), 2R addition line, as well as the 2R(2D) substitution line studied by Zhuang et al. (2011) showed increased resistance to PM. In the first case, it is a favorable effect of the *PmJZHM2RLb* gene introgression into wheat, whereas, in the second case, the genetic determinants of the obtained resistance have not been identified (An et al. 2006). Furthermore, the addition of $2R^{afr}$ from *Secale africanum* Stapf. or the substitution of $2R^{afr}$(2D) as well as the addition of $2R^{afr}L$ or the translocation of 2DS.$2R^{afr}L$ significantly increased the resilience of wheat to YR (Lei et al. 2013). The 2R introgression to the wheat genome can also contribute to a better adaptation to drought stress through the improvement of water use and root system architecture (Ehdaie et al. 2003). It also improves weed suppression ability in winter wheat (Bertholdsson et al. 2012).

The 3R chromosome carries the *Sr27* and Sr^{Satu} resistance genes, which are effective against a wide spectrum of SR races (Rahmatov et al. 2016b). The 3RS.3AL translocation provided the ability to take advantage in wheat of the resistance derived from *Sr27* (Friebe et al. 1996). Some 3R(3D) substitution lines contain Sr^{Satu} and/or probably a novel SR resistance gene (s) (Rahmatov et al. 2016b).

The 4R chromosome contains resistance genes against pests and diseases (Lukaszewski et al. 2001; An et al. 2013, 2019). However, the very low recombination frequency of 4R with 4A, 4B, and 4D and the compensation problems, makes it extremely difficult to use the potential of this chromosome in wheat breeding (Lukaszewski et al. 2001). Nevertheless, the translocations 4BL.4RL + 7AS.4RS and 5DS-4RS.4RL have resulted in increased resistance to PM, although the genes that determine this resistance have not been identified (An et al. 2013; Fu et al. 2014). The $4R^{add}$ disomic addition line exhibits resistance to PM, YR, and sharp eyespot, which is most likely determined by novel genes (An et al. 2019).

Similarly, like the case of 4R, the use of 5R elite genes is hampered by low-frequency and the non-compensating recombination of the chromosome with its wheat homeologues (Lukaszewski et al. 2001; Lukaszewski 2015). The wheat lines containing 5AS.5RL + 1RS.1BL translocation were resistant to LR and PM (Chumanova et al. 2014). A set of $5R^{Ku}$ dissection lines carrying potentially new YR resistance gene(s) were recently obtained by Xi et al. (2019). They may be used for diversity enrichment in wheat breeding programs for YR resistance. Apart from the resistance to biotic agents, the 5R(5A) substitution increased the tolerance to excess copper in wheat (Schlegel et al. 1991; Bálint et al. 2003). The 5R chromosome also has the potential to increase low temperature tolerance of wheat. However, to date, no significant change in cold tolerance has been observed in a line containing the 5AS.5RL translocation (Rabanus-Wallace et al. 2019). Further research in this field is necessary.

The 6R introgressions into the wheat genetic background have provided resistance to fungal diseases, pests, nematodes, weeds, and drought. Resistance to PM determined by *Pm20* and *Pm56* has been established by 6BS.$6RL^{rec}$ and 6RS.6AL translocations, respectively (Friebe et al. 1994; Hao et al. 2018). Also, the 6R disomic addition line and 6RL monotelosomic or ditelosomic addition lines showed high levels of resistance to PM (An et al. 2015; Fu et al. 2014). Unfortunately, genetic imbalances caused by the long arm of 6R in wheat disturb the utilization of these genes. More stable lines are small-segment introgression lines, which can be used in wheat breeding (Li et al. 2016). Research on small-segment wheat-rye 6D.6RL^{Ku} translocation line with a minichromosome of rye showed the potential of this chromosome to counteract PM virulent pathotypes. However, further studies on the structure of this line and its agronomic traits are necessary prior to including it in breeding programs (Du et al. 2018). A new YR resistance gene (*Yr83*) was discovered through disomic substitution 6R(6D) (Li et al. 2020b). Both the 6DS.6RL translocation line and the 6R(6D) substitution line contain the cereal cyst nematode resistance gene (*CreR*). Thus, they can be used in resistance breeding against *Heterodera avenae* and *Heterodera filipjevi* (Dundas et al. 2001; Cui et al. 2012). The 4BS.4BL-6RL translocation also contains the *H25* gene determining resistance to Hessian fly (Friebe et al. 1996). The studies on 6R disomic addition lines showed the presence of genes controlling drought resistance (Farshadfar et al. 2013). Moreover, the additional 6R pair in the 1R(1D) multisubstitution wheat lines increased their allelopathic activity and weed suppressive ability (Bertholdsson et al. 2012).

The 7R chromosome is the only one that is not the source of resistance to biotic stresses (Crespo-Herrera et al. 2017), but it carries genes controlling zinc efficiency. The addition of 7R (as well as 1R) into wheat cv. 'Holdfast' reduced the severity of Zn deficiency symptoms (Cakmak et al. 1997).

As shown above, rye chromosome introgressions have provided multiple benefits to wheat breeding programs, which are not only related to the resistance to biotic factors but also result from increasing yield potential, biomass production (Carver and Rayburn 1994; Villareal et al. 1998; Shearman et al. 2005) and better response to abiotic stress, especially drought (Kim et al. 2004; Hoffmann 2008; Ehdaie et al. 2012; Farshadfar et al. 2013; Howell et al. 2014).

References

Abdullah S, Sehgal SK, Glover KD, Ali S (2017) Reaction of global collection of rye (*Secale cereale* L.) to tan spot and *Pyrenophora tritici-repentis* races in South Dakota. Plant Pathol J 33(3):229–237. https://doi.org/10.5423/PPJ.OA.12.2016.0265

Aguirre-Rojas LM, Khalaf LK, Garcés-Carrera S et al (2017) Resistance to wheat curl mite in arthropod-resistant rye-wheat translocation lines. Agronomy 7:74. https://doi.org/10.3390/agronomy7040074

Alam MD, Fei X, Wang C, Ji W (2011) Powdery mildew resistance genes in wheat: identification and genetic analysis. J Mol Biol Res 1:1–20. https://doi.org/10.5539/jmbr.v1n1p20

Alter S, Bader K, Spannagl M et al (2015) Drought DB: an expert-curated compilation of plant drought stress genes and their homologs in nine species. Database, bav046. https://doi.org/10.1093/database/bav046

An DG, Li LH, Li JM et al (2006) Introgression of resistance to powdery mildew conferred by chromosome 2R by crossing wheat nullisomic 2D with rye. J Integr Plant Biol 48(7):838–847. https://doi.org/10.1111/j.1744-7909.2006.00275.x

An DG, Zheng Q, Zhou YL et al (2013) Molecular cytogenetic characterization of a new wheat-rye 4R chromosome translocation line resistant to powdery mildew. Chromosome Res 21:419–432. https://doi.org/10.1007/s10577-013-9366-8

An D, Zheng Q, Luo Q et al (2015) Molecular cytogenetic identification of a new wheat-rye 6R chromosome disomic addition line with powdery mildew resistance. PLoS One 10(8):e0134534. https://doi.org/10.1371/journal.pone.0134534

An D, Ma P, Zheng Q et al (2019) Development and molecular cytogenetic identification of a new wheat-rye 4R chromosome disomic addition line with resistances to powdery mildew, stripe rust and sharp eyespot. Theor Appl Genet 132(1):257–272. https://doi.org/10.1007/s00122-018-3214-3

Anderson GR, Papa D, Peng J et al (2003) Genetic mapping of *Dn7*, a rye gene conferring resistance to the Russian wheat aphid in wheat. Theor Appl Genet 107(7):1297–1303. https://doi.org/10.1007/s00122-003-1358-1

Andersson SC, Johansson E, Baum M (2015) New resistance sources to Russian wheat aphid (*Diuraphis noxia*) in Swedish wheat substitution and translocation lines with rye (*Secale cereale*) and *Leymus mollis*. Czech J Genet Plant Breed 51(4):162–165. https://doi.org/10.17221/72/2015-CJGPB

Bakera B, Rakoczy–Trojanowska M (2017) Genetic basis of the biosynthesis of benzoxazinoids in rye (*Secale cereale* L.). Characterization of *ScBx* gene expression under increased salinity. In: 4th National Conference - "Natural plant substances - structural and application aspects". 6–8 September 2017. Puławy, Poland, pp 62–63

Bakera B, Rakoczy-Trojanowska M (2020) Isolation and structural analysis of the *Bx6* and *Bx7* genes controlling the biosynthesis of benzoxazinoids in rye (*Secale cereale* L.). Acta Physiol Plant 42:56. https://doi.org/10.1007/s11738-020-03046-8

Bakera B, Makowska B, Groszyk J et al (2015) Structural characteristics of *ScBx* genes controlling the biosynthesis of hydroxamic acids in rye (*Secale cereale* L.). J Appl Genet 56:287–298. https://doi.org/10.1007/s13353-015-0271-z

Bakera B, Święcicka M, Stochmal A et al (2020) Benzoxazinoid biosynthesis in rye (*Secale cereale* L.) is affected by low temperature. Agronomy 10(9):1260. https://doi.org/10.3390/agronomy10091269

Bálint A, Kovács G, Börner A et al (2003) Substitution analysis of seedling stage copper tolerance in wheat. Acta Agron Hung 51(4):397–404. https://doi.org/10.1556/AAgr.51.2003.4.4

Ball MC, Harris-Pascal D, Egerton JJG et al (2012) The paradoxical increase in freezing injury in a warming climate: frost as a driver of change in cold climate vegetation. In Storey KB, Tanino KK (eds) Temperature adaptation in a changing climate nature at risk. CABI, pp 179–185

Bauer E, Schmutzer T, Barilar I et al (2017) Towards a whole-genome sequence for rye (*Secale cereale* L.). Plant J 89(5):853–869. https://doi.org/10.1111/tpj.13436

Bertholdsson NO, Andersson SC, Merker A (2012) Allelopathic potential of *Triticum* spp., *Secale* spp. and *Triticosecale* spp. and use of chromosome substitutions and translocations to improve weed suppression ability in winter wheat. Plant Breed 131(1):75–80. https://doi.org/10.1111/j.1439-0523.2011.01895.x

Berzonsky WA, Ding H, Haley SD et al (2003) Breeding wheat for resistance to insects. Plant Breed Rev 22:221–296. https://doi.org/10.1002/9780470650202.ch5

Bienias A, Góralska M, Masojć P et al (2020) The *GAMYB* gene in rye: sequence, polymorphisms, map location, allele-specific markers, and relationship with α-amylase activity. BMC Genom 21:578. https://doi.org/10.1186/s12864-020-06991-3

Blum U, Gerig TM, Worsham AD et al (1992) Allelopathic activity in wheat-conventional and wheat-no-till soils: development of soil extract bioassays. J Chem Ecol 18:2191–2221. https://doi.org/10.1007/BF00984946

Borovskii GB, Stupnikova IV, Antipina AI et al (2002) Accumulation of dehydrin-like proteins in the mitochondria of cereals in response to cold, freezing, drought and ABA treatment. BMC Plant Biol 2:5. https://doi.org/10.1186/1471-2229-2-5

Braun EM, Tsvetkova N, Rotter B et al (2019) Gene expression profiling and fine mapping identifies a *gibberellin 2-oxidase* gene co-segregating with the dominant dwarfing gene *Ddw1* in rye (*Secale cereale* l.). Front Plant Sci 10:1–19. https://doi.org/10.3389/fpls.2019.00857

Cakmak I, Derici R, Torun B et al (1997) Role of rye chromosomes in improvement of zinc efficiency in wheat and triticale. Plant Soil 196:249–53. https://doi.org/10.1023/a:1004210309876

Campoli C, Matus-Cádiz MA, Pozniak CJ et al (2009) Comparative expression of *Cbf* genes in the Triticeae under different acclimation induction temperatures. Mol Genet Genom 282(2):141–152. https://doi.org/10.1007/s00438-009-0451-9

Carlsen SCK, Kudsk P, Lauren B etal (2009) Allochemicals in rye (*Secale cereale* L.): cultivar and tissue differences in the production of Benzoxazinoid and phenolic acids. Nat Prod Commun 4(2):199–208. https://doi.org/10.1177/1934578X0900400206

Carver BF, Rayburn AL (1994) Comparison of related wheat stocks possessing 1B or 1RS. 1BL chromosomes: agronomic performance. Crop Sci 34(6):1505–1510. https://doi.org/10.2135/cropsci1994.0011183X003400060017x

Chaves M, Martinelli J, Guterres C, Graichen F (2008) The cereal rusts: an overview. Pest Technol 2:38–55. https://doi.org/10.3389/fpls.2020.00667

Chinnusamy V, Zhu J, Zhu J-K (2007) Cold stress regulation of gene expression in plants. Trends Plant Sci 12(10):444–451. https://doi.org/10.1016/j.tplants.2007.07.002

Chumanova E, Efremova T, Trubacheeva N et al (2014) Chromosome composition of wheat-rye lines and the influence of rye chromosomes on disease resistance and agronomic traits. Russ J Genet 50(11):1169–1178. https://doi.org/10.1134/S1022795414110039

Collins NC, Shirley NJ, Saeed M et al (2008) An *ALMT1* gene cluster controlling aluminum tolerance at the *Alt4* locus of rye (*Secale cereale* L.). Genetics 179:669–682. https://doi.org/10.1534/genetics.107.083451

Crespo-Herrera LA, Smith CM, Singh RP, Åhman I (2013) Resistance to multiple cereal aphids in wheat–alien substitution and translocation lines. Arthropod Plant Interact 7(5):535–545. https://doi.org/10.1007/s11829-013-9267-y

Crespo-Herrera LA, Garkava-Gustavsson L, Åhman I (2017) A systematic review of rye (*Secale cereale* L.) as a source of resistance to pathogens and pests in wheat (*Triticum aestivum* L.). Hereditas 154(1):14. https://doi.org/10.1186/s41065-017-0033-5

Crespo-Herrera LA, Singh RP, Sabraoui A (2019) El-Bouhssini M (2019) Resistance to insect pests in wheat—rye and *Aegilops speltoides* Tausch translocation and substitution lines. Euphytica 215:123. https://doi.org/10.1007/s10681-019-2449-7 (0123456789

Cui L, Gao X, Wang X et al (2012) Characterization of interaction between wheat roots with different resistance and Heterodera filipjevi. Acta Agronomica Sin 38(6):1009–1017. https://doi.org/10.3724/SP.J.1006.2012.01009

Czyczyło-Mysza I, Myśków B (2017) Analysis of the impact of drought on selected morphological, biochemical and physiological traits of rye inbred lines.

Acta Physiol Plant 39:87. https://doi.org/10.1007/s11738-017-2385-x

De Sousa A, AbdElgawad H, Asard H et al (2016) Oxidative metabolism of rye (*Secale cereale* L.) after short term exposure to aluminum: uncovering the glutathione-ascorbate redox network. Front Plant Sci 7:685. https://doi.org/10.3389/fpls.2016.00685

Dhariwal R, Fedak G, Dion Y et al (2018) High density single nucleotide polymorphism (SNP) mapping and quantitative trait loci (QTL) analysis in a biparental spring triticale population localized major and minor effect Fusarium head blight resistance and associated traits QTL. Genes 9(1):1–26. https://doi.org/10.3390/genes9010019

Dong Z, Ma C, Tian X et al (2020) Genome-wide impacts of alien chromatin introgression on wheat gene transcriptions. Sci Rep 10(1):1–12. https://doi.org/10.1038/s41598-020-61888-1

Du H, Tang Z, Duan Q et al (2018) Using the 6RLKu minichromosome of rye (*Secale cereale* L.) to create wheat-rye 6D/6RLKu small segment translocation lines with powdery mildew resistance. Int J Mol Sci 19(12):3933. https://doi.org/10.3390/ijms19123933

Duan Q, Wang YY, Qiu L et al (2017) Physical location of new PCR-based markers and powdery mildew resistance gene(s) on rye (*Secale cereale* L.) chromosome 4 using 4R dissection lines. Front Plant Sci 8:1716. https://doi.org/10.3389/fpls.2017.01716

Dundas IS, Frappell DE, Crack DM, Fisher JM (2001) Deletion mapping of a nematode resistance gene on rye chromosome 6R in wheat. Crop Sci 41(6):1771–1778. https://doi.org/10.2135/cropsci2001.1771

Ehdaie B, Whitkus RW, Waines JG (2003) Root biomass, water-use efficiency, and performance of wheat-rye translocations of chromosomes 1 and 2 in spring bread wheat "Pavon." Crop Sci 43(2):710. https://doi.org/10.2135/cropsci2003.7100

Ehdaie B, Layne AP, Waines JG (2012) Root system plasticity to drought influences grain yield in bread wheat. Euphytica 186(1):219–232. https://doi.org/10.1007/s10681-011-0585-9

El Bouhssini M, Ogbonnaya FC, Chen M et al (2012) Sources of resistance in primary synthetic hexaploid wheat (*Triticum aestivum* L.) to insect pests: Hessian fly, Russian wheat aphid and Sunn pest in the Fertile Crescent. Genet Resour Crop Evol 60:621–627. https://doi.org/10.1007/s10722-012-9861-3

Erath W, Bauer E, Fowler DB et al (2017) Exploring new alleles for frost tolerance in winter rye. Theor Appl Genet 130:2151–2164. https://doi.org/10.1007/s00122-017-2948-7

Faris DJ, Zhaohui L, Xu SS (2013) Genetics of tan spot resistance in wheat. Theor Appl Genet 126:2197–2217. https://doi.org/10.1007/s00122-013-2157-y

Farshadfar E, Mohammadi R, Aghaee M et al (2003) Identification of QTLs involved in physiological and agronomic indicators of drought tolerance in rye using a multiple selection index. Acta Agron Hungarica 51 (4):419–428. https://doi.org/10.1556/AAgr.51.2003.4.7

Farshadfar E, Farshadfar M, Kiani M (2012) Involvement of chromosome 5R carrying the genes controlling yield and yield stability in rye (*Secale cereale* cv Imperial). IPCBEE 31:78–85. http://www.ipcbee.com/vol31/016-ICBBB2012-T30026.pdf

Farshadfar E, Mohammadi R, Farshadfar M et al (2013) Relationships and repeatability of drought tolerance indices in wheat-rye disomic addition lines. Aust J Crop Sci 7(1):130–138. https://api.semanticscholar.org/CorpusID:54979989

Fowler DB (2008) Cold acclimation threshold induction temperatures in cereals. Crop Sci 48(3):1147–1154. https://doi.org/10.2135/cropsci2007.10.0581

Fowler D, Limin A (1987) Exploitable genetic variability for cold tolerance in commercially grown cereals. Can J Plant Sci 67(1):278–278

Frey M, Schullehner K, Dick R et al (2009) (2009) Benzoxazinoid biosynthesis, a model for evolution of secondary metabolic pathways in plants. Phytochemistry 70:1645–1651. https://doi.org/10.1016/j.phytochem.2009.05.012

Friebe B, Hatchett J, Sears R, Gill B (1990) Transfer of Hessian fly resistance from 'Chaupon'rye to hexaploid wheat via a 2BS/2RL wheat-rye chromosome translocation. Theor Appl Genet 79(3):385–389. https://doi.org/10.1007/BF01186083

Friebe B, Heun M, Tuleen N et al (1994) Cytogenetically monitored transfer of powdery mildew resistance from rye into wheat. Crop Sci 34(3):621–625. https://doi.org/10.2135/cropsci1994.0011183X003400030003x

Friebe B, Jiang J, Raupp W et al (1996) Characterization of wheat-alien translocations conferring resistance to diseases and pests: current status. Euphytica 91(1):59–87. https://doi.org/10.1007/BF00035277

Fu S, Ren Z, Chen X et al (2014) New wheat-rye 5DS-4RS.4RL and 4RS-5DS.5DL translocation lines with powdery mildew resistance. J Plant Res 127:743–753. https://doi.org/10.1007/s10265-014-0659-6

Gaikpa DS, Lieberherr B, Maurer HP et al (2020) Comparison of rye, triticale, durum wheat and bread wheat genotypes for Fusarium head blight resistance and deoxynivalenol contamination. Plant Breed 139:251–262. https://doi.org/10.1111/pbr.12779

Galiano-Carneiro AL, Boeven PHG, Maurer HP et al (2019) Genome-wide association study for an efficient selection of Fusarium head blight resistance in winter triticale. Euphytica 215(1):4. https://doi.org/10.1007/s10681-018-2327-8

Galiba G, Quarrie SA, Sutka J et al (1995) RFLP mapping of the vernalization (*Vrn1*) and frost resistance (*Fr1*) genes on chromosome 5A of wheat. Theor Appl Genet 90(7–8):1174–1179. https://doi.org/10.1007/BF00222940

Galiba G, Vágújfalvi A, Li C et al (2009) Regulatory genes involved in the determination of frost tolerance in temperate cereals. Plant Sci 176(1):12–19. https://doi.org/10.1016/j.plantsci.2008.09.016

Gavazzi C, Schulz M, Marocco A, Tabaglio V (2010) Sustainable weed control by allelochemicals from rye cover crops: from the greenhouse to field evidence.

Allelopathy J 25(1):259–274. https://hdl.handle.net/10807/5007

Geiger HH, Miedaner T (2009) Rye breeding. In: Carena MJ (ed) Cereals (Handbook of Plant Breeding), vol 3. Springer. US, New York, pp 157–181

Golkari S, Hasaniani A (2017) Molecular evaluation of 1RS-rye translocation distribution in the Iranian dryland wheat cultivars and elite promising lines. Iran J Field Crop Sci 48:119–128. https://doi.org/10.22059/IJFCS.2017.216253.654181

Griffith M, Yaish MWF (2004) Antifreeze proteins in overwintering plants: a tale of two activities. Trends Plant Sci 9(8):399–405. https://doi.org/10.1016/j.tplants.2004.06.007

Griffith M, Ala P, Yang DS et al (1992) Antifreeze protein produced endogenously in winter rye leaves. Plant Physiol 100(2):593–596. https://doi.org/10.1104/pp.100.2.593

Gruner P, Schmitt AK, Flath K et al (2020) Mapping stem rust (*Puccinia graminis* f. sp. *secalis*) resistance in self-fertile winter rye populations. Frontiers Plant Sci 11:667. https://doi.org/10.3389/fpls.2020.00667

Gusta L, Wisniewski M, Trischuk R (2009) Patterns of freezing in plants: the influence of species, environment and experiential procedures. In: Gusta LV, Wisniewski ME, Tanino KK (eds) Plant cold hardiness: from the laboratory to the field. CABI, pp 214–225. https://doi.org/10.1079/9781845935139.0214

Haffke S, Wilde P, Schmiedchen B et al (2015) Toward a selection of broadly adapted germplasm for yield stability of hybrid rye under normal and managed drought stress conditions. Crop Sci 55(3):1026–1034. https://doi.org/10.2135/cropsci2014.08.0532

Haley SD, Peairs FB, Walker CB et al (2004) Occurrence of a new Russian wheat aphid biotype in Colorado. Crop Sci 44(5):1589–1592. https://doi.org/10.2135/cropsci2004.1589

Hao M, Liu M, Luo J et al (2018) Introgression of powdery mildew resistance gene *Pm56* on rye chromosome arm 6RS into wheat. Front Plant Sci 9:1040. https://doi.org/10.3389/fpls.2018.01040

Harry S, Graham R (1981) Tolerance of triticale, wheat and rye to copper deficiency and low and high soil pH. J Plant Nutr 3:721–730. https://doi.org/10.1080/01904168109362874

Harvey TL, Seifers DL, Martin TJ, Brown-Guedira GL, Gill BS (1999) Survival of wheat curl mites on different sources of resistance in wheat. Crop Sci 39:1887–1889. https://doi.org/10.2135/cropsci1999.3961887x

Haseneyer G, Schmutzer T, Seidel M et al (2011) From RNA-seq to large-scale genotyping—genomics resources for rye (*Secale cereale* L.). BMC Plant Biol 11:1–13. https://doi.org/10.1186/1471-2229-11-131

Hatchett JH, Sears RG, Cox TS (1993) Inheritance of resistance to Hessian fly in rye and in wheat-rye translocation lines. Crop Sci 33:730–734. https://doi.org/10.2135/cropsci1993.0011183X003300040019x

Hawliczek A, Bolibok L, Borzęcka E et al (2019) A search for genetic determinants of phosphorus deficiency tolerance in rye (*Secale cereale* L.). In: 5th Conference on Cereal Biotechnology and Breeding, Budapest, Hungary, November 4–7, 2019, pp 72–73

Hawliczek A, Bolibok L, Tofil K et al (2020) Deep sampling and pooled amplicon sequencing reveals hidden genic variation in heterogeneous rye accessions. BMC Genom 21:845. https://doi.org/10.1186/s12864-020-07240-3

Hincha DK, Espinoza C, Zuther E (2012) Transcriptomic and metabolomic approaches to the analysis of plant freezing tolerance and cold acclimation. In: Improving crop resistance to abiotic stress, pp 255–287. https://doi.org/10.1002/9783527632930.ch11

Hlavinka P, Trnka M, Semera D et al (2009) Effect of drought on yield variability of key crops in Czech Republic. Agric for Meteorol 149:431–442. https://doi.org/10.1016/j.agrformet.2008.09.004

Hoffmann B (2008) Alteration of drought tolerance of winter wheat caused by translocation of rye chromosome segment 1RS. Cereal Res Commun 36(2):269–278. https://doi.org/10.1556/CRC.36.2008.2.7

Howell T, Hale I, Jankuloski L et al (2014) Mapping a region within the 1RS. 1BL translocation in common wheat affecting grain yield and canopy water status. Theor Appl Genet 127(12):2695–2709. https://doi.org/10.1007/s00122-014-2408-6

Hsam SLK, Mohler V, Hartl L et al (2000) Mapping of powdery mildew and leaf rust resistance genes on the wheatrye translocated chromosome T1BL·1RS using molecular and biochemical markers. Plant Breed 119:87–89. https://doi.org/10.1046/j.1439-0523.2000.00444.x

Hübner M, Wilde P, Schmiedchen B et al (2013) Hybrid rye performance under natural drought stress in Europe. Theor Appl Genet 126(2):475–482. https://doi.org/10.1007/s00122-012-1994-4

Hysing S-C, Hsam SL, Singh RP et al (2007) Agronomic performance and multiple disease resistance in t2bs.2rl wheat-rye translocation lines. Crop Sci 47(1):254–260. https://doi.org/10.2135/cropsci2006.04.0269

Jang JH, Jung WJ, Kim DY, Seo YW (2017) cDNA-AFLP analysis of 1BL.1RS under water-deficit stress and development of wheat-rye translocation-specific markers. New Zealand J Crop Hort Sci 45(2):150–164. https://doi.org/10.1080/01140671.2016.1269018

Janmohammadi M, Sabaghnia N, Mahfoozi S (2018) Frost tolerance and metabolite changes of rye (*Secale cereale*) during the cold hardening and overwintering. Acta Physiol Plant 40(3):42. https://doi.org/10.1007/s11738-018-2620-0

Jia Y, Ding Y, Shi Y et al (2016) The *cbfs* triple mutants reveal the essential functions of *CBFs* in cold acclimation and allow the definition of *CBF* regulons in *Arabidopsis*. New Phytol 212(2):345–353. https://doi.org/10.1111/nph.14088

Johansson E, Henriksson T, Prieto-Linde ML et al (2020) Diverse wheat-alien introgression lines as a basis for durable resistance and quality characteristics in bread wheat. Front Plant Sci 11:1067. https://doi.org/10.3389/fpls.2020.01067

Jung WJ, Seo YW (2019) Identification of novel C-repeat binding factor (*CBF*) genes in rye (*Secale cereale* L.) and expression studies. Gene 684:82–94. https://doi.org/10.1016/j.gene.2018.10.055

Jurkowski A, Bujak H (2019) A search for 4 specific markers linked to *Pm3* alleles for resistance to powdery mildew (*Blumeria graminis*) in rye (*Secale cereale*). Biological Lett 54(1):37–45. https://doi.org/10.2478/biolet-2019-0004

Jurkowski A, Bujak H, Nowosad K (2014) Analysis of presence of markers STS638, Xgwm356, ResPm4 in rye (*Secale cereale* L.). Jokull 64:22–33

Kalih R, Maurer HP, Hackauf B, Miedaner T (2014) Effect of a rye dwarfing gene on plant height, heading stage, and Fusarium head blight in triticale (X *Triticosecale* Wittmack). Theor Appl Genet 127(7):1527–1536. https://doi.org/10.1007/s00122-014-2316-9

Kalih R, Maurer HP, Miedaner T (2015) Genetic architecture of Fusarium head blight resistance in four winter triticale populations. Phytopathology 105 (3):334–341. https://doi.org/10.1094/PHYTO-04-14-0124-R

Karki D, Wyant III W, Berzonsky WA et al (2014) Investigating physiological and morphological mechanisms of drought tolerance in wheat (*Triticum aestivum* L.) lines with 1RS translocation. Am J Plant Sci 05(13):1936–1944. https://cyberleninka.org/viewer_images/1016883/f/1.png

Kim W, Johnson J, Baenziger P et al (2004) Agronomic effect of wheat-rye translocation carrying rye chromatin (1R) from different sources. Crop Sci 44 (4):1254–1258. https://doi.org/10.2135/cropsci2004.1254

Klocke B (2004) Virulenzstruktur und dynamik des Roggenbraunrostes (*Puccinia recondita* f. sp. *secalis*) in der Bundesrepublik Deutschland. PhD thesis, Martin Luther University, Halle-Wittenberg, Germany, pp 1–141

Knox AK, Li C, Vágújfalvi A et al (2008) Identification of candidate *CBF* genes for the frost tolerance locus *Fr-Am* 2 in *Triticum monococcum*. Plant Mol Biol 67 (3):257–270. https://doi.org/10.1007/s11103-008-9316-6

Kong CH, Xuan TD, Khanh TD et al (2019) Allelochemicals and signaling chemicals in plants. Molecules 24(15):2737. https://doi.org/10.3390/molecules24152737

Kong Y, Zhang T, Guan Y et al (2020) Comparative transcriptome analysis reveals the responses of winter rye to cold stress. Acta Physiol Plant 42(5):77. https://doi.org/10.1007/s11738-020-03059-3

Kosová K, Prášil IT, Vítámvás P (2008) The relationship between vernalization-and photoperiodically-regulated genes and the development of frost tolerance in wheat and barley. Biol Plant 52(4):601. https://doi.org/10.1007/s10535-008-0120-6

Kottmann L, Wilde P, Schittenhelm S (2016) How do timing, duration, and intensity of drought stress affect the agronomic performance of winter rye? Eur J Agron 75:25–32. https://doi.org/10.1016/j.eja.2015.12.010

La Hovary C (2012) Allelochemicals in *Secale cereale*: Biosynthesis and molecular biology of benzoxazinones. PhD dissertation. http://gradworks.umi.com/34/63/3463787.html

Lapitan NL, Peng J, Sharma V (2007) A high-density map and PCR markers for Russian wheat aphid resistance gene Dn7 on chromosome 1RS/1BL. Crop Sci 47 (2):811–818. https://doi.org/10.2135/cropsci2006.08.0529

Lei MP, Li GR, Zhou L et al (2013) Identification of wheat-*Secale africanum* chromosome 2R afr introgression lines with novel disease resistance and agronomic characteristics. Euphytica 194(2):197–205. https://doi.org/10.1007/s10681-013-0913-3

Li Y, Böck A, Haseneyer G et al (2011) Association analysis of frost tolerance in rye using candidate genes and phenotypic data from controlled, semi-controlled, and field phenotyping platforms. BMC Plant Biol 11 (1):1–14. https://doi.org/10.1186/1471-2229-11-146

Li G, Wang Y, Chen MS et al (2015) Precisely mapping a major gene conferring resistance to hessian fly in bread wheat using genotyping-by-sequencing. BMC Genom 16:108. https://doi.org/10.1186/s12864-015-1297-7

Li M, Tang Z, Qiu L et al (2016a) Identification and physical mapping of new PCR-based markers specific for the long arm of rye (*Secale cereale* L.) chromosome 6. J Genet Genom 43(4):199–206. https://doi.org/10.1016/j.jgg.2015.11.005

Li G, Tang L, Yin Y et al (2020a) Molecular dissection of *Secale africanum* chromosome $6R^{afr}$ in wheat enabled localization of genes for resistance to powdery mildew and stripe rust. BMC Plant Biol 20:1–11. https://doi.org/10.1186/s12870-020-02351-1

Li J, Dundas I, Dong C et al (2020b) Identification and characterization of a new stripe rust resistance gene *Yr83* on rye chromosome 6R in wheat. Theor Appl Genet 133:1095–1107. https://doi.org/10.1007/s00122-020-03534-y

Lu H, Rudd JC, Burd J, Weng Y (2010) Molecular mapping of greenbug resistance genes Gb2 and Gb6 in T1AL. 1RS wheat-rye translocations. Plant Breed 129 (5):472–476. https://doi.org/10.1111/j.1439-0523.2009.01722.x

Lukaszewski AJ (2015) Introgressions between wheat and rye. In: Molnár-Láng M, Ceoloni C, Doležel J (eds) Alien introgression in wheat. Springer, pp 163–189. https://doi.org/10.1007/978-3-319-23494-6_7

Lukaszewski AJ, Porter DR, Baker CA et al (2001) Attempts to transfer Russian wheat aphid resistance from a rye chromosome in Russian triticales to wheat. Crop Sci 41(6):1743–1749. https://doi.org/10.2135/cropsci2001.1743

Luo P, Zhang H, Shu K et al (2008) Stripe rust (*Puccinia striiformis* f. sp. *tritici*) resistance in wheat with the wheat-rye 1BL/1RS chromosomal translocation. Can J

Plant Pathol 30(2):254–259. https://doi.org/10.1080/07060661.2008.10540540
Macias FA, Marin D, Oliveros-Bastidas A et al (2005) Structure-activity relationship (SAR) studies of benzoxazinones, their degradation products, and analogues. Phytotoxicity on standard target species (STS). J Agric Food Chem 53:538–548. https://doi.org/10.1021/jf050903h
Mago R, Miah H, Lawrence GJ et al (2005) High-resolution mapping and mutation analysis separate the rust resistance genes *Sr31*, *Lr26* and *Yr9* on the short arm of rye chromosome 1. Theor Appl Genet 112:41–50. https://doi.org/10.1007/s00122-005-0098-9
Maighany F, Khalghani J, Baghestani MA, Najafpour M (2007) Allelopathic potential of *Trifolium resupinatum* L. Persian clover and *Trifolium alexandrium* L. Berseem clover. Weed Biol Menag 73:178–183. https://doi.org/10.1111/j.1445-6664.2007.00250.x
Makowska B, Bakera B, Rakoczy-Trojanowska M (2015) The genetic background of benzoxazinoid biosynthesis in cereals. Acta Physiol Plant 37:176. https://doi.org/10.1007/s11738-015-1927-3
Mago R, Zhang P, Vautrin S et al (2015) The wheat *Sr50* gene reveals rich diversity at a cereal disease resistance locus. Nature Plants 1(12):1–3. https://doi.org/10.1038/nplants
Mahajan S, Tuteja N (2005) Cold, salinity and drought stresses: an overview. Arch Biochem Biophys 444 (2):139–158. https://doi.org/10.1016/j.abb.2005.10.018
Mahmood K, Orabi J, Kristensen PS et al (2020) *De novo* transcriptome assembly, functional annotation, and expression profiling of rye (*Secale cereale* L.) hybrids inoculated with ergot (*Claviceps purpurea*). Sci Rep 10:13475. https://doi.org/10.1038/s41598-020-70406
Makleit P (2005) Changes in cyclic hydroxamic acid content of various rye varieties for the effect of abiotic stress. Acta Biol Szeged 49(1–2):103–104
Malik R, Brown-Guedira GL, Smith CM et al (2003) Genetic mapping of wheat curl mite resistance genes *Cmc3* and *Cmc4* in common wheat. Crop Sci 43:644–650. https://doi.org/10.2135/cropsci2003.0644
McIntosh RA, Friebe B, Jiang J et al (1995) Cytogenetical studies in wheat. XVI. Chromosome location of a new gene for resistance to leaf rust in a Japanese wheat-rye translocation line. Euphytica 82:141–147. https://doi.org/10.1007/BF00027060
McKenzie RIH, Lamb RJ, Aung T et al (2002) Inheritance of resistance to wheats midge, *Sitodiplosis mosellana*, in spring wheat. Plant Breed 121:383–388. https://doi.org/10.1046/j.1439-0523.2002.745267.x
Meyer SLF, Rice CP, Zasada IA (2009) DIBOA: Fate in soil and effects on root-knot nematode egg numbers. Soil Biol Biochem 41:1555–1560. https://doi.org/10.1016/j.soilbio.2009.04.016
Miedaner T, Geiger HH (2015) Biology, genetics, and management of ergot (*Claviceps spp.*) in rye, sorghum, and pearl millet. Toxins 7:659–678. https://doi.org/10.3390/toxins7030659
Miedaner T, Gey A-KM, Sperling U et al (2002) Quantitative-genetic analysis of leaf-rust resistance in seedling and adult-plant stages of inbred lines and their testcrosses in winter rye. Plant Breed 121:475–479. https://doi.org/10.1046/j.1439-0523.2002.00765.x
Miedaner T, Klocke B, Flath K et al (2012) Diversity, spatial variation, and temporal dynamics of virulences in the German brown rust (*Puccinia recondita* f. sp. *secalis*) population in winter rye. Eur J Plant Pathol 132:23–35. https://doi.org/10.1007/s10658-011-9845-8
Mohler V, Hsam S, Zeller F, Wenzel G (2001) An STS marker distinguishing the rye-derived powdery mildew resistance alleles at the *Pm8/Pm17* locus of common wheat. Plant Breed 120(5):448–450. https://doi.org/10.1046/j.1439-0523.2001.00622.x
Mondal S, Rutkoski JE, Velu G et al (2016) Harnessing diversity in wheat to enhance grain yield, climate resilience, disease and insect pest resistance and nutrition through conventional and modern breeding approaches. Front Plant Sci 7:991. https://doi.org/10.3389/fpls.2016.00991
Musa GLC, Duck PL, Samborski DI (1984) The inheritance of resistance in rye to *Puccinia recondita* f. sp. *secalis* and f. sp. *tritici*. Can J Plant Sci 64(3):511–519. https://doi.org/10.4141/cjps84-074
Myśków B, Czyczyło-Mysza I, Sokołowska S et al (2018) QTL analysis of moderate drought response in a rye recombinant inbred line (RIL) population shows co-localization of QTLs for morphological and physiological traits. Acta Biol Cracoviensia Ser Bot 60 (2):59–67 http://journals.pan.pl/dlibra/docmetadata?id=102667
Niculaes C, Abramov A, Hannemann L, Frey M (2018) Plant protection by benzoxazinoids—recent insights into biosynthesis and function. Agronomy 8:143. https://doi.org/10.3390/agronomy8080143
Nyvall RF(1989) Diseases of rye. In: Field crop diseases handbook. Springer, Boston, MA. https://doi.org/10.1007/978-1-4757-5221-2_15
Oeser B, Kind S, Schurack S et al (2017) Cross-talk of the biotrophic pathogen *Claviceps purpurea* and its host *Secale cereale*. BMC Genom 18:273. https://doi.org/10.1186/s12864-017-3619-4
Ostrowska A, Tyrka M, Dziurka M et al (2019) Participation of wheat and rye genome in drought induced senescence in winter triticale (X *Triticosecale* Wittm.). Agronomy 9(4):195. https://doi.org/10.3390/agronomy9040195
Owuoche JO, Briggs KG, Taylor GJ (1996) The efficiency of copper use by 5A/5RL wheat-rye translocation lines and wheat (*Triticum aestivum* L.) cultivars. Plant Soil 180:113–120. https://doi.org/10.1007/BF00015417
Pandey R (2006) Root exuded acid phosphatase and 32P uptake kinetics in wheat, rye and triticale under phosphorus starvation. J Nucl Agric Biol 35:168–179
Pandey R, Singh B, Nair TVR (2005) Phosphorus use efficiency of wheat, rye and triticale under deficient

and sufficient levels of phosphorus. Indian J Plant Physiol 10:292–296

Paponov IA, Lebediskai S, Koshkin EI (1999) Growth Analysis of solution culture-grown winter rye, wheat and triticale at different relative rates of nitrogen supply. Ann Bot 84:467–473. https://doi.org/10.1006/anbo.1999.0935

Pociecha E, Dziurka M, Oklestkova J, Janeczko A (2016) Brassinosteroids increase winter survival of winter rye (*Secale cereale* L.) by affecting photosynthetic capacity and carbohydrate metabolism during the cold acclimation process. Plant Growth Regul 80(2):127–135. https://doi.org/10.1007/s10725-016-0149-z

Rabanus-Wallace MT, Hackauf B, Mascher M et al (2021) Chromosome-scale genome assembly provides insights into rye biology, evolution and agronomic potential. Nat Genet 53:564-573

Rahmatov M, Rouse MN, Nirmala J et al (2016a) A new 2DS· 2RL Robertsonian translocation transfers stem rust resistance gene *Sr59* into wheat. Theor Appl Genet 129(7):1383–1392. https://doi.org/10.1007/s00122-016-2710-6

Rahmatov M, Rouse MN, Steffenson BJ et al (2016b) Sources of stem rust resistance in wheat-alien introgression lines. Plant Dis 100(6):1101–1109. https://doi.org/10.1094/PDIS-12-15-1448-RE

Rakoczy-Trojanowska M, Krajewski P, Bocianowski J et al (2017) Identification of single nucleotide polymorphisms associated with brown rust resistance, α-amylase activity and pre-harvest sprouting in rye (*Secale cereale* L.). Plant Mol Biol Rep 35:366–378. https://doi.org/10.1007/s11105-017-1030-6

Rakoczy-Trojanowska M, Bakera B, Wlazło A et al (2019) Benzoxazinoids—an effective weapon of rye against biotic and abiotic stresses. Bulletin Plant Breeding Acclimatization Inst 287:103–104

Rakoczy-Trojanowska M, Święcicka M, Bakera B et al (2020) Co-cultivating rye with Berseem clover affects Benzoxazinoid production and expression of related genes. Crop Sci 60(6):3228–3346. https://doi.org/10.1002/csc2.20263

Ren T, Tang Z, Fu S et al (2017) Molecular cytogenetic characterization of novel wheat-rye T1RS. 1BL translocation lines with high resistance to diseases and great agronomic traits. Front Plant Sci 8:799. https://doi.org/10.3389/fpls.2017.00799

Ritonga FN, Chen S (2020) Physiological and molecular mechanism involved in cold stress tolerance in plants. Plants 9(5):560. https://doi.org/10.3390/plants9050560

Roux SR, Wehling P (2010) Nature of mixed infection type 2(5) observed in rye (*Secale cereale* L.) plants carrying the *Pr1* leaf-rust resistance gene. J Kulturpflanzen 62:29–34. https://doi.org/10.5073/JfK.2010.02.01

Roux SR, Ruge B, Linz A et al (2000) Leaf rust resistance in rye—evaluation, genetic analysis and molecular mapping. Acta Phytopathol Entomol Hung 35:65–73

Roux SR, Hackauf B, Linz A et al (2004) Leaf rust resistance in rye (*Secale cereale* L.). 2. Genetic analysis and maping of resistance genes *Pr3*, *Pr4*, and *Pr5*. Theor Appl Genet 110:192–201. https://doi.org/10.1007/s00122-004-1807-5

Roux SR, Hackauf B, Ruge-Wehling B et al (2007). Exploitation and comprehensive characterization of leaf-rust resistance in rye. In: International Symposium on Rye Breeding & Genetics, Groß Lüsewitz, Germany, June 28–30, 2006, vol 71. Vortr Pflanzenzüchtg, pp 144–150

Ruge B, Roux SR, Linz A et al (1999) Erschließung und molekulare Charakterisierung von Resistenzen gegen Braunrost bei Roggen (*Secale cereale* L.). Vortr Pflanzenzüchtg 46:169–176

Ruge B, Roux SR, Hackauf B et al (2001) Leaf rust resistance in rye: genetic analysis and mapping with molecular markers. Eucarpia Rye Meet. Radzikow, Poland, July 4–7, 2001, p 30

Salvador Moreno N, Ryan PR, Holguin I et al (2018) Transcriptional profiling of wheat and wheat-rye addition lines to identify candidate genes for aluminum tolerance. Biol Plant 62:741–749. https://doi.org/10.1007/s10535-018-0804-5

Santos E, Matos M, Silva P et al (2016) Molecular diversity and genetic relationships in *Secale*. J Genet 95:273–281. https://doi.org/10.1007/s12041-016-0632-3

Santos E, Benito C, Gallego FJ, Figueiras AM (2018) Characterization, genetic diversity, phylogenetic relationships, and expression of the aluminum tolerance *MATE1* gene in *Secale species*. Biol Plant 62:109–120. https://doi.org/10.1007/s10535-017-0749-0

Schlegel RH (2013) Rye: genetics, breeding, and cultivation. CRC Press, pp 49–52. https://doi.org/10.1201/b15553

Schlegel R Current list of wheats with rye and alien introgression (2019). http://www.rye-gene-map.de/rye-introgression. Accessed 19 Dec 2020

Schlegel R, Werner T (1991) Hülgenhof E (1991) Confirmation of a 4BL/5RL Wheat-rye chromosome translocation line in the wheat cultivar 'Viking' showing high copper efficiency. Plant Breed 107:226–234. https://doi.org/10.1111/j.1439-0523.1991.tb01210.x

Schlegel R, Melz G, Korzun V (1998) Genes, marker and linkage data of rye (*Secale cereale* L.): 5th updated inventory. Euphytica 101:23–67. https://doi.org/10.1023/A:1018351008889

Schulz M (2013) Benzoxazinoids in rye allelopathy - from discovery to application in sustainable weed control and organic farming. J Chem Ecol 39:154–174. https://doi.org/10.1007/s10886-013-0235-x

Sharma S (2009) Characterization of quantitative loci for morphological and anatomical root traits on the short arm of chromosome 1 of rye in bread wheat. Electronic theses and dissertations UC Riverside. https://escholarship.org/uc/item/1mr5p3z0

Shearman V, Sylvester-Bradley R, Scott R, Foulkes M (2005) Physiological processes associated with wheat yield progress in the UK. Crop Sci 45(1):175–185. https://doi.org/10.2135/cropsci2005.0175a

Shi B-J, Gustafson JP, Button J et al (2009) Physical analysis of the complex rye (*Secale cereale* L.) *Alt4* aluminium (aluminum) tolerance locus using a whole-genome BAC library of rye cv Blanco. Theor Appl Genet 119:695–704. https://doi.org/10.1007/s00122-009-1080-8

Sidhu JS, Ramakrishnan SM, Ali S et al (2019) Assessing the genetic diversity and characterizing genomic regions conferring tan spot resistance in cultivated rye. PLoS One 14(3):e0214519. https://doi.org/10.1371/journal.pone.0214519

Silva-Navas J, Benito C, Téllez-Robledo B et al (2012) The *ScAACT1* gene at the *Qalt5* locus as a candidate for increased aluminum tolerance in rye (*Secale cereale* L.). Mol Breed 30:845–856. https://doi.org/10.1007/s11032-011-9668-5

Smith CM, Clement SL (2012) Molecular bases of plant resistance to arthropods. Annu Rev Entomol 57:309–328. https://doi.org/10.1146/annurev-ento-120710-100642

Smolik M (2013) Discrimination of population of recombinant IL of rye (*Secale cereale* L.) for different responses to nitrogen-potassium stress assessed at the seedling stage under in vitro conditions. Electron J Biotechnol 16. https://doi.org/10.2225/vol16-issue1-fulltext-5

Solodukhina OV (2002) Genetic characterization of rye accessions with regard to leaf rust resistance. Russian J Genet 38:399–407. https://doi.org/10.1023/A:1015202303392

Solodukhina OV, Kobylyanski VD (2003) Problems of winter rye breeding for resistance to leaf and stem rusts. Plant Breed Seed Sci 48:87–97

Solodukhina OV, Kobylyanski VD (2007) Possibility of rye breeding for long-term resistance to leaf and stem rust. Vortr Pflanzenzüchtg 71:151–157

Sperling U, Lessner B, Scholz M et al (1996) Qualitative and quantitative variation for resistance of winter rye to leaf rust. Vortr Pflanzenzüchtg 35:175

Stitt M, Hurry V (2002) A plant for all seasons: alterations in photosynthetic carbon metabolism during cold acclimation in *Arabidopsis*. Curr Opin Plant Biol 5 (3):199–206. https://doi.org/10.1016/s1369-5266(02)00258-3

Sue M, Nakamura C, Nomura T (2011) Dispersed benzoxazinone gene cluster: Molecular characterization and chromosomal localization of glucosyltransferase and glucosidase genes in wheat and rye. Plant Physiol 157:985–997. https://doi.org/10.1104/pp.111.182378

Święcicka M, Siedlecka M, Bakera B, Rakoczy-Trojanowska M (2019) Analysis of rye *Lr* gene orthologue expression in seedlings inoculated with *Puccinia recondita* f.sp. *secalis*. Bulletin Plant Breeding Acclimatization Inst 287:337

Tabaglio V, Marocco A, Schulz M (2013) Allelopathic cover crop of rye for integrated weed control in sustainable agroecosystems. IJA 81:5. https://doi.org/10.4081/ija.2013.e5

Tanwir F, Dionisio G, Adhikari KB et al (2017) Biosynthesis and chemical transformation of benzoxazinoids in rye during seed germination and the identification of a rye Bx6-like gene. Phytochemistry 140:95–107. https://doi.org/10.1016/j.phytochem.2017.04.020

Thakur P, Nayyar H (2013) Facing the cold stress by plants in the changing environment: sensing, signaling, and defending mechanisms. In: Tuteja N, Singh GS (eds) Plant acclimation to environmental stress. Springer, pp 29–69. https://doi.org/10.1007/978-1-4614-5001-6_2

Villareal RL, Bañuelos O, Mujeeb-Kazi A, Rajaram S (1998) Agronomic performance of chromosomes 1B and T1BL.1RS near-isolines in the spring bread wheat Seri M82. Euphytica 103(2):195–202. https://doi.org/10.1023/A:1018392002909

Wang D, Zhuang L, Sun L et al (2010) Allocation of a powdery mildew resistance locus to the chromosome arm 6RL of *Secale cereale* L. cv.'Jingzhouheimai'. Euphytica 176(2):157–166. https://doi.org/10.1007/s10681-010-0199-7

Wehling P, Linz A, Hackauf B et al (2003) Leafrust resistance in rye (*Secale cereale* L.). 1. Genetic analysis and mapping of resistance genes *Pr1* and *Pr2*. Theor Appl Genet 107:432–438. https://doi.org/10.1007/s00122-003-1263-7

Weidner S, Amarowicz R, Karamać M et al (2000) Changes in endogenous phenolic acids during development of *Secale cereale* caryopses and after dehydration treatment of unripe rye grains. Plant Physiol Biochem 38(7–8):595–602. https://doi.org/10.1016/S0981-9428(00)00774-9

Wheeler T, von Braun J (2013) Climate change impacts on global food security. Science 134:508. https://doi.org/10.1126/science.1239402

Wise IL, Fox LS, Smith MAH (2015) Seed damage by *Sitodiplosis mosellana* (*Diptera: Cecidomyiidae*) to spring wheat cultivars with the *Sm1* gene 2015. Can Entomol 147(6):754–765. https://doi.org/10.4039/tce.2014.75

Wlazło A, Święcicka M, Koter MD et al (2010) Genes *ScBx1* and *ScIgl*—competitors or cooperators? Genes 11(2):223. https://doi.org/10.3390/genes11020223

Wouters FC, Blanchett B, Gershenzo J, Vassão DG (2016) Plant defense and herbivore counter-defense: Benzoxazinoids and insect herbivores. Phytochem Rev 15:1127–1151. https://doi.org/10.1007/s11101-016-9481-1

Wu H, Kang Z, Li X et al (2020) Identification of wheat leaf rust resistance genes in Chinese wheat cultivars and the improved germplasms. Plant Dis 104:2669–2680. https://doi.org/10.1094/PDIS-12-19-2619-RE

Würschum T, Longin CFH, Hahn V et al (2017) Copy number variations of *CBF* genes at the *Fr-A2* locus are essential components of winter hardiness in wheat. Plant J 89:764–773. https://doi.org/10.1111/tpj.13424

Xi W, Tang Z, Luo J, Fu S (2019) Physical location of new stripe rust resistance gene (s) and PCR-based markers on rye (*Secale cereale* L.) chromosome 5 using 5R dissection lines. Agronomy 9(9):498. https://doi.org/10.3390/agronomy9090498

Yang E, Li G, Li L et al (2016) Characterization of stripe rust resistance genes in the wheat cultivar Chuanmai45. Int J Mol Sci 17(4):601. https://doi.org/10.3390/ijms17040601

Yokosho K, Yamaji N, Ma JF (2010) Isolation and characterisation of two *MATE* genes in rye. Funct Plant Biol 37:296–303. https://doi.org/10.1071/FP09265

Yu XM, Griffith M (2001) Winter rye antifreeze activity increases in response to cold and drought, but not abscisic acid. Physiol Plant 112(1):78–86. https://doi.org/10.1034/j.1399-3054.2001.1120111.x

Yu XM, Griffith M. Wiseman SB (2001) Ethylene induces antifreeze activity in winter rye leaves. Plant Physiol 126:1232–1240. https://doi.org/10.1104/pp.126.3.1232

Zasada IA, Meyer SLF, Halbrendt JM, Rice C (2005) Activity of hydroxamic acids from *Secale cereale* against the plant-parasitic nematodes *Meloidogyne incognita* and *Xiphinema americanum*. Phytopathol 95:1116–1121. https://doi.org/10.1094/PHYTO-95-1116

Zasada IA, Rice CP, Meyer SLF (2007) Improving the use of rye (*Secale cereale*) for nematode management: potential to select cultivars based on *Meloidogyne incognita* host status and benzoxazinoid content. Nematology 9:53–60. https://doi.org/10.1163/156854107779969745

Zeller FJ (1973) 1B/1R wheat-rye chromosome substitutions and translocations. In: Proceedings 4th international wheat genetics symposium, Mo agricultural experiment station. Columbia, Mo, pp 209–222

Zeller FJ, Hsam SLK (1983) Broadening the genetic variability of cultivated wheat by utilizing rye chromatin. In: Proceedings of 6th international wheat genetics symposium. Kyoto, pp 161–173

Zhao ZH, Sun HG, Song W et al (2013) Genetic analysis and detection of the gene *MlLX99* on chromosome 2BL conferring resistance to powdery mildew in the wheat cultivar Liangxing 99. Theor Appl Genet 126:3081–3089. https://doi.org/10.1007/s00122-013-2194-6

Zhuang L, Sun L, Li A et al (2011) Identification and development of diagnostic markers for a powdery mildew resistance gene on chromosome 2R of Chinese rye cultivar Jingzhouheimai. Mol Breed 27(4):455–465. https://doi.org/10.1007/s11032-010-9443-z

The manufacturer's authorised representative in the EU is Springer Nature Customer Service Centre GmbH, Europaplatz 3, 69115 Heidelberg, Germany. If you have any concerns regarding our products, please contact ProductSafety@springernature.com

Printed and bound by CPI Group (UK) Ltd, Croydon, CR0 4YY
05/10/2026
02228339-0010